Structural and Failure Mechanics of Sandwich Composites

SOLID MECHANICS AND ITS APPLICATIONS

Volume 121

Aims and Scope of the Series

The fundamental questions arising in mechanics are: *Why?*, *How?*, and *How much?* The aim of this series is to provide lucid accounts written by authoritative researchers giving vision and insight in answering these questions on the subject of mechanics as it relates to solids.

The scope of the series covers the entire spectrum of solid mechanics. Thus it includes the foundation of mechanics; variational formulations; computational mechanics; statics, kinematics and dynamics of rigid and elastic bodies: vibrations of solids and structures; dynamical systems and chaos; the theories of elasticity, plasticity and viscoelasticity; composite materials; rods, beams, shells and membranes; structural control and stability; soils, rocks and geomechanics; fracture; tribology; experimental mechanics; biomechanics and machine design.

The median level of presentation is the first year graduate student. Some texts are monographs defining the current state of the field; others are accessible to final year undergraduates; but essentially the emphasis is on readability and clarity.

For other titles published in this series, go to
www.springer.com/series/6557

L.A. Carlsson • G.A. Kardomateas

Structural and Failure Mechanics of Sandwich Composites

Springer

L.A. Carlsson
Department of Mechanical Engineering
Florida Atlantic University
777 Glades Road
Boca Raton, Florida 33431
USA
carlsson@fau.edu

G.A. Kardomateas
School of Aerospace Engineering
Georgia Institute of Technology
270 Ferst Dr.
Atlanta, Georgia 30332
USA
george.kardomateas@aerospace.gatech.edu

ISSN 0925-0042
ISBN 978-1-4020-3224-0 e-ISBN 978-1-4020-3225-7
DOI 10.1007/978-1-4020-3225-7
Springer Dordrecht Heidelberg London New York

Library of Congress Control Number: 2011927000

Cover design: eStudio Calamar S.L.

Printed on acid-free paper

Springer is part of Springer Science+Business Media (www.springer.com)

Contents

Preface

The need for a text focusing on the structural behavior and failure characteristics of sandwich materials and structures has emerged during recent years as a consequence of the increased acceptance of this structural concept in several weight critical applications. Much new research in this subject area has been reported and a new journal devoted to this specific topic has been successfully launched.[1] In addition to the more or less traditional topics, this book covers two topics of sandwich structures which seem not to have been systematically presented thus far in the existing book literature, namely the failure mechanisms (face/core debonding, core cracking, etc.) and the associated fracture mechanics treatment and the benchmark elasticity solutions for several classical mechanics problems (sandwich plate under transverse loading, sandwich shell under internal pressure, etc.).

The subject matter is laid out in the following way: Chapter 1 defines the concept of structural sandwich, highlights some important structural applications, and introduces some of the more common core concepts. By simple analysis of sandwich beams, important stiffnesses and strengths of the constituent faces and core are identified. Typical face and core material properties are listed. Chapter 2 describes test methods for determination of mechanical properties of face and core. Chapter 3 provides the framework for first-order shear deformation plate theory for analysis of stresses and displacement of flat sandwich panels and beams. This analysis is applied to determine bending deflections, buckling loads, and twisting response of sandwich beams and panels. In Chapter 4, analysis and description of the three-point test method for sandwich beams is outlined and examples are provided. Chapter 5 outlines elasticity theory for sandwich plates and shells, and presents application of this theory to cylindrical shell structures under

[1] *Journal of Sandwich Structures and Materials* (J.R. Vinson, Editor).

pressure and torsion loading. In Chapter 6, high-order core shear theories are presented. Chapters 7 and 8 are devoted to analysis and experimental test methods for some very important failure modes of sandwich panels, viz. global buckling and face wrinkling. Another important failure mode of sandwich panels is face/core separation, and analysis and testing related to this failure mode is presented in the four final chapters (Chapters 9 through 12). At the end of the book we have cited the literature which includes several already published texts on sandwich structures.

This book does not cover all aspects of sandwich structures. We have not considered dynamic loading of a sandwich body nor have we touched upon questions of plasticity and large deformations or the behavior at elevated temperatures. Further, important topics such as failure of sandwich structures under cyclic loads (fatigue), and localized failures due to concentrated static and dynamic loads, and failure of core joints are not addressed. We have rather focused on some topics that have received much attention in the research and engineering community. We would thus like to acknowledge at this point the need for attention to topics not included in this text.

This work would not have been possible without help from our friends and colleagues. Both of us very much value the long-term support and friendship of Dr. Yapa D.S. Rajapakse of the Office of Naval Research (ONR). His vision and directions for research have been beneficial, not only for our research but also for stimulating a transition from metal structures to much more efficient composite structures for naval applications. The first author (LAC) would like to express his gratitude to the late Dr. Alf de Ruvo for introducing me to sandwich structures as a graduate student and the many years of fruitful cooperation on sandwich structures at SCA Research. Support from DIAB through Chris Kilbourn and Baltek through Kurt Feichtinger in the form of supply of core materials for experimental testing has been most beneficial. Several of our colleagues have provided encouragement and critique, and we would like to acknowledge Profs George Simitses, Victor Birman, Charles W. Bert, Shuki Frostig, Dan Zenkert, Karl-Axel Olsson, Hassan Mahfuz, and Christian Berggreen. In particular, Professor Frostig provided valuable input on Chapter 6 (High-Order Theories). Moreover, our graduate students have made invaluable contributions. We owe thanks to Francis Aviles, Catherine Ferrie, Haiying Huang, Valeria LaSaponara, Andrew Layne, Renfu Li, Xiaoming Li, Tomas Nordstrand, Poorvi Patel, Srinivas Prasad, Mina Pelegri, Catherine Phan, Felipe Ramirez, Magnus Renman, Scott Sendlein, Vinod Vadakke, and Gilmer Viana. The 2004 summer visit at NAVSEA in Carderock by the first author enabled cooperation in research on sandwich specimens and panels with Douglas Loup, Diane

Hoyns, and Robert Matteson. Typing assistance by Alejandra Quesada, Riti Waghray, Trudy Jeffries, Nicole Carr, Laura Thornton, Anu Maharaj, and Megan Crosara, and graphics production by Shawn Pennell, Laurie Donahue, and Joshua Kahn have been most helpful. Finally, we would like to extend our thanks to Ms. Nathalie Jacobs of Springer (Dordrecht, the Netherlands) and Ms. Jolanda Karada of Karada Publishing Services (Slovenia) for their patience and flexibility in making this work possible.

We hope that this book will enable the specialists who encounter questions in the structural and failure mechanics of sandwich composites to have the basic material of this topic at their fingertips, including the recent advances, and to utilize it in their research or engineering practice.

Leif A. Carlsson, Boca Raton, Florida
George A. Kardomateas, Atlanta, Georgia

Chapter 1
Introduction

A structural sandwich typically consists of two thin "face sheets" made from stiff and strong relatively dense material such as metal or fiber composite bonded to a thick lightweight material called "core". This concept mimics an I-beam, but in two dimensions, where the face sheets support bending loads and the core transfers shear force between the faces in a sandwich panel under load. Figure 1.1 illustrates flat and curved elements from a sandwich structure.

Sandwich structures allow optimization of structures that are weight-critical such as parts of airplanes, space structures, sporting goods, naval structures, and blades for wind-power generation (see Figure 1.2).

In addition to providing a very efficient load-carrying structure, the sandwich concept enables design of multi-functional structures. Figure 1.3 shows an example of the enclosed mast of USS Radford, where stealth properties, i.e., invisibility to radar, is accomplished by embedding radar absorbing materials in the core.

In addition to advanced structural applications, the sandwich concept has long been utilized in packaging materials, such as corrugated paper board (Figure 1.4), and in natural materials and structures such as human and animal bones and skulls and wings of birds (see Gibson and Ashby, 1997).

Core materials are classified within two broad categories, i.e., "cellular" and "structural". Cellular implies that the material consists of "cells" containing open space enclosed by walls in a repetitive manner so that space-filling is achieved (see Figure 1.5). Cellular foams, e.g. polymer or metal foams, honeycomb core, and balsa wood, are very common in structural applications. Web core is a structural core that consists of a continuous web made from a solid material formed in such a way that it separates the faces and becomes effective in transferring shear forces.

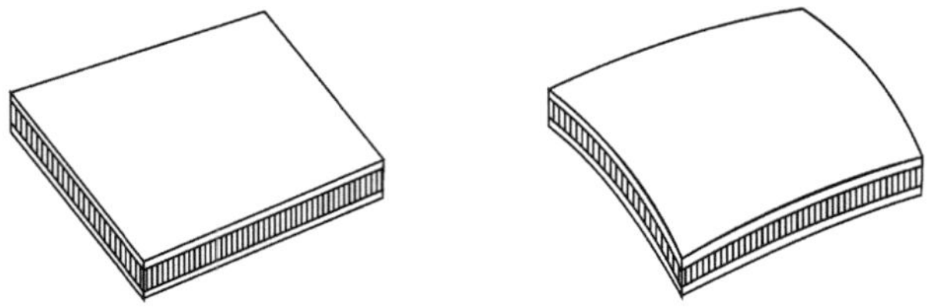

Figure 1.1 Flat and curved elements of a sandwich structure.

Because core materials are lightweight and stiffness and strength of materials scale with density (Gibson and Ashby, 1997), the core is commonly the weak constituent of a sandwich. In some instances, the bond between face and core may be critical for the integrity of the sandwich.

Proper selection of face and core materials requires understanding of the mechanics of sandwich structures. In this introductory chapter, we will examine some basic loading cases and failure modes of sandwich structures. Understanding of the contribution from the faces and core to important structural stiffnesses and strengths of a sandwich panel will guide the designer towards selection of appropriate materials and enable him or her to design a weight efficient and reliable structure. With almost no exceptions, sandwich structures utilize flat or curved panels (Figure 1.1). Still, much can be learned by consideration of a simpler sandwich structure, viz. a beam. This chapter will emphasize beams. Panels made from sandwich are examined in some detail in later chapters.

1.1 Bending Stiffness of a Sandwich Beam

The overall bending stiffness $E_x I$ of a sandwich beam is readily obtained from the parallel axis theorem (PAT) (Gere, 2004), which provides $E_x I$ in terms of the moduli and thicknesses of the constituents. For a symmetrical cross-section shown in Figure 1.6, PAT yields

$$E_x I = E_x^c I_c + 2E_x^f I_f, \tag{1.1}$$

where I_c and I_f are the moments of inertia of the core and each face sheet with respect to the neutral axis (y axis).

$$I_c = \frac{bh_c^3}{12}, \tag{1.2a}$$

Figure 1.2 Examples of sandwich structures.

$$I_f = \frac{bh_f^3}{12} + \frac{bh_f d^2}{4}. \tag{1.2b}$$

Hence, the bending stiffness per unit width of the sandwich beam becomes

$$\frac{E_x I}{b} = \frac{E_x^c h_c^3}{12} + E_x^f \left(\frac{h_f^3}{6} + \frac{h_f d^2}{2} \right). \tag{1.3}$$

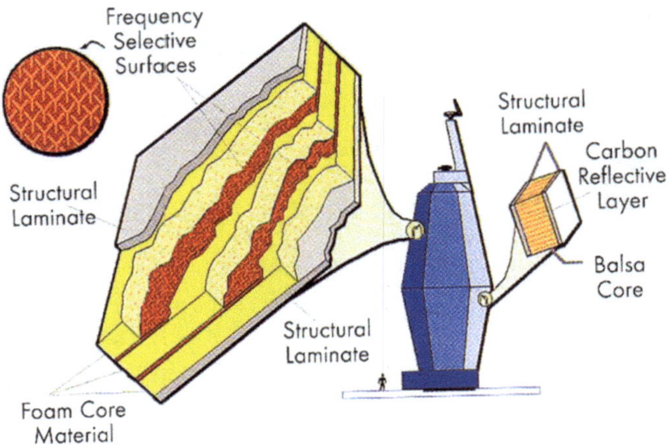

Figure 1.3 Cross-section of sandwich used to enclose the USS Radford mast.

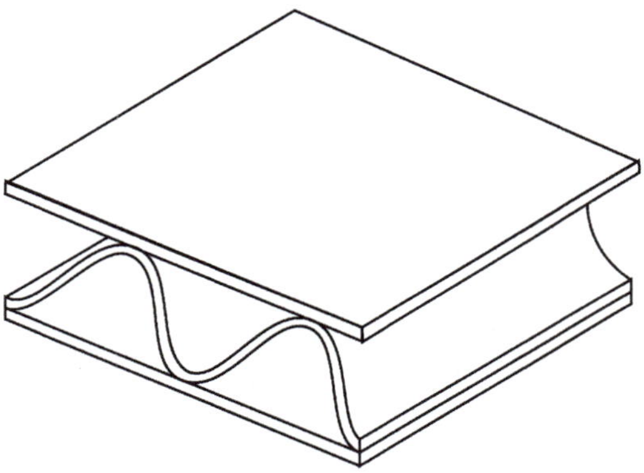

Figure 1.4 Corrugated core sandwich used in packaging boxes.

The quantity $E_x I/b$ is commonly referred to as "bending stiffness", D_x. Simplification of Equation (1.3) yields

$$D_x = E_x^f h_f d^2 \left[\frac{h_c^3}{12 h_f d^2} \left(\frac{E_x^c}{E_x^f} \right) + \frac{1}{6} \left(\frac{h_f}{d} \right)^2 + \frac{1}{2} \right]. \tag{1.4}$$

Sandwich structures are requested to be lightweight. Determination of optimal stiffness requires consideration of the density. The weight, W, of the sandwich beam, normalized by the beam width and length, is given by

$$\frac{W}{bl} = 2h_f \rho_f + h_c \rho_c, \tag{1.5}$$

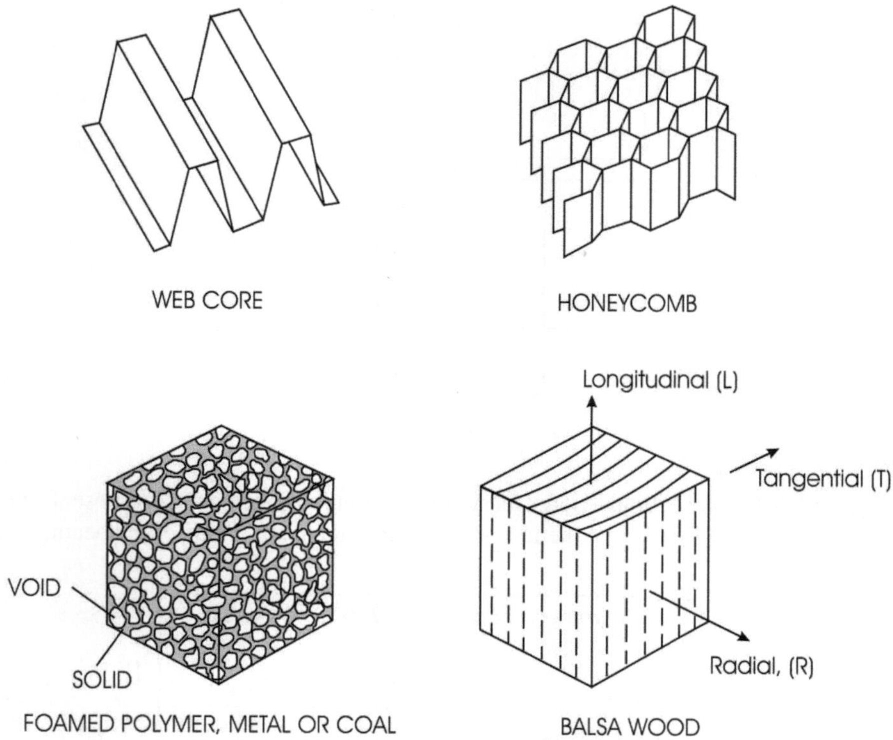

Figure 1.5 Core concepts utilized in sandwich structures.

where ρ_f and ρ_c are the densities (mass/unit volume) of the faces and core. The average (effective) density of the sandwich, ρ^*, becomes

$$\rho^* = 2\frac{h_f}{h}\rho_f + \frac{h_c}{h}\rho_c, \tag{1.6}$$

where h is the total thickness of the sandwich ($h = 2h_f + h_c$). Figure 1.7 shows D_x normalized by $E_x^f h_f d^2$, and ρ^* normalized by ρ_f, plotted vs. the core/face thickness ratio (h_c/h_f), for a typical sandwich consisting of aluminum face sheets and a H100 PVC foam core with: $E_x^f = 70$ GPa, $E_x^c = 0.1$ GPa, $\rho_f = 2.7$ g/cm^3, and $\rho_c = 0.1$ g/cm^3. Properties of typical face and core materials are provided in Tables 1.1 through 1.4. Inspection of the results in Figure 1.7 reveals that both the bending stiffness and density decrease with an increasing core-to-face thickness ratio. The normalized bending stiffness decreases rapidly at small thickness ratios and approaches 1/2 asymptotically, while the normalized density shows a continuous decrease with h_f/h_c.

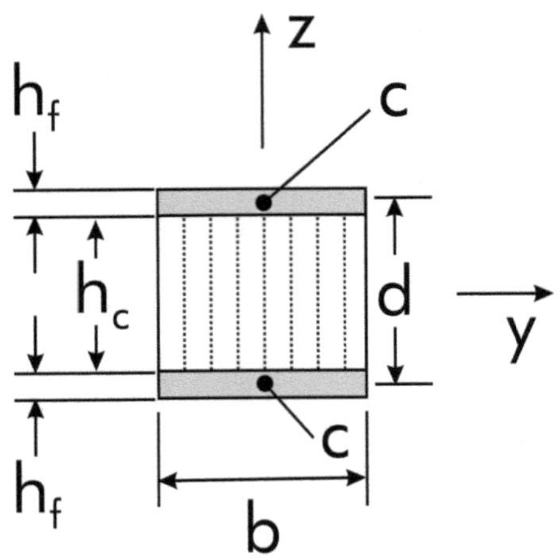

Figure 1.6 Cross-sectional view of a symmetric sandwich beam. "C" represents the centroid location for each of the face sheets, and y the neutral axis of the beam.

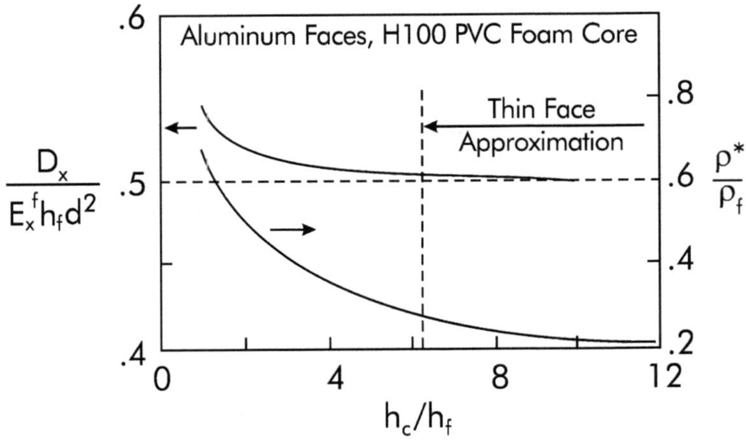

Figure 1.7 Bending stiffness and density of a sandwich beam vs. core-to-face thickness ratio. Face sheets are aluminum and the core is a H100 PVC foam.

The first term within the brackets of Equation (1.4) represents the bending stiffness contribution from the core, which is small by virtue of the small core-to-face modulus ratio (0.00143) for this combination and remains small for most other combinations of face and core materials, see Tables 1.1 through 1.4. The second term within the brackets in Equation (1.4) makes a significant contribution only for very thick faces (see Figure 1.7). Most practically used sandwich structures utilize thin face sheets, and the strictly geometry-dependent second term can be neglected in comparison to the third

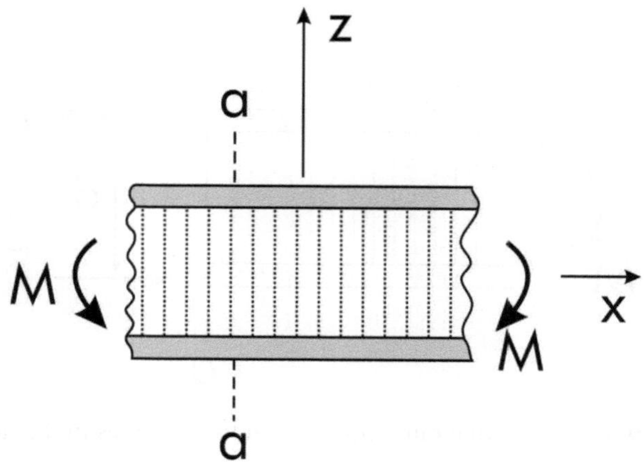

Figure 1.8 Sandwich element under pure bending.

term (1/2). We may at this point establish a limit on core/face thickness ratio above which the contribution from the second term to the bending stiffness is below 1%. Equation (1.4) yields

$$h_c/h_f \geq 5.35. \tag{1.7}$$

If this inequality is satisfied, the faces may be considered "thin", and the bending stiffness becomes

$$D_x = \frac{E_x^f h_f d^2}{2}. \tag{1.8}$$

This equation identifies the two most important factors for achieving high bending stiffness, i.e., high face sheet modulus, E_x^f, and a large distance, d, between the face sheet's centroids. A large value of the face sheet thickness, h_f, however, seemingly beneficial, will not be favorable from a weight point of view, see Figure 1.7. Consequently, from a bending stiffness and weight point of view, the most favorable sandwich design utilizes thin, high-modulus face sheets over a low-density core.

1.2 Stresses in the Face Sheets and Core

Consider the element of a sandwich under pure bending loads in Figure 1.8. Most core materials are compliant and do not significantly contribute to the bending rigidity. For such a case, and if the faces are thin compared to the

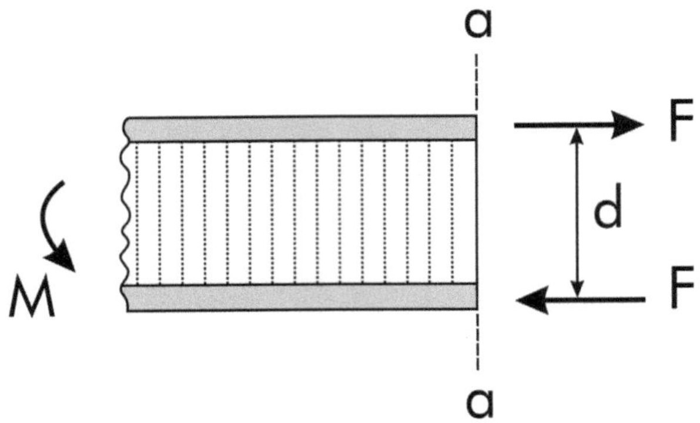

Figure 1.9 Free body diagram illustrating internal forces in the face sheets.

core, it is recognized that the bending moment, M, is equilibrated by internal tension and compression forces of equal magnitude acting at the centroids of the face sheets ("a couple"), as illustrated in the free body diagram in Figure 1.9.

If the bending stresses in the core are neglected, equilibrium of the element in Figure 1.9 yields an average bending stress in the face sheets

$$\sigma = \frac{M}{bdh_f}, \tag{1.9}$$

where d is the distance between the centroids of the faces, $d = h_c + h_f$, where h_c and h_f are the core and face thicknesses, respectively, and b is the width of the element (Figure 1.6). Notice that σ is tensile (positive) in the top face and compressive (negative) in the bottom face for the loading considered. Consequently, the face sheets need to be strong in tension and compression to be able to support the bending load.

If a sandwich beam is loaded by a bending moment that varies along the length of the beam, equilibrium analysis (Gere, 2004) shows that there will be a shear force, V, acting transversely to the beam axis (Figure 1.10).

$$V = \frac{dM}{dx}. \tag{1.10}$$

The shear stress, τ_{xz}, acting on the core, is obtained from equilibrium consideration of the element mm_1ab shown in the lower part of Figure 1.11. The horizontal (x axis) force due to the stress, σ, acting on the left side of the element is

$$F_1 = \sigma bh_f = \frac{M}{d}. \tag{1.11}$$

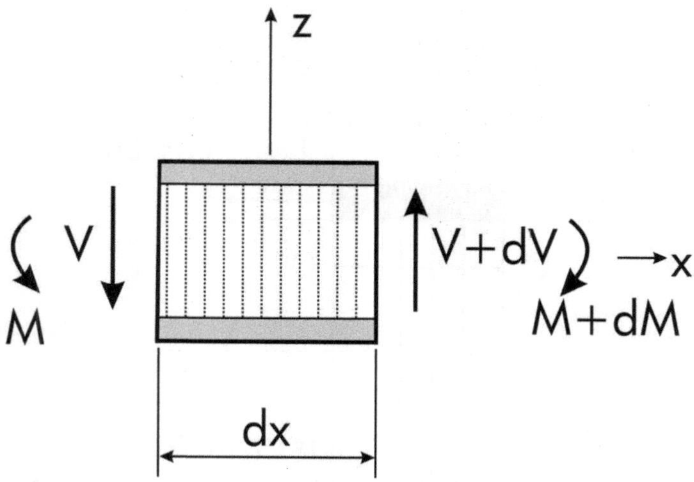

Figure 1.10 Element of a sandwich beam under variable bending moment.

The corresponding horizontal force acting on the right side of the element is

$$F_2 = \frac{M + dM}{d}. \tag{1.12}$$

The horizontal force due to the shear stress acting on the core surface at section *ab* is

$$F_3 = \tau_{xz}\, b dx. \tag{1.13}$$

Notice that the top surface (mm_1) is free from shear stress. Equilibrium yields

$$\tau_{xz} = \frac{dM}{dx}\frac{1}{bd} = \frac{V}{bd}. \tag{1.14}$$

This equation shows that the shear stress in the core, calculated based on the thin face/compliant core assumptions, is uniform (independent of the z coordinate). Exact analysis (Zenkert, 1997) reveals that the shear stress decreases almost linearly from the value, $V/(bd)$, at the face/core interfaces, to zero at the outer face surfaces. Equation (1.14) highlights the need for selecting a core material that is strong in shear. Further, as will be discussed later, for low modulus core material (Tables 1.2–1.4), shear deformation in the core may be excessive and may govern the overall deformation of the sandwich. Therefore, to avoid extensive shear deformation in sandwich structures, a core material with sufficiently high shear modulus must be used.

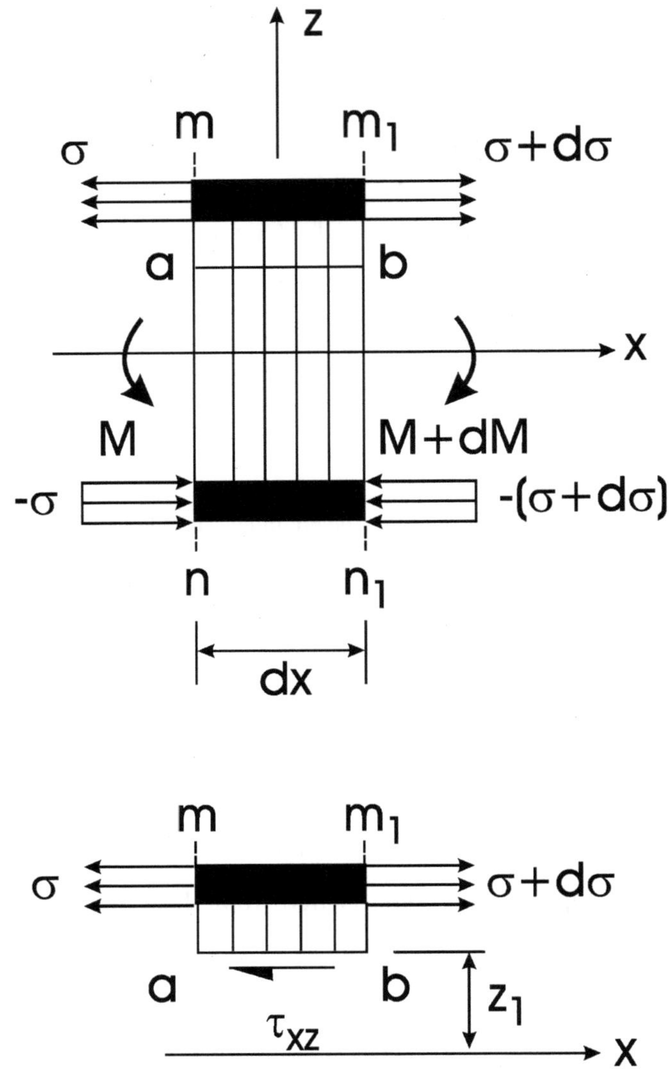

Figure 1.11 Sandwich elements considered in the calculations of core shear stress.

1.3 Local Failures

In addition to face failure in tension or compression, and core failure in shear, sandwich panels may fail locally through a host of failure modes to be discussed in detail in subsequent chapters. One such failure mode is "face wrinkling", sketched in Figure 1.12. Such a failure mode may occur in sand-

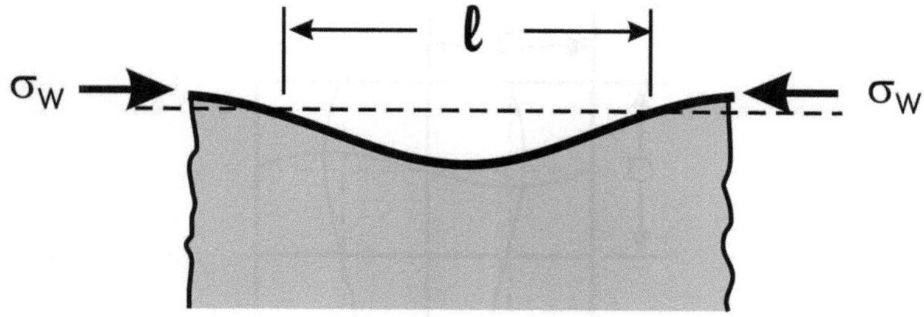

Figure 1.12 Wrinkling of the face sheets in compression loaded sandwich elements.

wich beams and panels with a soft homogeneous core (e.g. polymer foam or balsa wood core) under in-plane uniaxial compression loading. Wrinkling may also occur on the compression side of a sandwich panel or beam under bending. It manifests itself as a short wave-length buckling (local buckling) instability of the faces.

The wrinkling failure mode has been the subject of much research, as will be further discussed in Chapter 7. Such analysis shows that a high core stiffness will prevent such failures, in particular the out-of-plane extensional and shear stiffness.

For honeycomb-cored sandwich panels with thin faces, it is possible that the face sheets buckle between the supporting cells, as illustrated for square cells in Figure 1.13. Such a failure mode is called "intracell buckling" or "face dimpling" and this will be discussed later in this text. For the purpose of this chapter it is noticed that the local face buckling stress is proportional to the product of face modulus and face thickness squared ($E_f h_f^2$).

Sandwich panels with a web core (Figure 1.5) loaded in compression perpendicular to the corrugations (see Figure 1.14), may fail by local buckling of unsupported segments of both face and web core (Plantema, 1966).

Figure 1.15 shows that local buckling of a web-cored sandwich is a possible failure mode also when the panel is loaded in compression parallel to the corrugations.

The critical load for buckling of the face or web is proportional to the product of modulus and the ratio of thickness to unsupported length squared, i.e. $E(h_f/\lambda)^2$ for face buckling of the sandwich loaded perpendicular to the corrugations (Figure 1.14).

Sandwich panels with honeycomb or web-cores may also buckle locally when the sandwich is loaded in shear. As a guideline, such failures are cir-

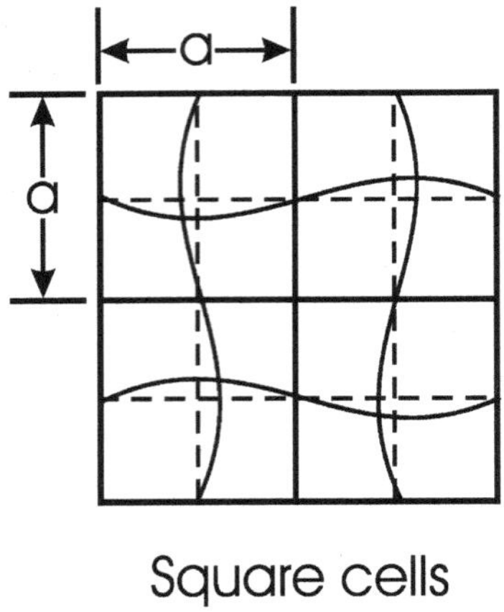

Figure 1.13 Local face buckling in a honeycomb-cored sandwich.

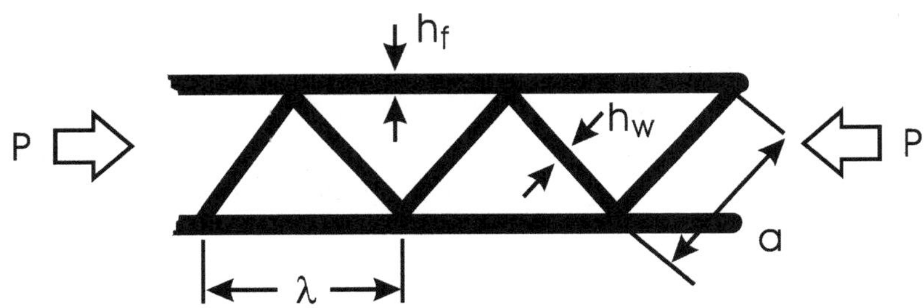

Figure 1.14 Web-cored sandwich loaded perpendicular to the corrugations.

cumvented by using short segments of high local bending stiffness (Eh^3) where h is the wall thickness of the honeycomb or web core.

Sandwich structures may suffer from failure due to concentrated loads acting normal to the plane of the sandwich panel, see Figure 1.16. Localized loads may occur due to hard object impact loading (dropped tools or hull/log collision for example), and at fittings and joints between panel sections.

Failure of sandwich beams due to localized loads have been analyzed by, e.g., Thomsen (1977), Ashby et al. (2000), and Steeves and Fleck (2004). Concentrated loads acting transverse to the plane of the sandwich may produce substantial local deformations of the faces and core, and induce a complex state of stress in the affected regions of the face and core. For the purposes of this chapter, it suffices to mention that the analysis of Ashby et al. (2000), provides an expression where the indentation load is directly proportional to the out-of-plane compression strength of the core. Consequently,

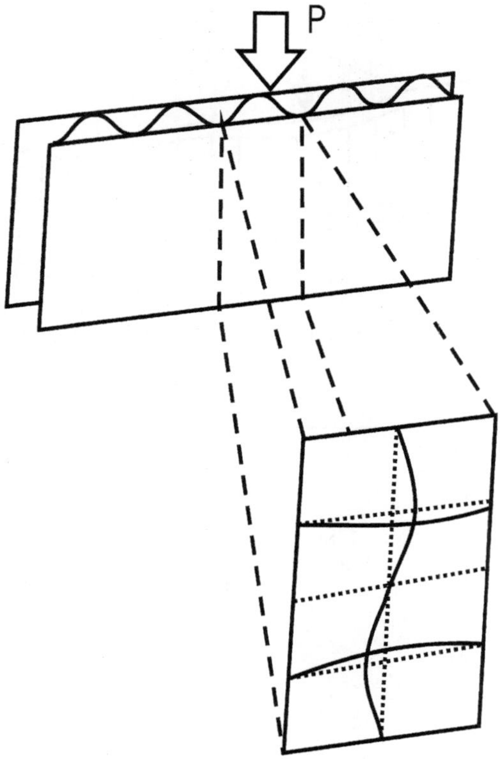

Figure 1.15 Local face buckling in corrugated core sandwich loaded parallel to the corrugations.

the out-of-plane compression stiffnes and strength of the core are important for the ability of the sandwich to resist localized loads.

1.4 Face and Core Materials

The preceding analysis of the stiffness and strength of sandwich has identified several important properties of the face sheets and core. The face sheets need to be stiff and strong in tension and compression to resist the bending and wrinkling loads. The core needs to be stiff and strong under shear and extension in the thickness direction to provide resistance to wrinkling and local indentation failure. At the same time, the core should be of low density in order to minimize the structural weight. Such demands are conflicting since, in general, low density materials are less stiff and strong than materi-

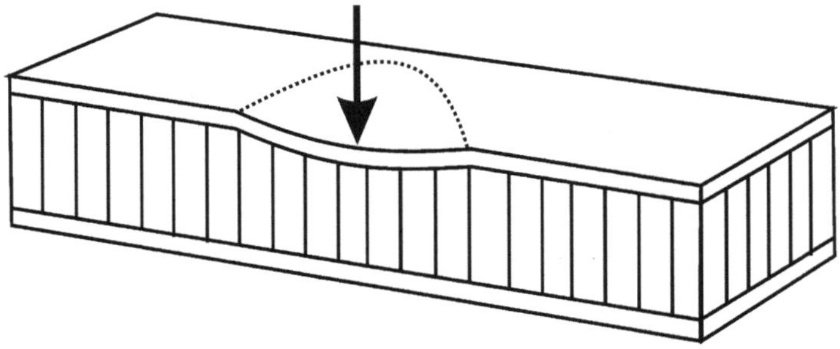

Figure 1.16 Local indentation failure due to concentrated load acting on a sandwich panel.

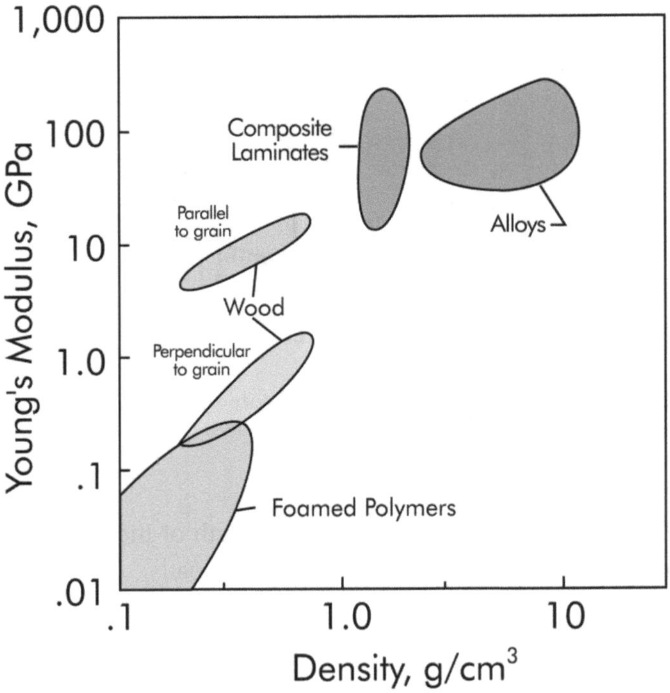

Figure 1.17 Modulus-density chart for various classes of materials. After Ashby (1999).

als of higher density. The selection of face and core materials may be guided by Ashby's materials property charts (Ashby, 1999). An example of such a chart is shown in Figure 1.17.

According to the guidelines outlined above, the face sheets should be made from high modulus materials, i.e., composite laminates or light-weight alloys (see Figure 1.17). The core should be of low density. Consequently, foamed polymers or balsa wood are common selections. In addition to the modulus-density chart shown above, Ashby (1999) presents material property charts for several other important mechanical as well as thermal insulation properties. Such graphs are extremely useful procedures for the proper selection of materials for given structural and thermal requirements.

Some typical mechanical properties of face and core materials will be provided. The mechanical properties of heterogeneous materials such as honeycomb core, foams, and wood are average properties, representative for a large volume of material, being part of a sandwich structure. The face materials may be made from isotropic metals or anisotropic composite laminates. Typically, however, the laminates are symmetric and balanced (same number of plies at positive and negative angles), which simplifies their mechanical description. It must be recognized that the mechanical properties of face laminates vary depending on type of fiber, ply orientation, and volume fractions of fiber and matrix. The type of matrix material will also influence the mechanical properties of the composite. In most applications of sandwich structures, however, the matrix is a thermoset resin such as epoxy, polyester, or vinylester, with much less stiffness and strength than the fibers. Consequently, for fiber-dominated lay-ups, the influence of matrix on the static, short-term mechanical properties is quite small.

Table 1.1 lists density and mechanical properties of some typical face materials. The properties represent short-term, room temperature values, as determined by standard test methods. It must be pointed out that such properties should not be used for actual design purposes since the properties may vary depending on temperature and humidity, and several other controlled and uncontrolled factors. The metal properties were obtained from Daniel and Ishai (2006) and Gere (2004). The S-glass/EP properties were determined by Aviles (2005), while the E-glass/EP and AS4-Carbon/EP properties were determined by Alif and Carlsson (1997).

Cores for sandwich panels are grouped in web core, honeycomb core, foams, and end-grain balsa wood (see Figure 1.5). It should again be pointed out that the most important core properties are the out-of-plane extensional and shear stiffnesses and strengths. It is not always possible or meaningful to test the core isolated without the presence of the faces since the faces tend to stabilize the core, especially for web and honeycomb cores. Furthermore, the mechanical stiffnesses of web cores are highly dependent on the geometry and the material of the web and, for these reasons, it is very difficult to list properties for such cores. For honeycomb cores, the most common materials are Nomex, which is an aramid fiber paper impregnated with a

Table 1.1 Mechanical properties of face materials. ρ = density, E = Young's modulus, G = shear modulus, ν = Poisson's ratio, X = strength, T = tension, C = compression.

Material	ρ g/cm^3	E GPa	G GPa	ν	X_T MPa	X_c MPa
Aluminum (2024-T3)	2.80	73	27.4	0.33	414	414
Steel (AISI 1025)	7.80	207	80.0	0.30	394	394
Titanium	4.40	108	42.4	0.30	550	475
S-Glass/EP[1]	1.73	20.6	3.10	0.12	261	177
E-Glass/EP[1]	2.00	26.6	4.63	0.144	422	410
AS4-Carbon/EP[1]	1.63	59.5	4.96	0.047	584	491

[1]The composites consist of woven 0 and 90°fibers in an epoxy (EP) matrix.

Table 1.2 Mechanical properties of honeycomb core. ρ = density, G = shear modulus, S = shear strength, W = width direction, L = length direction. From Zenkert (1997).

Material	ρ g/cm^3	G_L MPa	G_W MPa	S_L MPa	S_W MPa
Paper	0.056	141	38	1.3	0.48
Aluminum	0.070	460	200	2.2	1.50
Nomex	0.080	69	44	2.2	1.00
Nomex	0.129	112	64	3.2	1.70

polymer resin, usually phenolic, or aluminum alloy. The method of manufacturing of honeycomb core provides a structure with double walls in one direction and single walls in the other. As a result, the mechanical properties are different in the two in-plane principal directions (width W and length L). Mechanical properties of honeycomb cores are considered in great detail by Gibson and Ashby (1997). Product literature sometimes reports on modulus and strength in compression and shear, see, e.g., Hexcel product information www.hexcel.com), while other sources of data, e.g., Vinson (1999), reports only shear moduli. It is not practical to reproduce the very large amount of data on honeycomb cores available in a publication of this nature. Here we will only reproduce some typical data provided by Zenkert (1997), see Table 1.2.

Foams are very common core materials. Most commercial foams are made from polymers, although there is much interest in metallic foams (Ashby et al., 2000), and more recently carbon foams (Sihn and Rice, 2002). The

Table 1.3 Mechanical properties of various polymer foams. ρ = density, G = shear modulus, S = shear strength. Data obtained from Zenkert (1997), DIAB* and Rohacell**.

Material	ρ g/cm^3	G MPa	S MPa
Polyurethane	0.04	4	0.25
PVC H100	0.10	40	1.40
PVC HD130	0.13	40	1.50
PMI 110IG	0.11	50	2.40

*www.diabgroup.com; **www.roehm.com

Table 1.4 Mechanical properties of balsa wood core. ρ = density, G = shear modulus (out-of-plane), S = shear strength (out-of-plane). From www.alcanairex.com.

Product Designation	ρ g/cm^3	G MPa	S MPa
SB50	0.100	110	1.91
SB100	0.151	157	2.94
SB150	0.244	302	4.85

most common polymers used are polyurethane, polyvinylchloride (PVC), and polymethacrylimide (PMI). Such foams are closed-cell structures, making them isotropic and resistant to water penetration.

Balsa wood core is used as the core in structural sandwich panels because of its low density combined with good mechanical properties and a closed-cell structure. As a result of the unidirectional orientation of the fibers along the longitudinal direction of the wood (Figure 1.5), balsa wood is highly anisotropic, with much higher stiffness and strength in the longitudinal (along the grain) than in the radial and tangential directions. Balsa wood utilized as the core in a sandwich structure, is delivered in the desired thickness in the form of small square blocks with the L-direction (fiber direction) in the through-thickness direction, assembled in a panel held together with a scrim cloth on the top and bottom. The blocks are randomly oriented in the plane of the sandwich making the effective properties of the core in-plane isotropic. End-grain balsa wood core is available over a range of densities between about 0.1 to 0.3 g/cm^3. Typical mechanical properties of end grain balsa wood, obtained from Baltek (www.alcanairex.com or www.alcanbaltek.com) are provided in Table 1.4.

Chapter 2
Characterization of the Mechanical Properties of Face Sheet and Core Materials

Determination of mechanical properties of face sheets and core materials is important for analysis and design of sandwich structures. In many cases, especially for metals, mechanical property data exists in handbooks and textbooks on materials science and strength of materials (e.g., Gere, 2004). For composites, the large variety of fibers, matrix materials, ply lay-ups, and fiber volume fractions makes mechanical testing a necessity. The core may be corrugated, honeycomb, or foam as shown in Figure 1.5. The determination of the mechanical properties may be very challenging, especially for corrugated and honeycomb cores. For balsa wood, the anisotropy of the material causes further complications.

2.1 Face Mechanical Properties

2.1.1 Isotropic Face Sheets

For isotropic materials such as metal alloys, it is normally sufficient to perform a tensile test, which provides Young's modulus, E, Poisson's ratio, ν, the yield strength, σ_{ys}, and the ultimate strength, σ_{ult}, defined according to

$$E = \frac{\sigma}{\varepsilon_L} \quad (\varepsilon_L \ll 1), \tag{2.1}$$

$$\nu = \frac{-\varepsilon_T}{\varepsilon_L} \quad (\varepsilon_L \ll 1), \tag{2.2}$$

where σ is the tensile stress (force per unit cross-section area) and ε_L and ε_T are the longitudinal and transverse strains. The yield strength, σ_{ys}, is defined as the constant flow stress observed in some metal alloys after the linear elastic region or from the intersection of a line parallel to the initial linear response curve (with slope of E), but with a 0.2% strain offset. The ultimate strength, σ^{ult}, is defined as the maximum stress the material can endure. Further details are provided in Gere (2004), and the ASTM E8 standard (1998).

A complete characterization of the material demands the shear modulus, G, defined according to

$$G = \frac{\tau}{\gamma} \quad (\gamma \ll 1), \tag{2.3}$$

where τ is the shear stress and γ is the (engineering) shear strain. For isotropic materials, however, the shear modulus is not an independent quantity. With knowledge of E and ν it is possible to establish the shear modulus, G, according to Gere (2004)

$$G = \frac{E}{2(1+\nu)}. \tag{2.4}$$

Hence, there is no need for an independent shear test. The yield strength in shear, τ_{ys}, may also be estimated from the tensile yield strength (Norton, 2003), as

$$\tau_{\mathrm{ys}} \cong 0.57\, \sigma_{\mathrm{ys}}. \tag{2.5}$$

Because face alloys are available in the form of sheet materials, a flat test coupon is machined as illustrated in Figure 2.1. Prior to testing, the cross-sectional dimensions of the gage section should be measured. The tensile test coupon should be instrumented with strain gages oriented axially and transversely at the center of the gage region for measurement of the longitudinal and transverse strains, ε_L, and ε_T. ASTM E8 (1998) stipulates the use of extensometers for strain measurement, but electrical resistance strain gages are more accurate and demanded if Poisson's ratio, ν, is also to be determined.

Several details of test procedures are common to all types of mechanical testing. These details include, for example, correctly installing the specimen in the grips of the test frame or in the test fixture and ensuring that alignment is proper. Machine settings, such as load range, displacement speed, as well as calibrations of the load cell and strain gage instrumentation, and data acquisition systems, should be checked. The ASTM collection of standards includes several standards specific to sandwich structures.

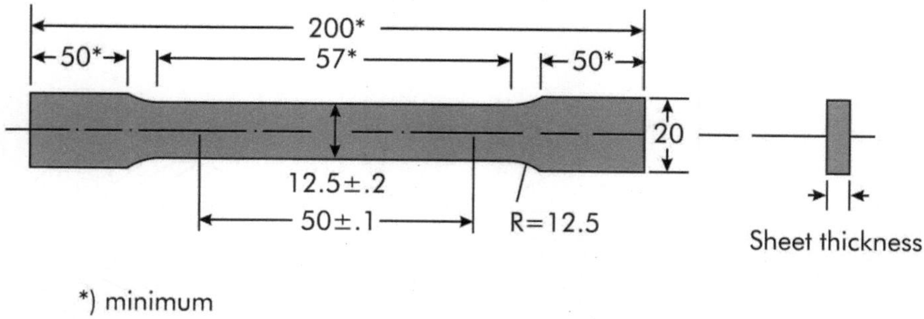

Figure 2.1 Tensile test specimen for metal face sheets (ASTM E8, 1998).

2.1.2 Composite Face Sheets

Composite face sheets for aerospace sandwich structures typically consist of laminates formed by stacking unidirectional plies of pre-impregnated unidirectional glass or carbon fibers embedded in a polymer matrix (epoxy, bismaleimid, etc.). Such laminates require processing in autoclaves at elevated temperatures and pressures over extended periods of times which leads to excellent quality (negligible void content) and high fiber volume fraction. The face sheets may be produced separately and subsequently adhesively bonded to the core or the sandwich may be co-cured. In the boating industry, wet lay-up procedures and room temperature cure of polyester or vinylester resins are quite common. Such procedures are cost-efficient for reasonable volumes but do not produce laminates with very high fiber volume fractions. Furthermore, lack of vacuum bagging leads to high void contents. A method that has become popular in ship structures consists of injection of resin in dry fiber mats that are placed on both sides of a core panel. This procedure saves time and often leads to improved bonding between face and core. The process is called vacuum assisted resin transfer molding (VARTM) and has been patented by Seeman (1990). Further details and comprehensive reviews of composite processing methods are provided by Advani and Sozer (2002).

Regardless of the way composites are manufactured, face laminates (with some rare exceptions) consist of fibers oriented in the plane of the laminate, making them anisotropic, especially when comparing in-plane, and out-of-plane stiffnesses, see Figure 2.2.

As there are no fibers oriented in the through-thickness direction (along the z axis in Figure 2.2), the out-of-plane extensional and shear stiffnesses of the laminate are of the same order as those for the polymer matrix, i.e. they are low (Hyer, 1998). Fortunately, as has been discussed, sandwich panels

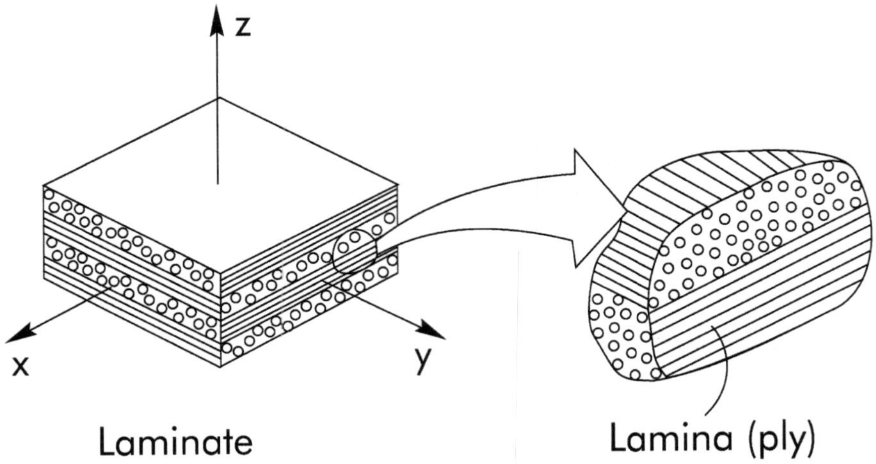

Figure 2.2 Fibers are oriented in the plane of the laminate (xy plane) resulting in anisotropic properties.

are loaded in such a way that the dominant loading of the face sheets is in the plane (membrane) and the low out-of-plane stiffnesses of the faces do not greatly influence the overall behavior of the structure. When performing three-dimensional (3D) stress analysis, or 2D plane stress or plane strain analysis, numerical values of out-of-plane stiffnesses are required as input. Although most standard test methods are designed for determining in-plane stiffnesses and strength of composite laminates, there are ASTM standard test methods and analytical schemes available for estimation of out-of-plane elastic stiffnesses (Hyer, 1998).

If the laminate is made from unidirectional prepreg it is common practice to make unidirectional laminates (fibers are oriented in one direction only) and characterize the mechanical stiffnesses and strengths of this basic building block of a composite laminate. A detailed description of test methods for characterization of the composite lamina is provided by Adams et al. (2003). With a full set of unidirectional lamina properties, it is possible to predict the overall stiffness and strength properties of the laminate. Such an analysis is based on classical laminated plate theory (CLPT), reviewed by Hyer (1998). This procedure is recommended for preliminary design analysis, but is no substitute for the mechanical characterization of the laminate used in an actual sandwich structure. Furthermore, this procedure does not readily apply to laminates formed by resin impregnation (VARTM). In this chapter, we will therefore describe test methods for mechanical characterization of composite laminates. The most important mechanical properties of the face sheets are the tensile, compressive and shear stiffnesses, and strengths.

Laminates from the face sheets should be symmetric and balanced. Symmetric means that the ply configurations above and below the geometrical midplane ($z = 0$ in Figure 2.2) are mirror-images of each other (notice that the laminate shown in Figure 2.2 is not symmetric). Symmetry thus requires that, for each ply above the midplane, there is an identical ply (material, thickness, and fiber orientation) at the same distance below the midplane. Balanced means that the number of off-axis plies oriented in a certain positive angle (θ) should be balanced by the same number of plies oriented in the negative angle ($-\theta$). As discussed for example by Hyer (1998), such laminates should be free from extension/bending and extension/shear couplings. Such couplings may promote undesired deformation shapes and complicate the analysis of the mechanical response quite substantially.

Many tests require end tabs to prevent crushing of the regions of the specimen in contact with gripping surfaces. End tabs are designed to achieve a gradual transfer of load into the test specimen. Most common are glass fabric/epoxy end tabs bonded adhesively to the test specimen's gripping regions. A detailed description of end-tab attachment procedures is presented in Adams et al. (2003). End tabs are typically bonded to the composite panel before specimens are machined in the orientations desired (x and y in Figure 2.2). Specimens may be cut using a thin abrasive disk in a tile cutter or surface grinder.

Tension Test Procedures

The tension test for composite materials is described in the ASTM D3039 Standard (2000) and by Adams et al. (2003). A laminate tensile specimen is typically about 225 mm long and 25 mm wide, see Figure 2.3. The specimen may be tabbed as shown in Figure 2.3, although lower strength composites may be tested without tabs (Adams et al., 2003). For a tensile test, two strain gages are mounted at the center of each specimen, one in the longitudinal direction and the other in the transverse direction, so that both the axial stiffness, E_x, and major Poisson's ratio, ν_{xy}, can be determined. Notice that the x and 0° directions coincide for laminates. Multiple strain gages may be used for increased confidence in the results.

The specimen should be mounted in the grips of the test machine and tested at a cross-head rate of about 0.5 to 1 mm/min. Avoid unprotected eyes in the test area. The strain readings may be recorded continuously or

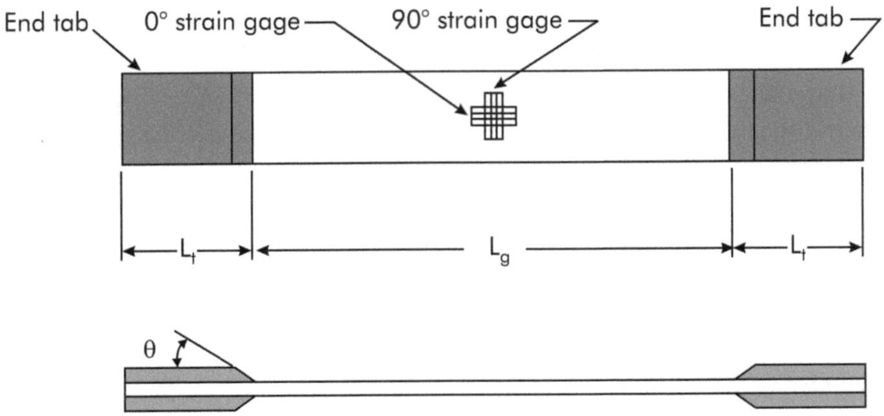

Figure 2.3 Straight-sided tension test specimen with adhesively bonded end tabs and strain gages. Tab angle θ may be between 5 and 90°.

at discrete load intervals to provide the required number of data points. The stress-strain data should allow determination of Young's modulus, E_x, Poisson's ratio, ν_{xy}, and the ultimate strength in tension, σ_x^{ult}, of the laminate

$$E_x = \frac{\sigma_x}{\varepsilon_x} \quad (\varepsilon_x \ll 1), \tag{2.6a}$$

$$\nu_{xy} = \frac{-\varepsilon_y}{\varepsilon_x} \quad (\varepsilon_x \ll 1), \tag{2.6b}$$

where σ_x is the axial stress (load per unit cross-section area), and ε_x and ε_y are the longitudinal and transverse strains, respectively, i.e., the strains in the 0 and 90° directions, see Figure 2.3. The tensile strength, X_t, is defined as the maximum stress measured during the test.

Compression Test Procedures

Because face laminates are thin, global buckling of a compression loaded test specimen is a possibility that must be prevented. Thus, if the specimen is long, some type of lateral restraint fixturing is required. It is important to ensure that this fixturing does not create a redundant load path.

Compression test fixtures are classified as "shear-loaded" or "end-loaded", depending on how load is introduced into the gage section. Compression testing of shear and end-loaded specimens requires attention and care since the failure mechanism is very sensitive to misalignment and imperfections. Any misalignment of the test specimen can induce bending of the specimen and promote failure by global buckling. The specimen must be checked for proper alignment of the loading surfaces. The ends of end-loaded specimens must be flat, mutually parallel, and perpendicular to the specimen axis. Specified parallel and perpendicular tolerances are typically in the order of 0.03 mm. Specimen end flatness is particularly important, so that the loading is centered around the central axis of the specimen and local end crushing is avoided. Uniformity of specimen thickness from end to end within 0.06 mm is a common requirement. Variations in thickness of any of the tabbing strips, along with variations in thickness of the adhesive bond lines, must be checked. Specimen bending and buckling cannot usually be detected visually during the compression test or by microscopic examination of the failed specimen. The use of two strain gages, mounted "back-to-back" on opposite faces of the specimen, is the only reliable method of detecting bending and buckling, by a difference in the strains.

Certain test preparation and test procedures need to be followed and careful attention must be directed towards detection of eccentric loading promoting bending and buckling of the specimen, as well as end-crushing in the end-loaded fixtures (Adams et al., 2003; ASTM D3410, 2003; ASTM D6641, 2001).

The IITRI compression test fixture shown in Figure 2.4 has long been recognized as a reliable standardized test procedure (ASTM D3410, 2003), which employs a tabbed unsupported short gage length specimen with the load introduced through shear. The IITRI test specimen is typically between 140–155 mm long and 25 mm wide. The gage length varies between 10 and 25 mm depending on the specimen thickness (ASTM D3410, 2003).

End-loaded test fixtures are sometimes used (Adams et al., 2003), although none has been approved as ASTM standard. A "combined loading" compression (CLC) test method, where the load is introduced both by shear and end pressure is ASTM standard D6641 (2001), see Figure 2.5.

The shear-loading component is achieved by clamping pairs of lateral support blocks, which have high-friction contact surfaces, to each end of the specimen. The end-loading component is induced directly because each end of the specimen is flush with the outer surfaces of the support blocks. The fraction of shear loading can be increased by increasing the torque in the clamping screws. The target is to achieve just enough shear loading to avoid crushing of the ends of the specimen. It is possible to test an untabbed, straight-sided specimen. This simplifies specimen preparation. A typical specimen is 140 mm long and 12 mm wide with an unsupported gage length of 12 mm. Such a gage length and the recess opening in the fixture

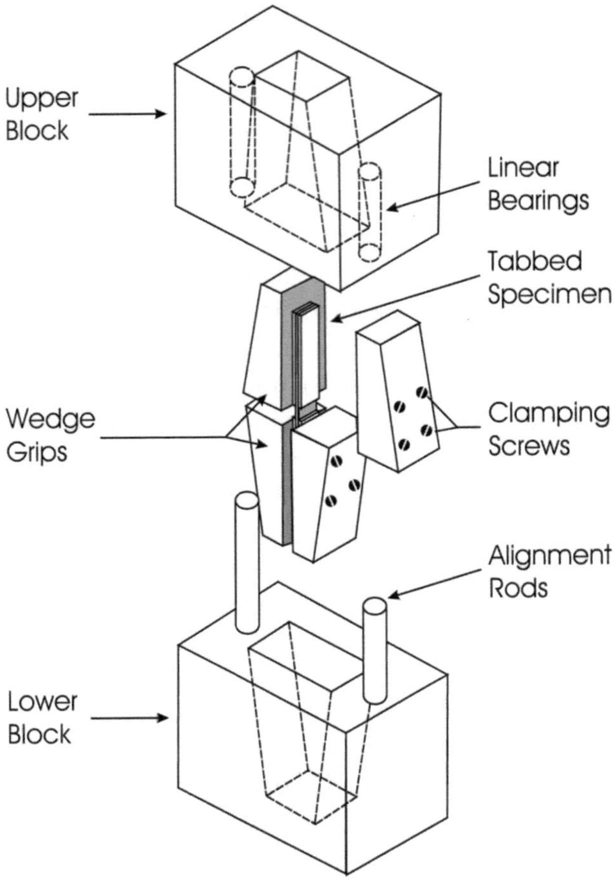

Figure 2.4 IITRI compression test fixture and tabbed test specimen.

(Figure 2.5) allows sufficient room to use bonded strain gages (see ASTM D6641, 2001, and Adams et al., 2003, for further details).

From a properly conducted test, it should be possible to determine Young's modulus, E_x, and the compressive strength, X_c. E_x is defined in Equation (2.1), with ε_x being the average of the front and backside strains if two gages are used. The compressive strength X_c is defined as the magnitude of the maximum stress recorded during the test.

Shear Test Procedures

Face laminates commonly contain 0, 90° and angle-plies, e.g., plies with the fibers oriented at ±45° to the axis of reference. For such laminates, the

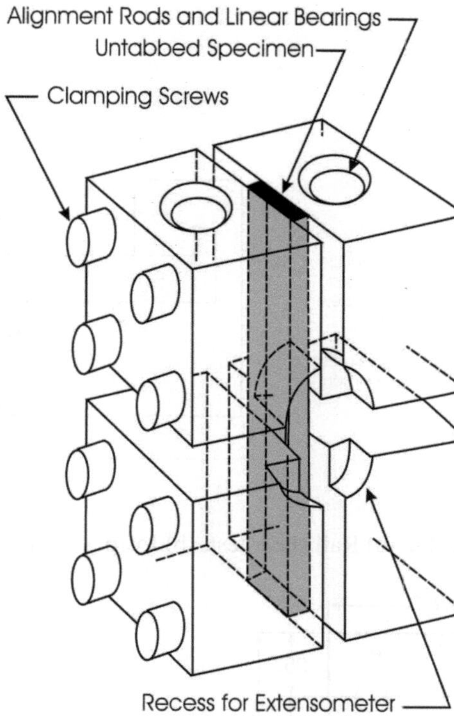

Figure 2.5 The Wyoming combined loading compression (CLC) fixture.

rail and Iosipescu shear test methods shown in Figure 2.6 are applicable for determination of the in-plane shear modulus and strength. Both the rail and Iosipescu shear tests are ASTM standards (ASTM D4255, 2002; ASTM D5379, 2001). Laminates containing angle-ply fibers do not have weak shear planes (which is structurally beneficial), leading to a tortuous failure path and an extensive zone of damage, which makes testing and interpretation of test results challenging.

The rail shear loading configuration in Figure 2.6a produces an essentially pure shear loading over the gage section of the specimen (the 12.7 mm wide portion of the specimen exposed between the rails). The rail shear test specimen is shown in Figure 2.7a. There are six clearance holes to allow clamping the specimen to the rails. A strain gage rosette, oriented at ±45° with respect to the specimen longitudinal axis should be bonded to the center of the test section. Sometimes only a single 45° gage is used. If added confidence is required, both sides of the gage section may be instrumented with strain gages.

A two-rail shear test is conducted by loading a test specimen in tension between the pairs of rails, as indicated in Figure 2.6a. It is important that

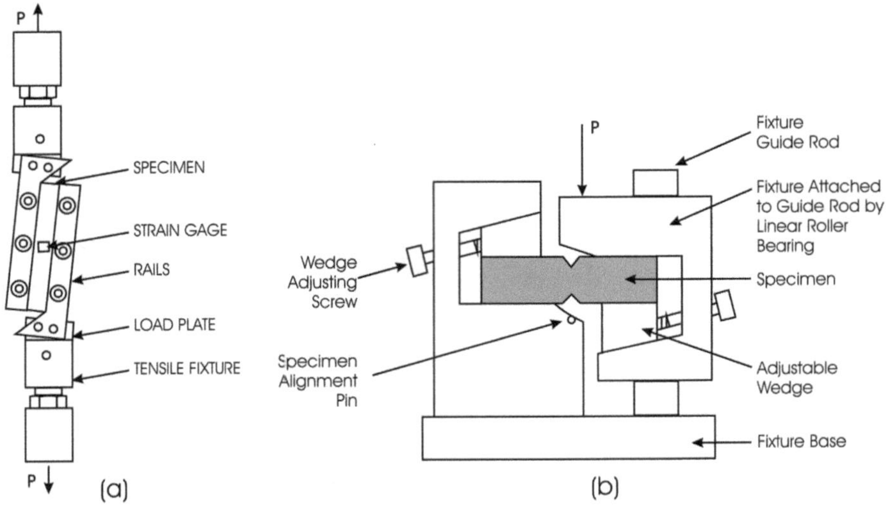

Figure 2.6 (a) Rail shear test, (b) Iosipescu shear test.

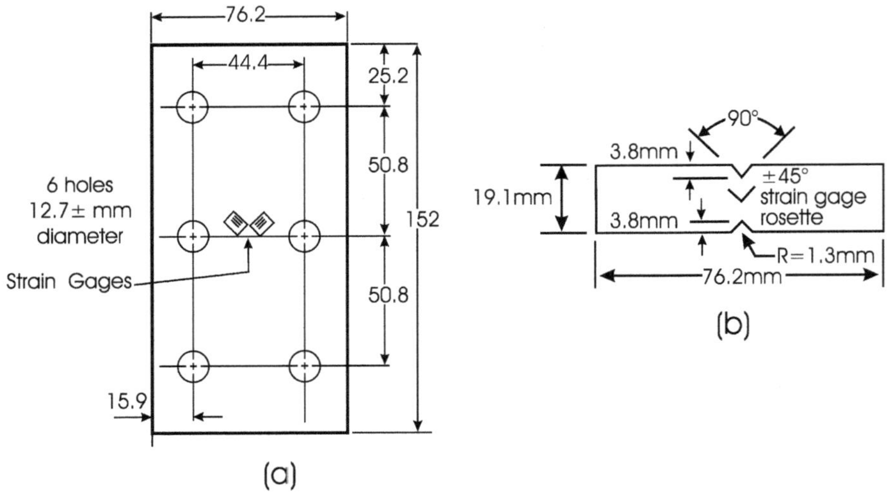

Figure 2.7 (a) Rail-shear specimen, (b) Iosipescu specimen. All dimensions in mm.

the rails do not slip during the test. If slipping occurs, the clamping bolts can bear against the clearance holes in the specimen, inducing local stress concentrations leading to premature failure. This, of course, results in an unacceptable test. For very high shear strength composites the clamping forces have to be very high to avoid slipping of the rails. A bolt torque of 100 Nm is specified in ASTM D4255 (2002), which is very high for a 9.5 mm diameter bolt. Commercially available fixtures have thermal-sprayed tungsten carbide particle gripping surfaces to improve the shear transfer of load.

The average shear stress is

$$\tau_{xy} = \frac{P}{A}, \tag{2.7}$$

where P is the applied load, and A is the area of the specimen cross-section parallel to the rails (length times thickness of the specimen). The shear strain, γ_{xy}, is determined from the ±45° strain gage readings according to Adams et al. (2003)

$$\gamma_{xy} = |\varepsilon(45^\circ)| + |\varepsilon(-45^\circ)| \tag{2.8}$$

or, if a single 45° gage is used,

$$\gamma_{xy} = 2|\varepsilon(45^\circ)|. \tag{2.9}$$

The specimen is loaded at a cross-head rate of 1–10 mm/min The shear stress-shear strain curve (τ_{xy} vs. γ_{xy}) may be constructed from load and strain data collected and reduced according to Equations (2.7)–(2.9). The in-plane shear modulus, G_{xy}, is determined from the slope of the initial linear region of the curve τ_{xy} vs. γ_{xy},

$$G_{xy} = \frac{\tau_{xy}}{\gamma_{xy}} \quad (\gamma_{xy} \ll 1). \tag{2.10}$$

Shear strength S_{xy} is the maximum shear stress value.

The Iosipescu test specimen shown in Figure 2.7b, is much smaller than the rail shear specimen (Figure 2.7a). The test fixture, shown in Figure 2.6b, allows a specimen of up to 12.7 mm in thickness, although a thickness of about 2.5 mm is more common. The top and bottom edges must be carefully machined to be flat, parallel to each other, and perpendicular to the faces of the specimen, to avoid out-of-plane bending and twisting when the load is applied. It is convenient to machine the V-shaped notches on a stack of specimen using a grinding wheel. The specimen should be instrumented with a ±45° strain gage rosette attached to one or both sides of the notched test region, as shown in Figure 2.7b. The dimensions of the specimen, including the length between the notches, should be measured.

The Iosipescu test fixture should be set up for compressive loading. The upper half of the test fixture should be attached to the moving cross-head of the testing machine. Failure should occur within 1–10 minutes, corresponding to a cross-head rate at about 1–10 mm/min.

The shear modulus and strength are determined as for the rail shear test discussed earlier. Notice that the shear stress, τ_{xy}, is defined as load divided by the cross-section area between the notches of the specimen.

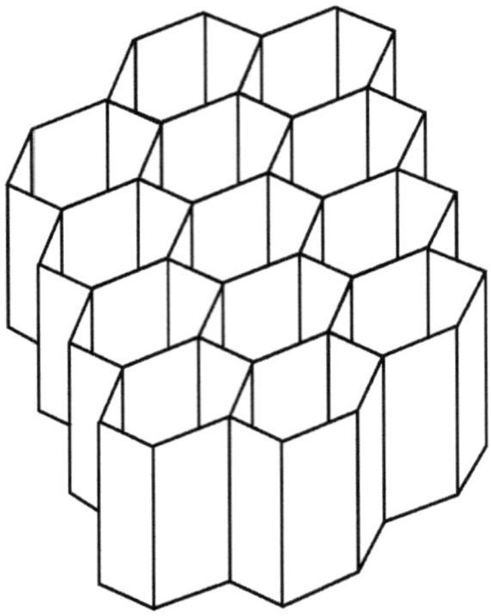

Figure 2.8 Representative element of core material.

2.2 Core Mechanical Properties

A volume of core material (Figure 2.8) large enough to represent the behavior of the core in a typical sandwich structure is selected. Testing of the element will provide average (homogenized) mechanical properties of the core. The properties of honeycomb and web cores, however, are very different with and without the support from the face sheets, and usually specimens from the core are adhesively bonded to a rigid test fixture to approximate the constraint from the faces. Notice, however, that if the faces are thin and of low bending stiffness, they may deform in bending under in-plane shear and bending loads, see, e.g., Libove and Hubka (1951) and Nordstrand et al. (1994). Such behavior substantially complicates the analysis of the response.

The actual size of the core element depends on the type of core and what kind of test that is to be conducted. For honeycomb and web cores (Figure 1.5) the size of core element should be quite large. For example, the out-of-plane tension test of a honeycomb core (to be discussed) requires at least 60 cells in the test specimen according to ASTM C297 (2004). Most foam cores, on the other hand, have quite small cells (order of mm), allowing the use of smaller test specimens.

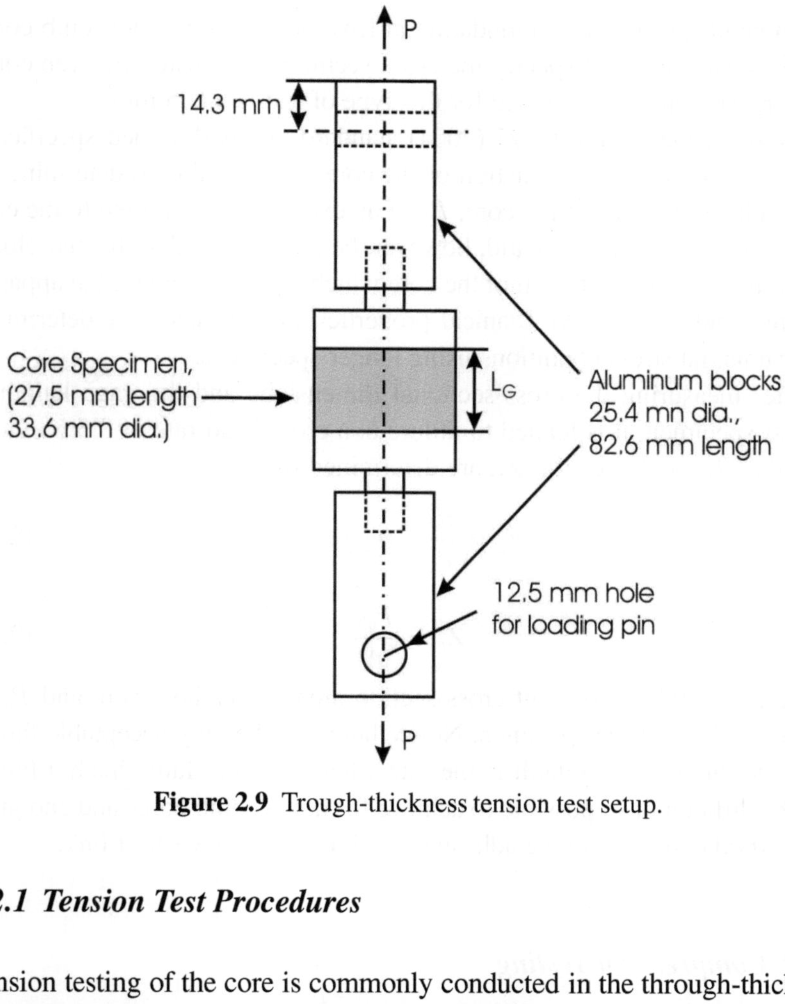

Figure 2.9 Trough-thickness tension test setup.

2.2.1 Tension Test Procedures

Tension testing of the core is commonly conducted in the through-thickness direction. ASTM C297 (2004) describes bonding of an element of the core (or sandwich) to metal loading blocks that are loaded in uniaxial tension until the specimen fails. Figure 2.9 shows the test configuration used by Viana and Carlsson (2002), for testing of PVC foam cores between adhesively bonded aluminum blocks of circular cross-sections. The loading fixture should be self-aligning, which may be accomplished by using a universal joint (Figure 2.9). The specimen gage length (L_G in Figure 2.9) is defined by the thickness of the foam panel. ASTM C297 (2004) specifies 40–50 mm thick end blocks with a square or circular cross-section in order to prevent non-uniform deformation (and stress concentrations) during loading. The minimum cross-sectional area is 625 mm^2 for foam and balsa wood cores, and depends on the cell size for honeycomb cores. The ASTM C297 standard recommends an area accommodating at least 60 cells for honeycomb cores. This standard does not specify the cross-sectional dimension for web cores. The largest area recommended for this type of test is 5,625 mm^2.

Although the ASTM C297 (2004) standard test is designed specifically for tensile strength determination of the core, it is possible to determine the out-of-plane modulus of the core, E_z, if an extensometer is fitted to the edge of the core specimen. It should, however, be pointed out that the end blocks constrain lateral determation of the core which tends to increase the apparent modulus and strength. Mechanical properties should ideally be determined under uniaxial stress conditions using longer specimens.

After measuring the cross-sectional dimensions and the gage length of the test specimen, it is loaded to failure at a cross-head rate of 0.5 mm/min. Modulus, E_z, and strength, Z_t, are determined from

$$E_z = \frac{\sigma_z}{\varepsilon_z} \quad (\varepsilon_z \ll 1), \tag{2.11}$$

$$Z_t = \frac{P_{\mathrm{ult}}}{A}, \tag{2.12}$$

where $\sigma_z = P/A$ (load/unit cross-section area), ε_z is the strain, and P_{ult} is the failure load of the specimen. Notice here that the only acceptable failure mode defining core strength is the one where the core fails. Such a failure may be difficult or impossible to achieve for honeycomb cores and end grain balsa wood cores where the adhesive bond may be the weakest link.

2.2.2 Compression Testing

The out-of-plane compression modulus and strength of the core are fundamental properties that govern wrinkling failure and resistance to localized loading. A compression test method consisting of loading a short specimen between parallel metal platens is outlined in ASTM Standard C365 (2003). Such a test may be performed on a core or sandwich specimen. The test specimen should be machined from a core or sandwich panel and should have a circular or square cross-section. The minimum cross-sectional area should be at least 625 mm^2 for foam and balsa cores. For web-cores, ASTM C365 does not provide guidance on the specimen size, but the instructions for honeycomb cores are helpful: for honeycomb cores with cell size below 6 mm, the cross-sectional area should be 2,500 mm^2. For cells greater than 6 mm, the area should be 5,800 mm^2. To avoid crushing of the ends of honeycomb

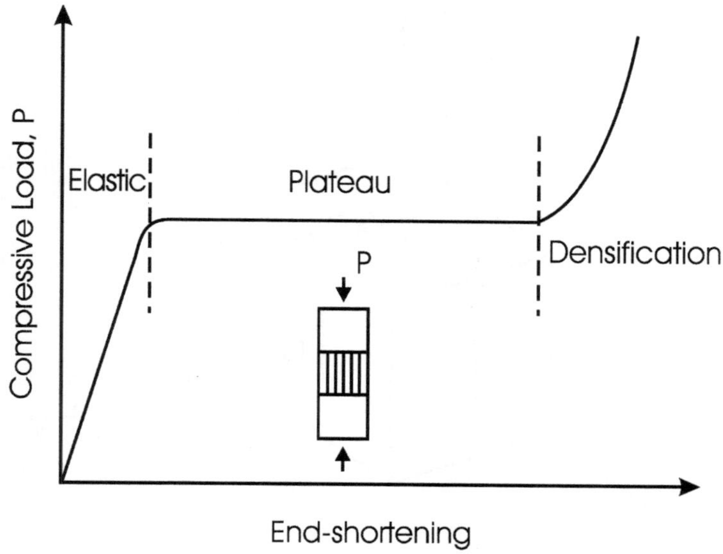

Figure 2.10 Stress-strain curve for compression loading of cellular materials.

cores, the ends may be reinforced with a layer of resin, or thin faces may be bonded to the core. With reinforced ends the test is called "stabilized compression test", while tests on unreinforced honeycomb core is called a "bare compression test".

The specimen should be tested between two parallel platens at 0.5 mm/min. The loading platens should be self-aligned through a spherical ball joint to promote a uniform distribution of compressive stress over the end surfaces. Deflection of the specimen during loading may be monitored using the cross-head travel or, more accurately, by an extensometer fitted to the top and bottom ends of the gage section of the specimen.

As discussed by Gibson and Ashby (1997), compression loading of cellular materials tends to produce a stress-strain curve of the type shown in Figure 2.10, where the stress reaches a constant plateau level after the initial linear region of response and a steeply increasing region at large strains.

The initial linear region corresponds to elastic extensional and bending deformation of the cell walls, while the plateau region occurs due to the formation of a band where localized buckling and collapse of the cell walls occurs. Upon further deformation, this band progresses through the gage region until all the empty space within the core is consumed and the cell walls begin to impinge on each other and take the direct load in compression. At this point (or before) the compression test should be stopped to protect the

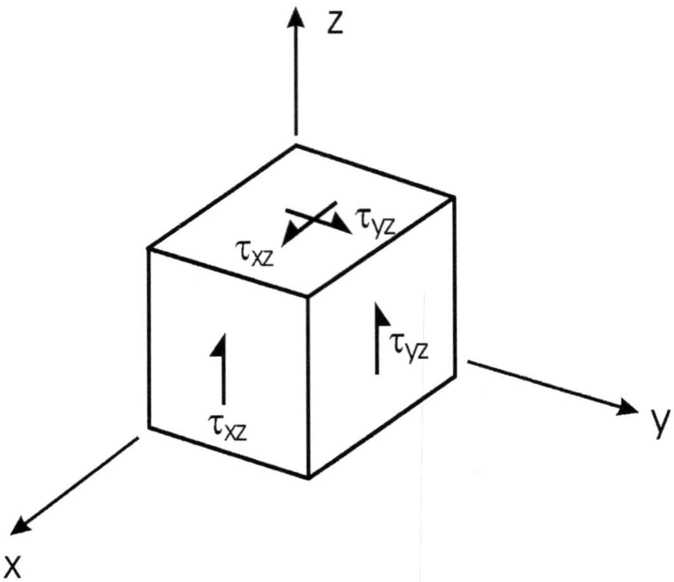

Figure 2.11 Core element under shear stresses τ_{xz} and τ_{yz}.

test equipment. The modulus and strength, E_z are Z_c, are reduced from the compressive stress strain curve (Figure 2.10).

2.2.3 Shear Testing

For core materials, the out-of-plane shear response is of major importance, see Section 1.2. The out-of-plane shear moduli G_{xz} and G_{yz} and shear strengths S_{xz} and S_{yz} are determined by subjecting a core element to shear stresses τ_{xz} and τ_{yz} as defined in Figure 2.11.

The ASTM standard C273 (2000), i.e. the plate shear test (Figure 2.12), considers a sandwich or core specimen adhesively bonded to two steel blocks that are loaded in tension to produce out-of-plane shear loading of the core.

The plate shear test does not produce a pure uniform state of shear in the entire test specimen. By using long specimens and adjusting the specimen length so that the line of load passes through the diagonally opposite corners of the specimen (see Figure 2.13), it is possible to minimize the influence of secondary stresses on the response. Notice that the line of load action is defined by the centers of the pins at A and B in Figure 2.13.

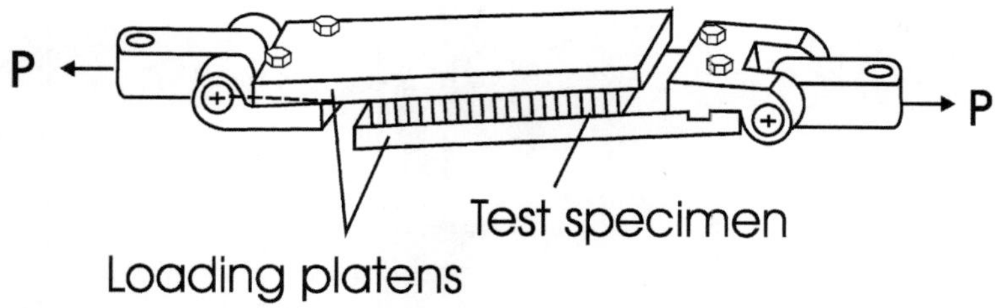

Figure 2.12 Plate shear specimen.

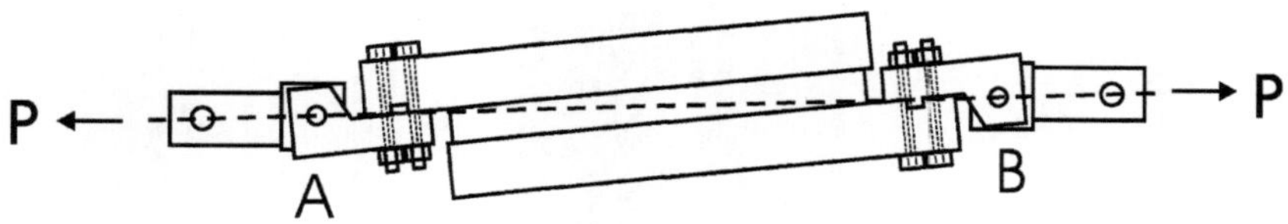

Figure 2.13 Specimen length is adjusted so that the line of load is passing through opposite corners of the shear specimen.

The test specimen should have a width of 50 mm or more, and a length of more than 12 times the thickness, while the loading line condition discussed above should be targeted as closely as possible. The dimensions of the test specimen (length, thickness, and width) should be measured before adhesively bonding the specimen to the loading platens. In order to determine shear strength of strong core materials, a very strong aerospace grade epoxy adhesive is required. Notice that a thick and flexible bond layer may deform in shear and add to the measured compliance. Such effects will reduce the apparent shear modulus of the core. Alignment of the specimen during bonding is important but may be difficult to achieve without proper care since the uncured, low viscosity adhesive is slippery.

After mounting the fixture assembly in the test frame, a fixture that supports a LVDT or an extensometer may be attached to each of the steel plates in order to allow direct measurement of the shear displacement during the test. Figure 2.14 shows a photograph of such an arrangement. The specimen should be loaded at a cross-head rate of 0.5 mm/min, while recording load vs. shear displacement until failure of the test specimen occurs. If the specimen fails at the specimen/steel plate interface, the test will not provide the shear strength of the core, while the shear modulus determined from the test at low stress and strain levels may still be valid unless the adhesive is very flexible.

The shear stress (τ_{xz} or τ_{yz}) is determined from

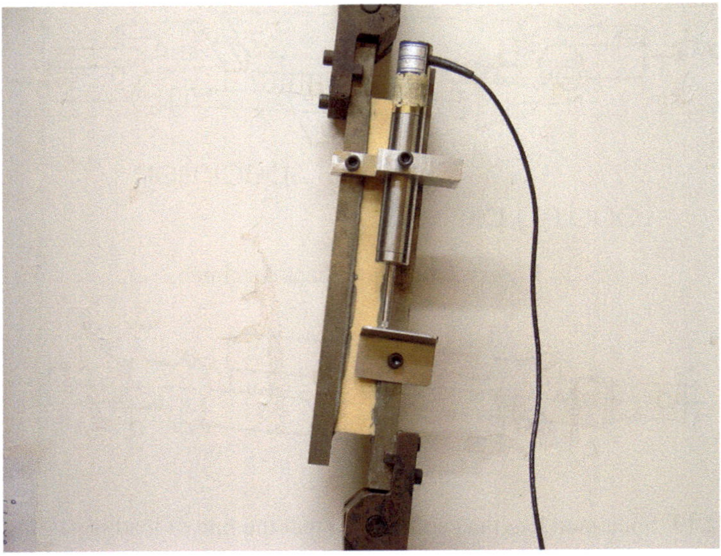

Figure 2.14 LVDT is attached to the plate shear test for measurement of shear displacement.

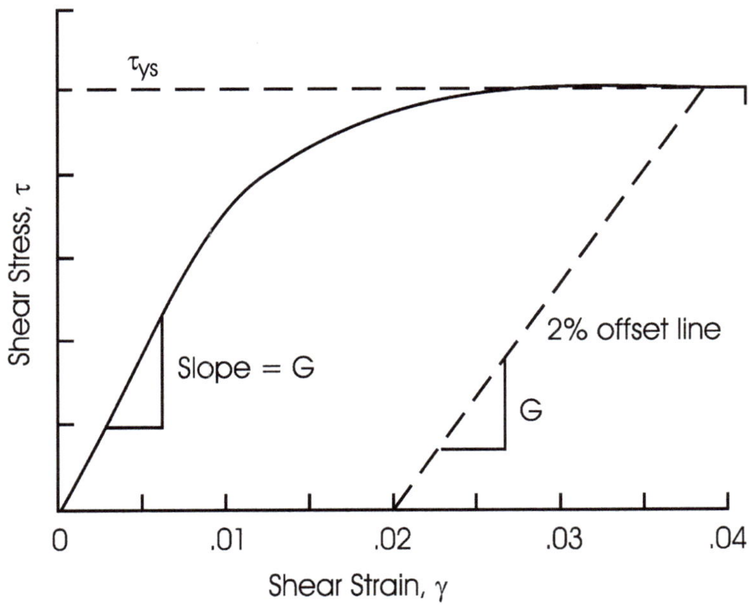

Figure 2.15 Schematic shear stress vs. shear strain curve for core material.

$$\tau = \frac{P}{Lb}, \tag{2.13}$$

where L and b are the length and width of the test specimen. The shear modulus, G_{xz} or G_{yz}, is determined from

$$G = \frac{\tau}{\gamma}, \tag{2.14}$$

where $\gamma = u/h_c$, with u being the shear displacement of the loading plates and h_c the core thickness. The shear modulus is determined using the initial linear-elastic region of response (see Figure 2.15). The shear strength is determined using the maximum load. Sometimes a yield strength in shear, τ_{ys}, is evaluated using a 2% offset shear strain (ASTM C273, 2000). Figure 2.15 illustrates schematically a shear stress vs. shear strain curve, and the determination of the yield stress.

Chapter 3
Classical and First-Order Shear Deformation Analysis of Sandwich Plates

This chapter will present classical laminated plate theory (CLPT) analysis of composite face sheets and sandwich plates. It is recognized that the transverse shear deformation is not incorporated in CLPT. Shear deformation of sandwich plates is important and first-order shear deformation analysis will be outlined. Applications of CLPT and first-order shear deformation analysis to sandwich panels will be presented. Two experimental sandwich plate tests, viz. bending under transverse pressure load and twisting, will be described. Experimental data generated from such tests will be compared to predictions from plate theory analysis and finite elements.

3.1 Classical Laminated Plate Theory Analysis

Classical laminated plate theory (CLPT) aims to relate the mechanical response of a layered plate to that of the individual constituent piles. This theory is an extension of the theory for homogeneous isotropic plates presented by Timoshenko and Woinowsky-Krieger (1959) to thin laminated plates. The analysis is most appropriate for thin plates since, as will be shown, this theory does not accommodate transverse shear deformation. Hence, CLPT is of limited applicability to sandwich panels since they often possess a thick, shear deformable core. CLPT, however, is applicable to the analysis of the in-plane response of face sheets and, furthermore, constitutes an important reference for sandwich panels with in-plane dimensions much greater than the thickness.

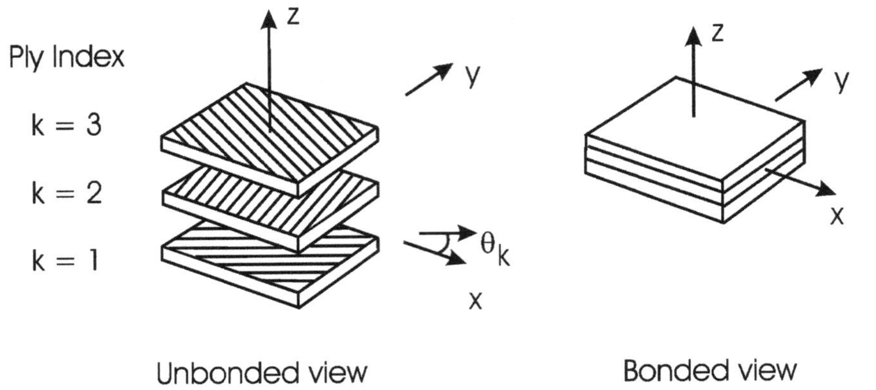

Figure 3.1 Unbonded and bonded views of a multi-ply laminate consisting of N plies.

3.1.1 Laminate Nomenclature

Figure 3.1 shows unbonded and bonded views of a laminate consisting of N orthotropic plies with the fibers oriented at any in-plane direction, θ.

The plies may be unidirectional (as shown) or fabric weaves. For fabric weaves, the angle θ refers to the warp or weft fiber directions. Figure 3.1 also illustrates the global xyz laminate coordinate system and the local (ply) coordinate system 123, where the 3 axis is parallel to the thickness coordinate (z) of the laminated plate. The plies in the laminate are numbered $1, 2, \ldots, N$ from bottom and up. The ply index, k, identifies the particular ply considered, and θ_k denotes the orientation of ply k.

For the purpose of subsequent analysis, the "ply coordinates", z_k, are defined in Figure 3.2. The origin of the z coordinate is located at the mid-plane of the laminate. Hence, $z_0 = -h/2$ and $z_N = h/2$, where h is the total thickness of the laminate (Figure 3.2). The ply coordinates indicate the location of the ply interfaces, and ply k is bound by z_{k-1} and z_k.

The lay-up sequence of a laminate is standardized, see Adams et al. (2003). The ply orientations in degrees are listed within brackets starting with the first ply laid up, followed by a slash (/) and then the next ply, and so on until the top ply. For symmetric laminates, only the bottom half of the plies are shown, and a subscript capital S follows the right closing bracket. For example, a six-ply symmetric laminate with plies oriented at 45°, 0°, −30°, −30°, 0° and 45° would be expressed as $[45/0/-30]_s$. For symmet-

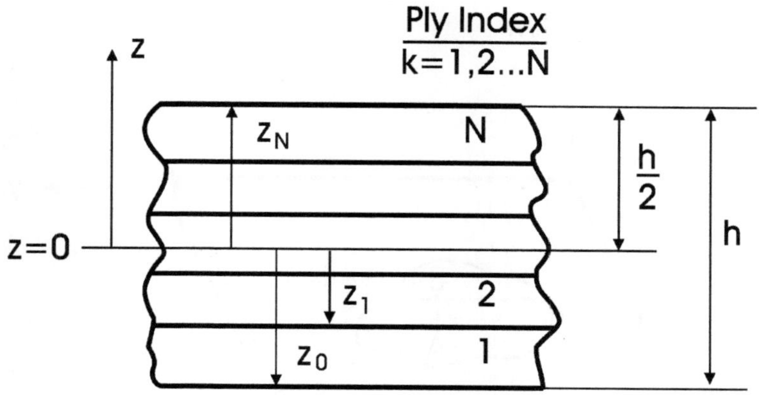

Figure 3.2 Definition of the ply coordinates, z_k.

ric laminates with an odd number of plies, the center ply is designated with an overbar.

3.1.2 Kinematics of Deformation

Figure 3.3 shows a flat composite laminate plate before loading and the xyz coordinate system. Deformation of the laminate plate is expressed using the displacement vector

$$\mathbf{u} = u\mathbf{i} + v\mathbf{j} + w\mathbf{k}, \tag{3.1}$$

where (u, v, w) are the components of the displacement vector (Figure 3.3) and **i**, **j**, **k** are the unit vectors for the x, y, z coordinates. From the displacement components, we can determine the extensional and shear strains,

$$\varepsilon_x = \frac{\partial u}{\partial x}, \tag{3.2a}$$

$$\varepsilon_y = \frac{\partial v}{\partial y}, \tag{3.2b}$$

$$\varepsilon_z = \frac{\partial w}{\partial z}, \tag{3.2c}$$

$$\gamma_{yz} = \frac{\partial v}{\partial z} + \frac{\partial w}{\partial y}, \tag{3.2d}$$

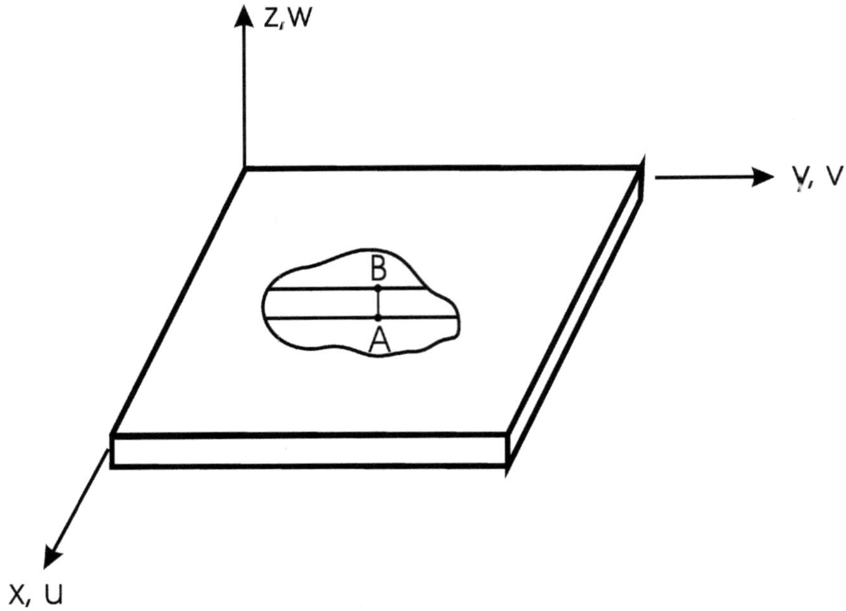

Figure 3.3 Composite laminate plate before deformation and definition of displacement components.

$$\gamma_{xz} = \frac{\partial u}{\partial z} + \frac{\partial w}{\partial x}, \tag{3.2e}$$

$$\gamma_{xy} = \frac{\partial u}{\partial y} + \frac{\partial v}{\partial x}. \tag{3.2f}$$

Notice here that the strains are assumed to be small ($\ll 1$), since Equations (3.2) include only first-order derivatives.

Figure 3.3 shows an element of the interior of the laminated plate cut parallel to the y axis. The deformation of this element will be examined in detail. Figure 3.4 illustrates the element before and after deformation.

When the laminate plate is loaded, the cross-section defined by the line AB, originally straight and perpendicular to the reference plane ($z = 0$) will translate and rotate as shown in Figure 3.4. The y and z axis displacements of the point O, on the mid-plane, are v_0 and w_0 (Figure 3.4). It is assumed that the line segment AB remains straight and normal to the deformed mid-plane. Further, the segment is assumed to maintain its length during deformation, implying

$$w(x, y) = w_0(x, y), \tag{3.3}$$

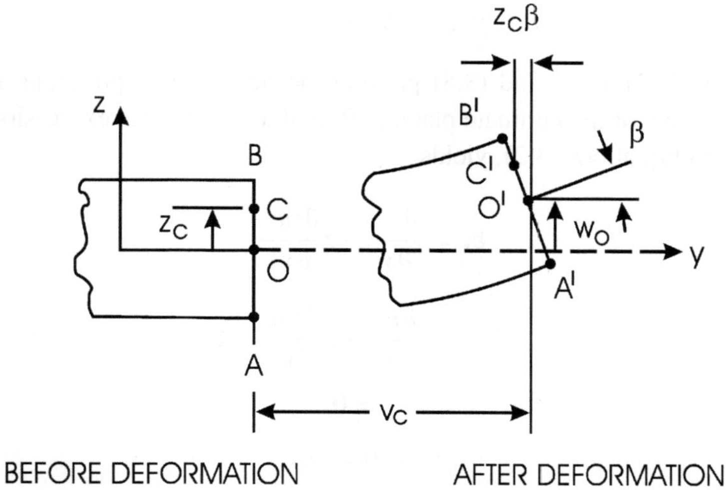

Figure 3.4 Element of plate cut parallel to the x axis before and after deformation.

where subscript zero refers to the mid-plane ($z = 0$). The assumptions constitutes the famous Kirchhoff hypothesis for plates, and implies that flat sections originally oriented normal to the mid-plane remain flat and normal to the deformed mid-plane after loading.

Further, because the line AB remains straight and perpendicular to the deformed mid-plane, the slope of the cross-section, β, in Figure 3.4, is equal to the slope of the panel, i.e.,

$$\beta = \frac{\partial w_0}{\partial y}. \tag{3.4}$$

The y axis displacement of point C, at a distance z_c from the mid-plane, becomes

$$v_c = v_0 - \beta z_c. \tag{3.5}$$

For any point on the line segment we will get

$$v = v_0 - \beta z. \tag{3.6}$$

Substitution of Equation (3.4) into (3.6) yields

$$v(x, y, z) = v_0(x, y) - z\frac{\partial w_0}{\partial y}. \tag{3.7}$$

Consideration of a cross-section of the plate cut parallel to the x axis similarly yields

$$u(x, y, z) = u_0(x, y) - z\frac{\partial w_0}{\partial x}. \tag{3.8}$$

Equations (3.3), (3.7) and (3.8) provide the necessary expressions for the displacements of the laminate plate. Differentiation of these expressions, according to Equations (3.2), yields

$$\varepsilon_x = \frac{\partial u_0}{\partial x} - z\frac{\partial^2 w_0}{\partial x^2}, \tag{3.9a}$$

$$\varepsilon_y = \frac{\partial v_0}{\partial y} - z\frac{\partial^2 w_0}{\partial y^2}, \tag{3.9b}$$

$$\varepsilon_z = 0, \tag{3.9c}$$

$$\gamma_{yz} = 0 = \gamma_{xz}, \tag{3.9d}$$

$$\gamma_{xy} = \frac{\partial u_0}{\partial y} + \frac{\partial v_0}{\partial x} - 2z\frac{\partial^2 w_0}{\partial x \partial y}. \tag{3.9e}$$

Hence, the CLPT does not accommodate transverse shear deformation and thickness stretch. The only non-zero strains are the in-plane strains ε_x, ε_y and γ_{xy}. These strains are commonly expressed in the following form

$$\begin{bmatrix} \varepsilon_x \\ \varepsilon_y \\ \gamma_{xy} \end{bmatrix} = \begin{bmatrix} \varepsilon_x^0 \\ \varepsilon_y^0 \\ \gamma_{xy}^0 \end{bmatrix} + z \begin{bmatrix} \kappa_x \\ \kappa_y \\ \kappa_{xy} \end{bmatrix}, \tag{3.10}$$

where $[\varepsilon_x^0, \varepsilon_y^0, \gamma_{xy}^0]$ and $[\kappa_x, \kappa_y, \kappa_{xy}]$ are the strains and curvatures of the reference mid-plane, defined according to Equations (3.9) as

$$\varepsilon_x^0 = \frac{\partial u_0}{\partial x}, \tag{3.11a}$$

$$\varepsilon_y^0 = \frac{\partial v_0}{\partial y}, \tag{3.11b}$$

$$\gamma_{xy}^0 = \frac{\partial u_0}{\partial y} + \frac{\partial v_0}{\partial x}, \tag{3.11c}$$

$$\kappa_x = -\frac{\partial^2 w_0}{\partial x^2}, \tag{3.12a}$$

$$\kappa_y = -\frac{\partial^2 w_0}{\partial y^2}, \tag{3.12b}$$

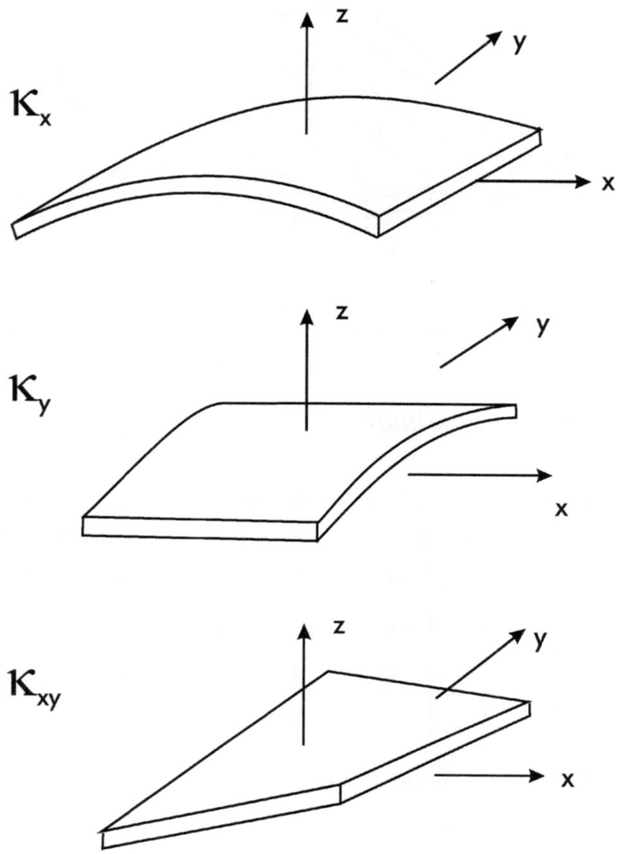

Figure 3.5 Illustration of bending and twisting deformation of a laminated plate. All curvatures shown are positive.

$$\kappa_{xy} = -2\frac{\partial^2 w_0}{\partial x \partial y}. \tag{3.12c}$$

The mid-plane strains signify membrane loading, since they represent extension and shear deformation of the mid-plane. The curvatures, κ_x and κ_y, represent bending deformation, while the curvature, κ_{xy}, represents twisting of the laminated plate, see Figure 3.5.

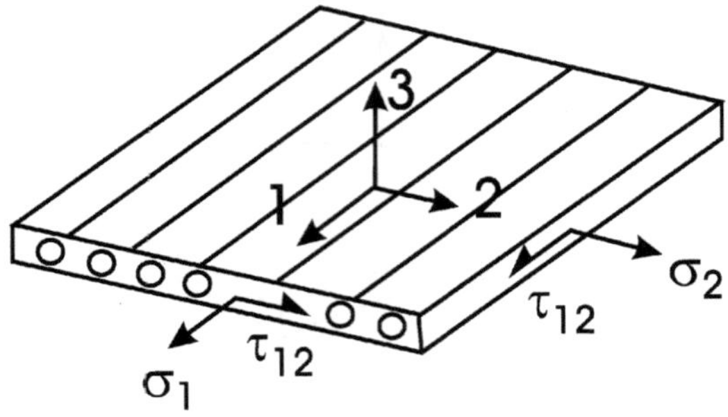

Figure 3.6 Orthotropic ply under in-plane stresses σ_1, σ_2 and τ_{12}.

3.1.3 Stresses in the Laminate

Consider a thin orthotropic ply with the principal material directions 1-2-3 loaded in the 1-2 plane of the ply as shown in Figure 3.6. The relation between stresses $[\sigma_1, \sigma_2, \tau_{12}]$ and strains $[\varepsilon_1, \varepsilon_2, \gamma_{12}]$ becomes (Hyer, 1998)

$$\begin{bmatrix} \sigma_1 \\ \sigma_2 \\ \tau_{12} \end{bmatrix} = \begin{bmatrix} Q_{11} & Q_{12} & 0 \\ Q_{12} & Q_{22} & 0 \\ 0 & 0 & Q_{66} \end{bmatrix} \begin{bmatrix} \varepsilon_1 \\ \varepsilon_2 \\ \gamma_{12} \end{bmatrix}, \tag{3.13}$$

where the stiffness, Q_{ij}, can be expressed

$$Q_{11} = E_1/(1 - \nu_{112}\nu_{21}), \tag{3.14a}$$

$$Q_{12} = \nu_{12}E_2/(1 - \nu_{12}\nu_{21}) = \nu_{21}E_1/(1 - \nu_{12}\nu_{21}), \tag{3.14b}$$

$$Q_{22} = E_2/(1 - \nu_{12}\nu_{21}), \tag{3.14c}$$

$$Q_{66} = G_{12}. \tag{3.14d}$$

Here, E_1 and E_2 represent the principal moduli in the fiber direction (1 in Figure 3.6) and the transverse direction (2 in Figure 3.6). ν_{12} and ν_{21} are the associated Poisson ratios, and G_{12} is the in-plane shear modulus.

For a ply within the laminate where the fibers are oriented at an angle, θ, to the global x coordinate of the laminate, Figure 3.7, the stresses, strains, and stiffnesses must be transformed to the new x–y axes. It may be shown that transformed relation between in-plane stresses and strains for the "off-axis" ply, Figure 3.7, will take the following form (Hyer, 1998):

$$\begin{bmatrix} \sigma_x \\ \sigma_y \\ \tau_{xy} \end{bmatrix}_k = \begin{bmatrix} \bar{Q}_{11} \bar{Q}_{12} \bar{Q}_{16} \\ \bar{Q}_{12} \bar{Q}_{22} \bar{Q}_{26} \\ \bar{Q}_{16} \bar{Q}_{26} \bar{Q}_{66} \end{bmatrix}_k \begin{bmatrix} \varepsilon_x \\ \varepsilon_y \\ \gamma_{xy} \end{bmatrix}, \tag{3.15}$$

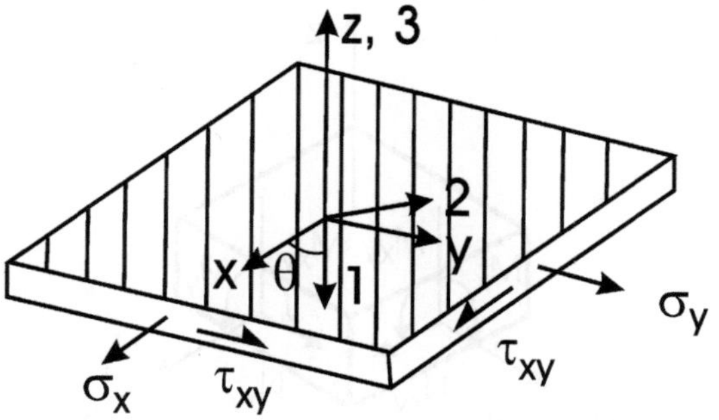

Figure 3.7 Off-axis ply element under in-plane loading.

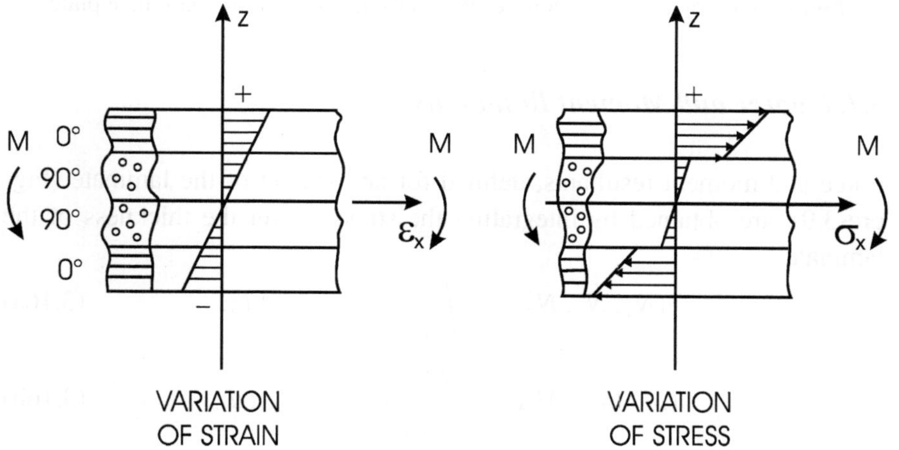

Figure 3.8 Variations of bending strain and stress in a $[0/90]_s$ laminate subject to pure bending loading.

where the elements, $\bar{Q}_{ij}$, of the stiffness matrix for ply k are defined in terms of the principal stiffnesses, Q_{ij}, and the ply orientation, θ_k, in Appendix A. The subscript k on the stresses signifies that the stresses may vary in a discontinuous manner from ply to ply, even if the strains vary in a continuous manner (Equations (3.10)).

Figure 3.8 illustrates the variations of strain, ε_x, and stress, σ_x, in a $[0/90]_s$ laminate under pure bending load. The stress, σ_x, is greater in the 0° plies than the 90° plies because the 0° plies have the fibers aligned with the longitudinal direction (x axis), while the 90° plies have the fibers aligned with the y axis. Typically, Young's modulus (E) is a factor 5–15 greater in the 0° direction than in the transverse direction.

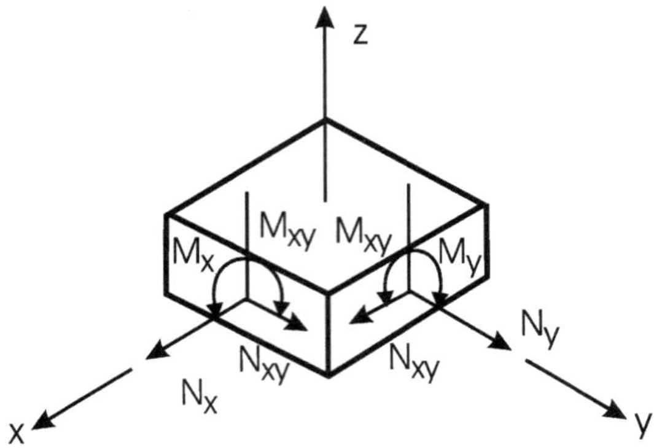

Figure 3.9 Force and moment resultants for an element of a laminate plate.

3.1.4 Force and Moment Resultants

Force and moment resultants, defined for an element of the laminate (Figure 3.9), are obtained by integrating the stresses over the thickness of the laminate

$$(N_x, N_y, N_{xy}) = \int_{-h/2}^{h/2} (\sigma_x, \sigma_y, \tau_{xy}) dz, \tag{3.16a}$$

$$(M_x, M_y, M_{xy}) = \int_{-h/2}^{h/2} (\sigma_x, \sigma_y, \tau_{xy}) z dz. \tag{3.16b}$$

Because the stresses vary in a continuous manner within a ply but may jump across the ply boundaries as shown in Figure 3.8, the integrations are conducted for each ply (k) defined by the ply coordinates z_{k-1} and z_k, and then the results are added.

$$\begin{bmatrix} N_x \\ N_y \\ N_{xy} \end{bmatrix} = \int_{-h/2}^{h/2} \begin{bmatrix} \sigma_x \\ \sigma_y \\ \tau_{xy} \end{bmatrix} dz = \sum_{k=1}^{N} \int_{z_{k-1}}^{z_k} \begin{bmatrix} \sigma_x \\ \sigma_y \\ \tau_{xy} \end{bmatrix}_k dz, \tag{3.17a}$$

$$\begin{bmatrix} M_x \\ M_y \\ M_{xy} \end{bmatrix} = \int_{-h/2}^{h/2} \begin{bmatrix} \sigma_x \\ \sigma_y \\ \tau_{xy} \end{bmatrix} z dz = \sum_{k=1}^{N} \int_{z_{k-1}}^{z_k} \begin{bmatrix} \sigma_x \\ \sigma_y \\ \tau_{xy} \end{bmatrix}_k z dz. \tag{3.17b}$$

The force and moment resultants have units of force per unit length and moment per unit length, and are generally dependent on the in-plane coordinates, x and y, but do not depend on the thickness coordinate (z) after integration.

Substitution of the in-plane stresses given by Equations (3.15) into (3.17) yields, after integration,

$$\begin{bmatrix} N_x \\ N_y \\ N_{xy} \end{bmatrix} = \begin{bmatrix} A_{11} A_{12} A_{16} \\ A_{12} A_{22} A_{26} \\ A_{16} A_{26} A_{66} \end{bmatrix} \begin{bmatrix} \varepsilon_x^0 \\ \varepsilon_y^0 \\ \gamma_{xy}^0 \end{bmatrix} + \begin{bmatrix} B_{11} B_{12} B_{16} \\ B_{12} B_{22} B_{26} \\ B_{16} B_{26} B_{66} \end{bmatrix} \begin{bmatrix} \kappa_x \\ \kappa_y \\ \kappa_{xy} \end{bmatrix}, \qquad (3.18)$$

$$\begin{bmatrix} M_x \\ M_y \\ M_{xy} \end{bmatrix} = \begin{bmatrix} B_{11} B_{12} B_{16} \\ B_{12} B_{22} B_{26} \\ B_{16} B_{26} B_{66} \end{bmatrix} \begin{bmatrix} \varepsilon_y^0 \\ \varepsilon_y^0 \\ \gamma_{xy}^0 \end{bmatrix} + \begin{bmatrix} D_{11} D_{12} D_{16} \\ D_{12} D_{22} D_{26} \\ D_{16} D_{26} D_{66} \end{bmatrix} \begin{bmatrix} \kappa_x \\ \kappa_y \\ \kappa_{xy} \end{bmatrix}. \qquad (3.19)$$

The $[A]$, $[B]$, and $[D]$ matrices in Equations (3.18) and (3.19) govern the response of a laminated plate to forces and moments. The $[A]$ matrix is called "extensional stiffness matrix", the $[B]$ matrix is called "coupling stiffness matrix", and the $[D]$ matrix is called "bending stiffness matrix" in accordance with their roles for the mechanical behavior of a laminated plate. The $[A]$ matrix relates extensional and shear strains to the force resultants, and the $[D]$ matrix relates bending and twisting curvatures to the moment resultants. The $[B]$ matrix appears both in the equations for the force resultants and moment resultants and acts to couple the responses in extension and bending.

The elements of the stiffness matrices are given by

$$A_{ij} = \sum_{k=1}^{N} (\bar{Q}_{ij})_k (z_k - z_{k-1}), \qquad (3.20a)$$

$$B_{ij} = \frac{1}{2} \sum_{k=1}^{N} (\bar{Q}_{ij})_k (z_k^2 - z_{k-1}^2), \qquad (3.20b)$$

$$D_{ij} = \frac{1}{3} \sum_{k=1}^{N} (\bar{Q}_{ij})_k (z_k^3 - z_{k-1}^3). \qquad (3.20c)$$

It may be shown that $B_{ij} = 0$ in Equations (3.18) and (3.19) for laminates with a symmetrical lay-up sequence. Hence, such laminates possess no coupling between the extensional and bending responses which greatly simplifies the analysis of the response. Face laminates are commonly laid-up in a symmetric manner. Furthermore, face laminates are usually "balanced", which means that for an off-axis ply with fiber orientation angle θ, there is a corresponding layer with orientation angle $-\theta$, which will have the consequence that $A_{16} = A_{26} = 0$ in Equations (3.18). These terms, if non-zero, signify coupling between extensional and shear response, which is undesirable.

3.1.5 Effective Engineering Elastic Constants of Laminates

When analyzing sandwich beams and panels, it is convenient to establish the effective engineering constants of the face laminates. Such constants are primarily the in-plane extensional and shear moduli, although sometimes the out-of-plane moduli are demanded. In this section we will present method to calculate the in-plane engineering constants based on the laminate extensional stiffness matrix.

Consider a symmetric and balanced laminated plate. According to the discussion in Section 3.1.4, the response to in-plane loading is given by

$$\begin{bmatrix} N_x \\ N_y \\ N_{xy} \end{bmatrix} = \begin{bmatrix} A_{11} & A_{12} & 0 \\ A_{12} & A_{22} & 0 \\ 0 & 0 & A_{66} \end{bmatrix} \begin{bmatrix} \varepsilon_x \\ \varepsilon_y \\ \gamma_{xy} \end{bmatrix}. \tag{3.21}$$

Notice here that the laminate strains coincide with the mid-plane strains due to the absence of bending curvatures (Equations (3.10)). For the purpose of establishing the effective engineering constants, it is more convenient to use the compliance (inverted) version of Equation (3.21), i.e.,

$$\begin{bmatrix} \varepsilon_x \\ \varepsilon_y \\ \gamma_{xy} \end{bmatrix} = \begin{bmatrix} a_{11} & a_{12} & 0 \\ a_{12} & a_{22} & 0 \\ 0 & 0 & a_{66} \end{bmatrix} \begin{bmatrix} N_x \\ N_y \\ N_{xy} \end{bmatrix}, \tag{3.22}$$

where

$$a_{11} = \frac{A_{22}}{A_{11}A_{22} - A_{12}^2}, \tag{3.23a}$$

$$a_{12} = \frac{-A_{12}}{A_{11}A_{22} - A_{12}^2}, \tag{3.23b}$$

$$a_{22} = \frac{A_{11}}{A_{11}A_{22} - A_{12}^2}, \tag{3.23c}$$

$$a_{66} = \frac{1}{A_{66}}. \tag{3.23d}$$

To further facilitate the deformation of the engineering constants, it is recognized that the average stresses $\bar{\sigma}_x$, $\bar{\sigma}_y$ and $\bar{\tau}_{xy}$ are given by

$$\bar{\sigma}_x = \frac{N_x}{h}, \tag{3.24a}$$

$$\bar{\sigma}_y = \frac{N_y}{h}, \tag{3.24b}$$

$$\bar{\tau}_{xy} = \frac{N_{xy}}{h}. \tag{3.24c}$$

With this, Equations (3.22) may be written as

$$\begin{bmatrix} \varepsilon_x \\ \varepsilon_y \\ \gamma_{xy} \end{bmatrix} = h \begin{bmatrix} a_{11} & a_{12} & 0 \\ a_{12} & a_{22} & 0 \\ 0 & 0 & a_{66} \end{bmatrix} \begin{bmatrix} \bar{\sigma}_x \\ \bar{\sigma}_y \\ \bar{\gamma}_{xy} \end{bmatrix}. \tag{3.25}$$

These equations may be compared to those for an orthotropic homogeneous material loaded in the principal system (Figure 3.6)

$$\begin{bmatrix} \varepsilon_1 \\ \varepsilon_2 \\ \gamma_{12} \end{bmatrix} = \begin{bmatrix} l/E_1 & -\nu_{12}/E_1 & 0 \\ -\nu_{21}/E_2 & 1/E_2 & 0 \\ 0 & 0 & 1/G_{12} \end{bmatrix} \begin{bmatrix} \sigma_1 \\ \sigma_2 \\ \tau_{12} \end{bmatrix}. \tag{3.26}$$

Direct comparison between Equations (3.25) and (3.26) yields the effective engineering constants of the laminated plate according to

$$E_x = \frac{1}{ha_{11}}, \tag{3.27a}$$

$$\nu_{xy} = \frac{-a_{12}}{a_{11}}, \tag{3.27b}$$

$$E_y = \frac{1}{ha_{22}}, \tag{3.27c}$$

$$\nu_{yx} = \frac{-a_{12}}{a_{22}}, \tag{3.27d}$$

$$G_{xy} = \frac{1}{ha_{66}}. \tag{3.27e}$$

Notice also that the Poisson ratios are not independent

$$\nu_{xy} = \nu_{xy}\frac{E_x}{E_y}, \tag{3.28}$$

which conforms to established orthotropic material behavior.

Equations (3.27) are very convenient for reducing a large set of ply mechanical properties and ply orientation angles into a set of four independent engineering constants.

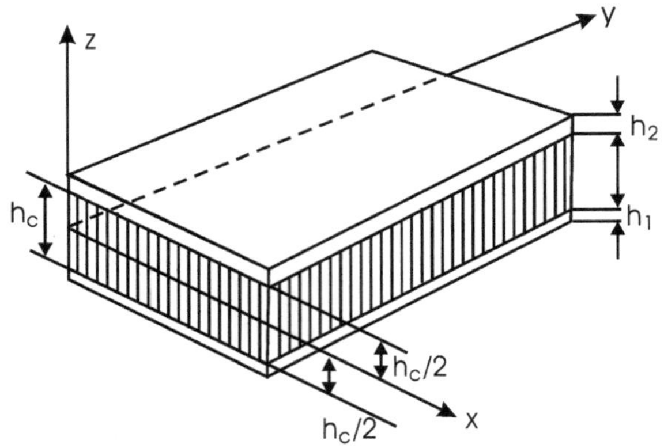

Figure 3.10 Definition of nomenclatures for sandwich plate.

For estimation of the effective out-of-plane stiffnesses E_z, G_{xz} and G_{yx} of the laminate several methods exists, see, e.g., Chou et al. (1972) and Bogetti et al. (2004). It should be pointed out that such methods are much more involved than the determination of the in-plane engineering constants. As a first estimate for a laminate consisting of transversely isotropic plies, it could be assumed that

$$E_z \cong E_3 = E_2, \tag{3.29a}$$

$$G_{xz} \cong G_{yz} \cong G_{13} = G_{12}. \tag{3.29b}$$

Such estimates are expected to be reasonable for laminates utilizing unidirectional transversely isotropic plies, where the properties in the out-of-plane direction should be close to those in the in-plane transverse direction.

3.2 First-Order Shear Deformation Analysis of a Sandwich Plate

Consider a sandwich plate consisting of face sheets of thicknesses h_1 and h_2, and a core of thickness h_c, see Figure 3.10. The faces and core may be isotropic or orthotropic with their principal directions along xyz, see Figure 3.10. The core may be corrugated (web), honeycomb, a foamed material, or balsa wood, (Figure 1.4). Such cores display macroscopic mechanical behavior that may be characterized as isotropic or orthotropic, i.e. having three mutually perpendicular planes of elastic symmetry (Hyer, 1998).

The early texts on sandwich structures forwarded by Plantema (1966) and Allen (1969), as well as the more recent text by Zenkert (1997), analyze the deflection, w, of sandwich panels using "partial deflections", i.e., deflections due to bending and shear separately, and obtain the total deflection by adding solutions for each mode of deformation. Analysis of sandwich plates and beams, however, is most conveniently performed extending the first-order shear deformation theory for homogeneous and isotropic plates developed by Reissner (1945) and Mindlin (1951) to sandwich plates with orthotropic face sheets. Such extensions were done by several authors, e.g. Libove and Batdorf (1948) and Whitney (1987). In this text we will forward the first-order shear deformation plate theory in a form similar to that presented for flat sandwich panels by Whitney (1987) who assumed that the in-plane displacements u and v of the faces are those at the face/core interfaces. Here we will modify the Whitney theory by assuming that the in-plane displacements of the faces are those at the centroids of the face sheets. This is consistent with traditional sandwich theory (Allen, 1969).

Figure 3.10 shows that the origin of the coordinate system xyz is placed at the center of the core, i.e., $z = 0$ in the core mid-plane. This is different from classical laminated plate theory where $z = 0$ in the geometrical mid-plane of the panel (Figure 3.10). The analysis is based on the following assumptions:

(i) The face sheets are thin compared to the core, i.e., $h_1, h_2 \ll h_c$ and in a state of plane stress ($\sigma_z = \tau_{xz} = \tau_{yz} = 0$).
(ii) The in-plane stresses, σ_x, σ_y, and τ_{xy}, in the core are negligible.
(iii) In-plane displacements, u and v, are uniform through the thickness of the face sheets and assume their mid-plane (centroidal) values.
(iv) The out-of-plane displacement, w, is independent of the z coordinate, i.e., the thickness strain, $\varepsilon_z = \partial w/\partial z = 0$.
(v) The in-plane displacements in the core, u and v, are linear in the thickness coordinate, z.

Based on the assumptions (iv) and (v), the displacements of the core are

$$u = u_0(x, y) + z\psi_x(x, y), \tag{3.30a}$$

$$v = v_0(x, y) + z\psi_y(x, y), \tag{3.30b}$$

$$w = w_0(x, y), \tag{3.30c}$$

where u_0, v_0 and w_0 are the displacements at the core mid-plane, see Figure 3.11, and ψ_x and ψ_y are the rotations of cross-sections originally perpendicular to the x and y axes, respectively. From continuity of displacements at the face/core interfaces ($z = \pm h_c/2$), and assumption (iii), the displacements of the bottom and top face sheets become (Figure 3.11)

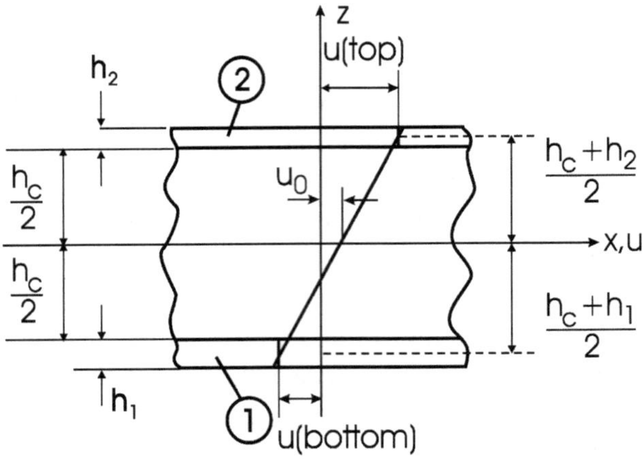

Figure 3.11 Illustration of displacement u for sandwich element oriented along the x axis.

$$u(\text{bottom}) = u_0 - \frac{(h_c + h_1)}{2}\psi_x, \tag{3.31a}$$

$$u(\text{top}) = u_0 + \frac{(h_c + h_2)}{2}\psi_x, \tag{3.31b}$$

$$v(\text{bottom}) = v_0 - \frac{(h_c + h_1)}{2}\psi_y, \tag{3.31c}$$

$$v(\text{top}) = v_0 + \frac{(h_c + h_2)}{2}\psi_y, \tag{3.31d}$$

$$w = w_0, \tag{3.31e}$$

where "bottom" and "top" refer to the lower and upper face sheets, respectively.

Figure 3.12 shows a section of a sandwich plate cut in the x–z plane. The core element ACFD represents a section of the core with the surfaces AC and DF perpendicular to the x axis before deformation. After deformation, point B displaces to assume a position at B′, and the vertical upwards displacement of point B (originally at $z = 0$) is w, and that of the adjacent point E assuming the new position E′ is $w + (\partial w/\partial w)dx$, where $\partial w/\partial x$ is the slope of the panel along the x axis.

Similar to classical plate theory for homogeneous and isotropic materials (Timoshenko and Woinowsky-Krieger, 1959), and classical laminated plate theory (see above), first-order shear deformation theory assumes that plane sections of the core, originally perpendicular to the plane of the sandwich panel remain plane after deformation. According to this theory, however,

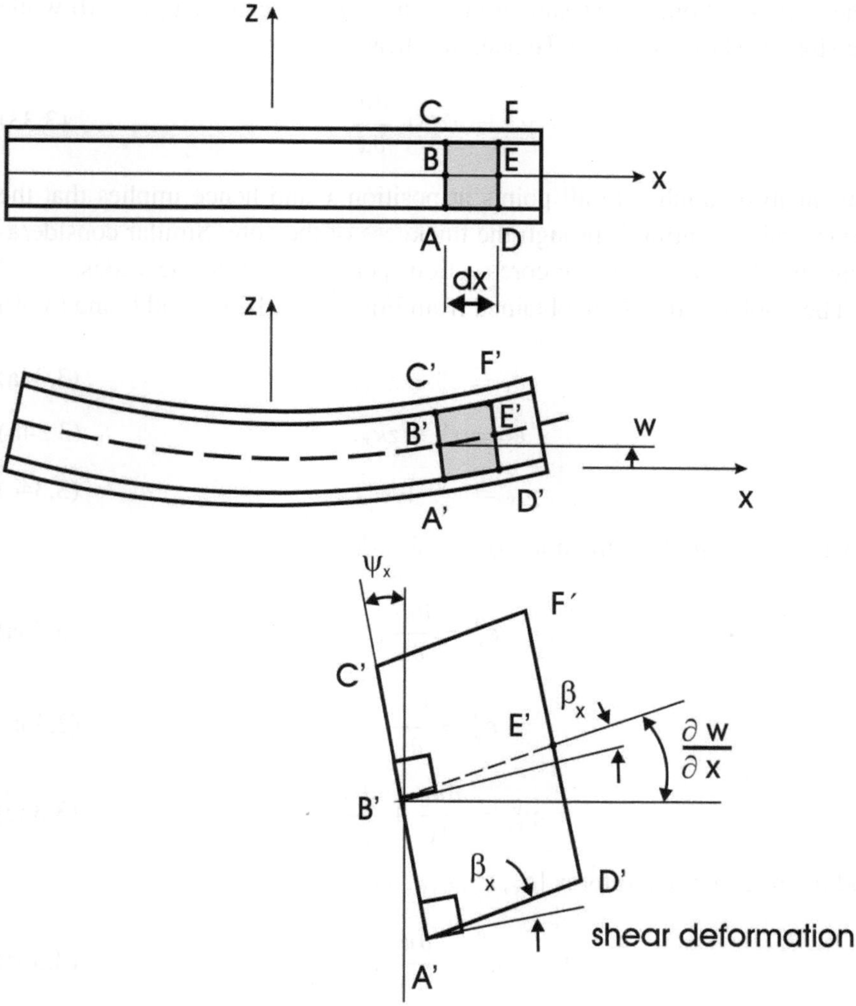

Figure 3.12 Deformation of core element in the x–z plane.

the cross-sections may not necessarily remain perpendicular to the deformed middle surface of the core, as shown in Figure 3.12. The slope of the middle surface, $\partial w/\partial x$ differs from the magnitude of the rotation of the cross-section, $|\psi_x|$, and the difference (β_x) constitutes the shear deformation. From Figure 3.12, it is recognized that

$$\beta_x = \frac{\partial w}{\partial x} - |\psi_x|, \tag{3.32}$$

where the rotation, ψ_x, as shown in Figure 3.12 is negative ($\psi_x < 0$) while the slope is shown positive. Hence, the shear strain is

$$\gamma_{xy} = \psi_x + \frac{\partial w}{\partial x}. \tag{3.33}$$

This analysis applies to all points at position x and hence implies that the shear strain is uniform through the thickness of the core. Similar considerations may be formulated for core sections perpendicular to the y axis.

The in-plane strains are obtained from Equations (3.2a, b and f) and (3.30)

$$\varepsilon_x = \varepsilon_x^0 + z\kappa_x, \tag{3.34a}$$

$$\varepsilon_y = \varepsilon_y^0 + z\kappa_x, \tag{3.34b}$$

$$\gamma_{xy} = \gamma_{xy}^0 + z\kappa_{xy}, \tag{3.34c}$$

where $[\varepsilon_x^0, \varepsilon_y^0, \gamma_{xy}^0]$ are the mid-core strains defined by

$$\varepsilon_x^0 = \frac{\partial u_0}{\partial x}, \tag{3.35a}$$

$$\varepsilon_y^0 = \frac{\partial v_0}{\partial y}, \tag{3.35b}$$

$$\gamma_{xy}^0 = \frac{\partial u_0}{\partial y} + \frac{\partial v_0}{\partial x} \tag{3.35c}$$

and the mid-core curvatures $[\kappa_x, \kappa_y, \kappa_{xy}]$ are

$$\kappa_x = \frac{\partial \psi_x}{\partial x}, \tag{3.36a}$$

$$\kappa_y = \frac{\partial \psi_y}{\partial y}, \tag{3.36b}$$

$$\kappa_{xy} = \frac{\partial \psi_x}{\partial y} + \frac{\partial \psi_y}{\partial x}. \tag{3.36c}$$

The out-of-plane shear strains, γ_{yz} and γ_{xz}, are defined in Equations (3.2d, e) which, combined with Equations (3.30), yield

$$\gamma_{xz} = \psi_x + \frac{\partial w}{\partial x}, \tag{3.37a}$$

$$\gamma_{yz} = \psi_y + \frac{\partial w}{\partial y}. \tag{3.37b}$$

The out-of-plane extensional strain in the core, ε_z, vanishes according to assumption (iv).

Labeling the bottom and top face sheets by (1) and (2), their in-plane strains become (Equations (3.36)),

Bottom

$$\varepsilon_x(1) = \varepsilon_x^0 - \frac{(h_c + h_1)}{2}\kappa_x, \tag{3.38a}$$

$$\varepsilon_y(1) = \varepsilon_y^0 - \frac{(h_c + h_1)}{2}\kappa_y, \tag{3.38b}$$

$$\gamma_{xy}(1) = \gamma_{xy}^0 - \frac{(h_c + h_1)}{2}\kappa_{xy}. \tag{3.38c}$$

Top

$$\varepsilon_x(2) = \varepsilon_x^0 + \frac{(h_c + h_2)}{2}\kappa_x, \tag{3.38d}$$

$$\varepsilon_y(2) = \varepsilon_y^0 + \frac{(h_c + h_2)}{2}\kappa_y, \tag{3.38e}$$

$$\gamma_{xy}(2) = \gamma_{xy}^0 + \frac{(h_c + h_2)}{2}\kappa_{xy}. \tag{3.38f}$$

The force and moment resultants for a sandwich element (Figure 3.13) are obtained by integrating the stresses over the element thickness,

$$(N_x, N_y, N_{xy}) = \int_{-(h_c/2+h_1)}^{-h_c/2} (\sigma_x(1), \sigma_y(1), \tau_{xy}(1))dz + \int_{h_c/2}^{h_c/2+h_2} (\sigma_x(2), \sigma_y(2), \tau_{xy}(2))dz, \tag{3.39a}$$

$$(M_x, M_y, M_{xy}) = \int_{-(h_c/2+h_1)}^{-h_c/2} (\sigma_x(1), \sigma_y(1), \tau_{xy}(1))z dz + \int_{h_c/2}^{h_c/2+h_2} (\sigma_x(2), \sigma_y(2), \tau_{xy}(2))z dz, \tag{3.39b}$$

$$(Q_x, Q_y) = \int_{-h_c/2}^{h_c/2} (\tau_{xz}, \tau_{yz})dz. \tag{3.39c}$$

Notice that the in-plane normal and in-plane shear stresses in the core are neglected by virtue of assumption (ii).

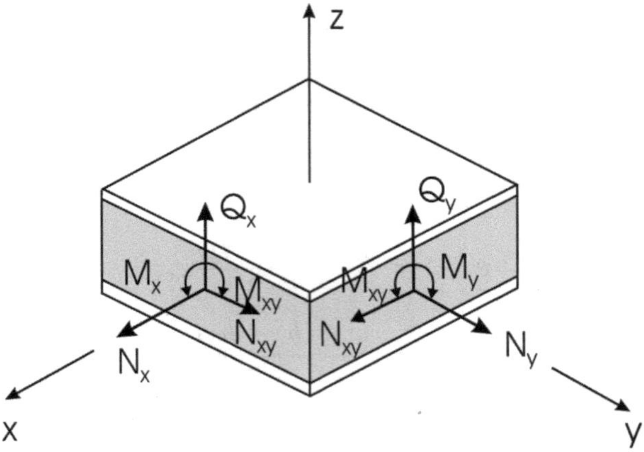

Figure 3.13 Force and moment resultants for a sandwich element.

The in-plane stresses in each ply (k) of the face sheets are given by Equation (3.15)

$$\begin{bmatrix} \sigma_x(i) \\ \sigma_y(i) \\ \tau_{xy}(i) \end{bmatrix}_k = \begin{bmatrix} \bar{Q}_{11} & \bar{Q}_{12} & \bar{Q}_{16} \\ \bar{Q}_{12} & \bar{Q}_{22} & \bar{Q}_{26} \\ \bar{Q}_{16} & \bar{Q}_{26} & \bar{Q}_{66} \end{bmatrix}_k \begin{bmatrix} \varepsilon_x(i) \\ \varepsilon_y(i) \\ \gamma_{xy}(i) \end{bmatrix}, \tag{3.40}$$

where k is the ply index (Figure 3.1) and $i = 1$ for the lower face, and $i = 2$ for the upper face. The matrix in Equation (3.40) is the transformed plane stress stiffness matrix as defined in Appendix A.

Substitution of the in-plane stresses given by Equations (3.40) into the expressions for the force and moment resultants (3.39a, b) yields

$$\begin{bmatrix} N_x \\ N_y \\ N_{xy} \end{bmatrix} = \begin{bmatrix} A_{11} & A_{12} & A_{16} \\ A_{12} & A_{22} & A_{26} \\ A_{16} & A_{26} & A_{66} \end{bmatrix} \begin{bmatrix} \varepsilon_x^0 \\ \varepsilon_y^0 \\ \gamma_{xy}^0 \end{bmatrix} + \begin{bmatrix} B_{11} & B_{12} & B_{16} \\ B_{12} & B_{22} & B_{26} \\ B_{16} & B_{26} & B_{66} \end{bmatrix} \begin{bmatrix} \kappa_x \\ \kappa_y \\ \kappa_{xy} \end{bmatrix}, \tag{3.41}$$

$$\begin{bmatrix} M_x \\ M_y \\ M_{xy} \end{bmatrix} = \begin{bmatrix} C_{11} & C_{12} & C_{16} \\ C_{12} & C_{22} & C_{26} \\ C_{16} & C_{26} & C_{66} \end{bmatrix} \begin{bmatrix} \varepsilon_x^0 \\ \varepsilon_y^0 \\ \gamma_{xy}^0 \end{bmatrix} + \begin{bmatrix} D_{11} & D_{12} & D_{16} \\ D_{12} & D_{22} & D_{26} \\ D_{16} & D_{26} & D_{66} \end{bmatrix} \begin{bmatrix} \kappa_x \\ \kappa_y \\ \kappa_{xy} \end{bmatrix}, \tag{3.42}$$

where

$$A_{ij} = A_{ij}(1) + A_{ij}(2), \tag{3.43a}$$

$$B_{ij} = \frac{(h_c + h_2)}{2} A_{ij}(2) - \frac{(h_c + h_1)}{2} A_{ij}(1), \tag{3.43b}$$

$$C_{ij} = C_{ij}(1) + C_{ij}(2), \tag{3.43c}$$

$$D_{ij} = \frac{(h_c + h_2)}{2} C_{ij}(2) - \frac{(h_c + h_1)}{2} C_{ij}(1), \tag{3.43d}$$

with

$$A_{ij}(1), C_{ij}(1) = \int_{-(h_c/2+h_1)}^{-h_c/2} \bar{Q}_{ij}(1, z) dz \quad \text{(bottom face)}, \tag{3.44a}$$

$$A_{ij}(2), C_{ij}(2) = \int_{h_c/2}^{h_c/2+h_2} \bar{Q}_{ij}(1, z) dz \quad \text{(top face)}. \tag{3.44b}$$

It may be shown that a symmetric sandwich plate, i.e., where the top and bottom faces are laid-up such that the mid-plane ($z = 0$) of the sandwich panel is a symmetry (mirror) plane; for example, the bottom face (#1) is a [0/90] laminate and the top face (#2) is a [90/0] laminate, fulfills the following equations:

$$A_{ij}(1) = A_{ij}(2), \tag{3.45a}$$

$$C_{ij}(1) = -C_{ij}(2). \tag{3.45b}$$

As a result, Equations (3.45) in (3.43) give

$$A_{ij} = 2A_{ij}(2), \tag{3.46a}$$

$$B_{ij} = 0 = C_{ij}, \tag{3.46b}$$

$$D_{ij} = (h_c + h_f) C_{ij}(2), \tag{3.46c}$$

where $h_f = h_1 = h_2$, is the face thickness.

Consequently, for such a panel the constitutive Equations (3.41) and (3.42) uncouple and become

$$\begin{bmatrix} N_x \\ N_y \\ N_{xy} \end{bmatrix} = \begin{bmatrix} A_{11} & A_{12} & A_{16} \\ A_{12} & A_{22} & A_{26} \\ A_{16} & A_{26} & A_{66} \end{bmatrix} \begin{bmatrix} \varepsilon_x^0 \\ \varepsilon_y^0 \\ \gamma_{xy}^0 \end{bmatrix}, \tag{3.47}$$

$$\begin{bmatrix} M_x \\ M_y \\ M_{xy} \end{bmatrix} = \begin{bmatrix} D_{11} & D_{12} & D_{16} \\ D_{12} & D_{22} & D_{26} \\ D_{16} & D_{26} & D_{66} \end{bmatrix} \begin{bmatrix} \kappa_x \\ \kappa_y \\ \kappa_{xy} \end{bmatrix}. \tag{3.48}$$

Another simplification of the analysis is obtained by treating the face sheets as homogeneous materials with average stiffnesses

$$\bar{Q}_{ij}(1) = A_{ij}(1)/h_1, \tag{3.49a}$$

$$\bar{Q}_{ij}(2) = A_{ij}(2)/h_2. \tag{3.49b}$$

For this case, the integrals in Equations (3.44) become

$$C_{ij}(1) = \frac{-A_{ij}(1)}{2}(h_c + h_1), \tag{3.50a}$$

$$C_{ij}(2) = \frac{A_{ij}(2)}{2}(h_c + h_2). \tag{3.50b}$$

Substitution in Equations (3.43) gives

$$A_{ij} = A_{ij}(1) + A_{ij}(2), \tag{3.51a}$$

$$B_{ij} = \frac{(h_c + h_2)}{2}A_{ij}(2) - \frac{(h_c + h_1)}{2}A_{ij}(1), \tag{3.51b}$$

$$C_{ij} = B_{ij}, \tag{3.51c}$$

$$D_{ij} = \left(\frac{h_c + h_1}{2}\right)^2 A_{ij}(2) + \left(\frac{h_c + h_2}{2}\right)^2 A_{ij}(2). \tag{3.51d}$$

For a symmetric sandwich with homogeneous faces: $A_{ij}(1) = A_{ij}(2) = \bar{Q}_{ij}(f)h_f$, and $h_1 = h_2 = h_f$, which leads to

$$A_{ij} = 2A_{ij}(2) = 2\bar{Q}_{ij}(f)h_f, \tag{3.52a}$$

$$B_{ij} = C_{ij} = 0, \tag{3.52b}$$

$$D_{ij} = 2\left(\frac{h_c + h_f}{2}\right)^2 \bar{Q}_{ij}(f)h_f. \tag{3.52c}$$

The expression for the flexural stiffness, Equation (3.52c), is the plate equivalent to the beam equation (1.8) for a sandwich beam with thin faces.

For calculation of the transverse shear resultants, Q_x and Q_y, defined in Equation (3.39c) and shown in Figure 3.13, the shear stresses in the core, τ_{xz} and τ_{yz}, are given in terms of the shear strains, γ_{xz} and γ_{yz} by

$$\tau_{xz} = G_{xz}\gamma_{xz}, \tag{3.53a}$$

$$\tau_{yz} = G_{yz}\gamma_{yz}, \tag{3.53b}$$

where G_{xz} and G_{yz} are the transverse shear moduli of the core.

In shear deformation theory for homogeneous plates, it is customary to use a correction factor (k), or correction factors (k_1 and k_2) for the transverse shear strains as introduced by Reissner (1945), Mindlin (1951), and Chow (1971), although a single factor is most commonly used for both transverse shear strains. The factor k is introduced as a multiplicative parameter in the constitutive relations between transverse shear forces and transverse shear strains (see, e.g., Whitney, 1987). The need for a correction factor in first-order theory for homogeneous plates originates from the fact that the transverse shear strains and shear stresses are uniform through the thickness instead of the classical parabolic shear stress distribution with zero shear stresses on the surfaces of the plate. The correction factor k is determined from exact solutions for the shear stresses at the center of the plate in terms of the transverse shear forces or from the total strain energy due to transverse shear forces (Whitney, 1987). Whitney (1972) determined two correction factors (k_1 and k_2) for the transverse shear strains in a sandwich panel by fitting the first-order shear results to an exact solution by Pagano (1970a). Most commonly, however, shear correction factors are not used for sandwich panels since the core shear stress indeed is fairly constant throughout the thickness of the core (see, e.g., Section 2.2). Furthermore, the faces are assumed free from shear stresses (assumption (i)). With no shear correction factor, integration of the (constant) shear stress given by Equations (3.37) into (3.53) over the core thickness (Equation (3.39c)) yields

$$Q_x = h_c G_{xz}\left(\psi_x + \frac{\partial w}{\partial x}\right), \tag{3.54a}$$

$$Q_y = h_c G_{yz}\left(\psi_y + \frac{\partial w}{\partial y}\right). \tag{3.54b}$$

3.2.1 Alternative Form of the Constitutive Equations for a Sandwich Plate Element

The force and moment resultants given by Equations (3.41) and (3.42) may be written in compressed form as

$$[N] = [N][\varepsilon^0] + [B][\kappa], \tag{3.55a}$$

$$[M] = [C][\varepsilon^0] + [D][\kappa], \tag{3.55b}$$

where $[N]$ and $[M]$ represent the 3×1 force and moment resultants, $[A]$, $[B]$, $[C]$ and $[D]$ are the 3×3 elastic stiffness matrices, and $[\varepsilon^0]$ and $[\kappa]$ are the

3×1 core mid-plane strains and curvatures, respectively. Equations (3.55) may also be written as

$$\begin{bmatrix} N \\ M \end{bmatrix} = \begin{bmatrix} A & B \\ C & D \end{bmatrix} \begin{bmatrix} \varepsilon^0 \\ \kappa \end{bmatrix}, \tag{3.56}$$

where the $ABCD$ matrix is of dimension 6×6.

Equations (3.56) are convenient when force and moment resultants are to be determined from known mid-core strains and curvatures. Often it is desirable to express the core mid-core strains and curvatures in terms of force and moment resultants, and this is achieved by inversion of the 6×6 $ABCD$ matrix in Equations (3.56)

$$\begin{bmatrix} \varepsilon^0 \\ \kappa \end{bmatrix} = \begin{bmatrix} a & b \\ c & d \end{bmatrix} \begin{bmatrix} N \\ M \end{bmatrix}, \tag{3.57}$$

where expressions for the 3×3 compliance matrices $[a]$, $[b]$, $[c]$ and $[d]$ in terms of $[A]$, $[B]$, $[C]$ and $[D]$ are provided in Appendix B.

3.2.2 Equilibrium Equations

The equilibrium equations for force and moment resultants are presented by Whitney (1987). These equations are

$$\frac{\partial N_x}{\partial x} + \frac{\partial N_{xy}}{\partial y} = 0, \tag{3.58a}$$

$$\frac{\partial N_{xy}}{\partial x} + \frac{\partial N_y}{\partial y} = 0, \tag{3.58b}$$

$$\frac{\partial M_x}{\partial x} + \frac{\partial M_{xy}}{\partial y} - Q_x = 0, \tag{3.58c}$$

$$\frac{\partial M_{xy}}{\partial x} + \frac{\partial M_y}{\partial y} - Q_y = 0, \tag{3.58d}$$

$$\frac{\partial Q_x}{\partial x} + \frac{\partial Q_y}{\partial y} + N_x \frac{\partial^2 w}{\partial x^2} + 2N_{xy} \frac{\partial^2 w}{\partial x \partial y} + N_y \frac{\partial^2 w}{\partial y^2} + q = 0, \tag{3.58e}$$

with

$$q = \sigma_z(h/2) - \sigma_z(-h/2), \tag{3.58f}$$

where h is the thickness of the sandwich plate.

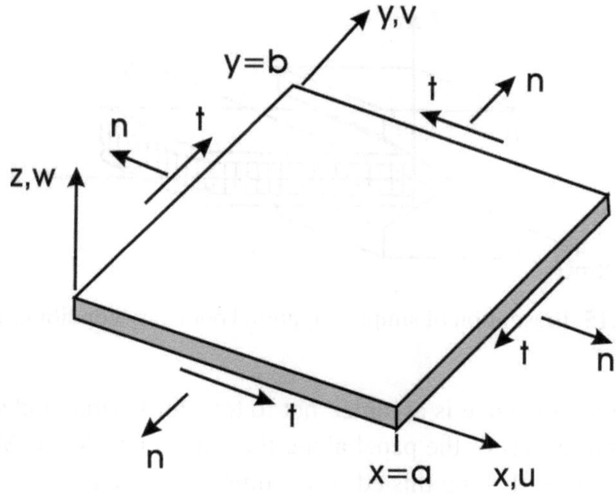

Figure 3.14 Definition of normal and tangential directions for a rectangular panel.

3.2.3 Boundary Conditions for a Rectangular Panel

We will consider a rectangular panel with the x and y coordinate axes parallel to the edges of the panel such as shown in Figure 3.14. The proper boundary conditions are necessary to guarantee achievement of a unique solution of the governing equations. Such conditions may be achieved by inspection of the problem which will reveal some of the more obvious boundary conditions. Other boundary conditions may be obtained by applying energy principles and calculus of variations (Whitney, 1987). Boundary conditions for rectangular plates refer to the normal and tangential in-plane directions as defined in Figure 3.14.

Boundary conditions specified for the present shear deformation theory requires specification of displacements, forces, and moments with respect to the normal and tangential directions of the panel (Figure 3.14). Five boundary conditions are generally required.

Simply-Supported

Figure 3.15 illustrates simply supported conditions for the edge $x = 0$, i.e., the deflection, w, is zero along the edge and at the same time, the edge can rotate freely with respect to a line along the edge ($x = 0$), i.e., $M_x = 0$ along this edge.

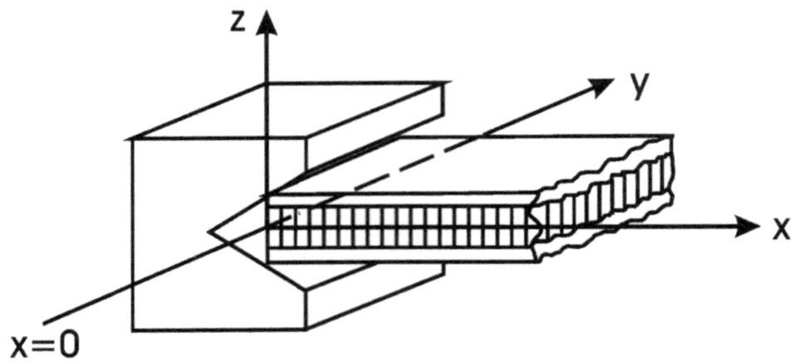

Figure 3.15 Illustration of simply-supported boundary conditions at $x = 0$.

Furthermore, the edge is assumed not to transfer normal and shear forces and prohibit rotations of the panel along the y axis, i.e., $N_x = N_{xy} = 0$, and the rotations $\psi_y = 0$ along this edge. In summary, we have

$$w(0, y) = M_x(0, y) = N_x(0, y) = N_{xy}(0, y) = \psi_y(0, y) = 0. \quad (3.59)$$

If all edges are simply supported, similar conditions apply for the edges $x = a$, $y = 0$, and $y = b$.

Hinged-Free Perpendicular to the Edge

If the edge, $x = 0$, is moment-free (hinged) and free to move in a direction normal to the edge (here along the x axis), the following conditions apply

$$N_x(0, y) = v_0(0, y) = M_x(0, y) = \psi_y(0, y) = w(0, y) = 0. \quad (3.60)$$

Hinged-Free Parallel to the Edge

If the edge, $x = 0$, is hinged and free to move tangentially, the following conditions apply:

$$u_0(0, y) = N_{xy}(0, y) = M_x(0, y) = \psi_y(0, y) = w(0, y) = 0. \quad (3.61)$$

Clamped

If the edge, $x = 0$, is clamped or "built-in", the deflection, w, along the edge is zero and the rotations, ψ_x and ψ_y, as well as mid-core displacements u_0 and v_0 (Equations (3.30)), are zero, i.e.

$$w(0, y) = \psi_x(0, y) = \psi_y(0, y) = u_0(0, y) = v_0(0, y) = 0. \quad (3.62)$$

Free Edge

If the edge, $x = 0$, is completely unconstrained, there are no resulting normal and shear forces in the plane, no bending and twisting moments, and no transverse shear force, i.e.,

$$N_x(0, y) = N_{xy}(0, y) = M_x(0, y) = M_{xy}(0, y) = Q_x(0, y) = 0. \quad (3.63)$$

3.3 Analysis of a Transversely Loaded Sandwich Plate

Sandwich panels are common structural elements in sandwich constructions such as boat hulls and containers. Analysis of flat sandwich panels of rectangular shape have been conducted by several investigators, e.g. Yen et al. (1951), Reissner (1948), Hoff (1950), and Riber (1997), and both linear small deflection analysis and geometrical nonlinear behavior have been addressed. In this section we will present analysis of the small deflection response of simply supported rectangular sandwich panels under transverse loading.

The strain-displacement relations for the core are those of Equations (3.30):

$$u = u_0 + z\psi_x, \quad (3.64a)$$

$$v = v_0 + z\psi_y, \quad (3.64b)$$

$$w = w_0. \quad (3.64c)$$

The face displacements are specified in Equations (3.31). The sandwich panel is assumed to be of symmetric construction and the face sheets are treated as homogeneous specially orthotropic materials. Hence, the constitutive relations for the force and moment resultants are given by (Equations (3.47) and (3.48))

$$\begin{bmatrix} N_x \\ N_y \\ N_{xy} \end{bmatrix} = \begin{bmatrix} A_{11} & A_{12} & 0 \\ A_{12} & A_{22} & A_{26} \\ 0 & 0 & A_{66} \end{bmatrix} \begin{bmatrix} \varepsilon_x^0 \\ \varepsilon_y^0 \\ \gamma_{xy}^0 \end{bmatrix}, \quad (3.65)$$

$$\begin{bmatrix} M_x \\ M_y \\ M_{xy} \end{bmatrix} = \begin{bmatrix} D_{11} & D_{12} & 0 \\ D_{12} & D_{22} & 0 \\ 0 & 0 & D_{66} \end{bmatrix} \begin{bmatrix} \kappa_x \\ \kappa_y \\ \kappa_{xy} \end{bmatrix}, \quad (3.66)$$

where, according to Equations (3.52),

$$A_{ij} = 2A_{ij}(f), \tag{3.67a}$$

$$D_{ij} = 2\left(\frac{h_c + h_f}{2}\right)^2 A_{ij}(f). \tag{3.67b}$$

For transverse shear loading, the constitutive equations (3.54) read

$$Q_x = h_c G_{xz}\left(\psi_x + \frac{\partial w}{\partial x}\right), \tag{3.68a}$$

$$Q_y = h_c G_{yz}\left(\psi_y + \frac{\partial w}{\partial y}\right). \tag{3.68b}$$

Substitution of the constitutive equations (3.65), (3.66) and (3.68) into the equilibrium equations (3.58) yields the following set of governing differential equations valid for symmetric sandwich panels with specially orthotropic or isotropic face sheets:

$$A_{11}\frac{\partial u_0}{\partial x^2} + (A_{12} + A_{66})\frac{\partial^2 v_0}{\partial x \partial y} + A_{66}\frac{\partial^2 u_0}{\partial y^2} = 0, \tag{3.69a}$$

$$A_{22}\frac{\partial^2 v_0}{\partial y^2} + (A_{12} + A_{66})\frac{\partial^2 u_0}{\partial x \partial y} + A_{66}\frac{\partial^2 v_0}{\partial x^2} = 0, \tag{3.69b}$$

$$D_{11}\frac{\partial^2 \psi_x}{\partial x^2} + (D_{12} + D_{66})\frac{\partial^2 \psi_y}{\partial x \partial y} + D_{66}\frac{\partial^2 \psi_x}{\partial y^2} - h_c G_{xz}\left(\psi_x + \frac{\partial w}{\partial x}\right) = 0, \tag{3.69c}$$

$$D_{22}\frac{\partial^2 \psi_y}{\partial y^2} + (D_{12} + D_{66})\frac{\partial^2 \psi_x}{\partial x \partial y} + D_{66}\frac{\partial^2 \psi_y}{\partial x^2} - h_c G_{yz}\left(\psi_y + \frac{\partial w}{\partial y}\right) = 0, \tag{3.69d}$$

$$h_c G_{xz}\left(\frac{\partial \psi_x}{\partial x} + \frac{\partial^2 w}{\partial x^2}\right) + H_c G_{yz}\left(\frac{\partial \psi_y}{\partial y} + \frac{\partial^2 w}{\partial y^2}\right) + q = 0. \tag{3.69e}$$

We will specifically apply these equations to a transversely loaded sandwich panel of planar dimension a and b with the edges hinged and unconstrained parallel to the edges, see Figure 3.16. Specifically, the conditions stipulated for the edge defined by $x = 0$ in Equation (3.61) apply to all edges.

The transverse loading is most commonly represented by a double Fourier series (see, e.g., Whitney, 1987)

$$q(x, y) = \sum_{m=1}^{\infty}\sum_{n=1}^{\infty} q_{mn} \sin\frac{m\pi x}{a} \sin\frac{n\pi y}{b}. \tag{3.70}$$

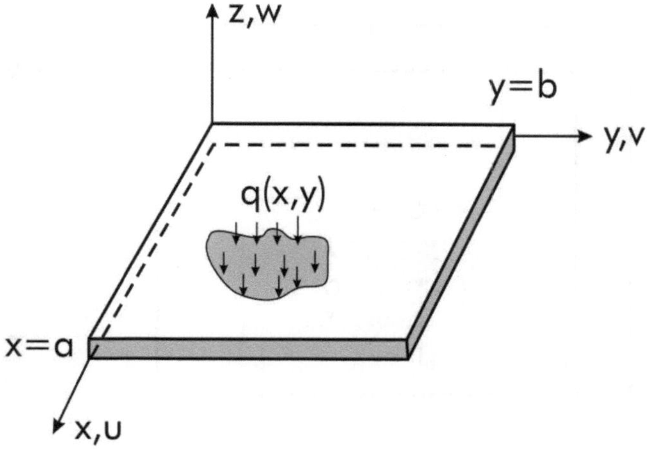

Figure 3.16 Rectangular sandwich panel under transverse loading.

The Fourier coefficients, q_{mn}, are determined from the actual load distribution, $q(x, y)$, using

$$q_{mn} = \frac{4}{ab} \int_0^a \int_0^b q(x, y) \sin \frac{m\pi x}{a} \sin \frac{n\pi y}{b} dxdy. \tag{3.71}$$

A common loading case studied is uniform pressure over the entire panel surface, $q = q_0 =$ constant. For this case, Equation (3.71) yields

$$q_{mn} = \frac{16q_0}{\pi^2 mn}, \quad m, n \text{ odd}, \tag{3.72a}$$

$$q_{mn} = 0, \quad m, n \text{ even}. \tag{3.72b}$$

Another important case is the rectangular area of uniform pressure shown in Figure 3.17. For this type of loading, Equation (3.71) yields

$$q_{mn} = \frac{16q_0}{\pi^2 mn} \sin \frac{m\pi\xi}{a} \sin \frac{m\pi\eta}{b} \sin \frac{m\pi c}{2a} \sin \frac{n\pi d}{2b}, \tag{3.73}$$

where ξ and η are the x and y coordinates of the center of the rectangle and c and d are the lengths of the rectangle along the x and y axes, see Figure 3.17.

A concentrated load, P, applied at $(x, y) = (\xi, \eta)$ may be represented by

$$q_{mn} = \frac{4P}{ab} \sin \frac{m\pi\xi}{a} \sin \frac{n\pi\eta}{b}. \tag{3.74}$$

By increasing the number of terms in the series, Equation (3.70), the exact solution will be approached asymptotically. The number of terms required

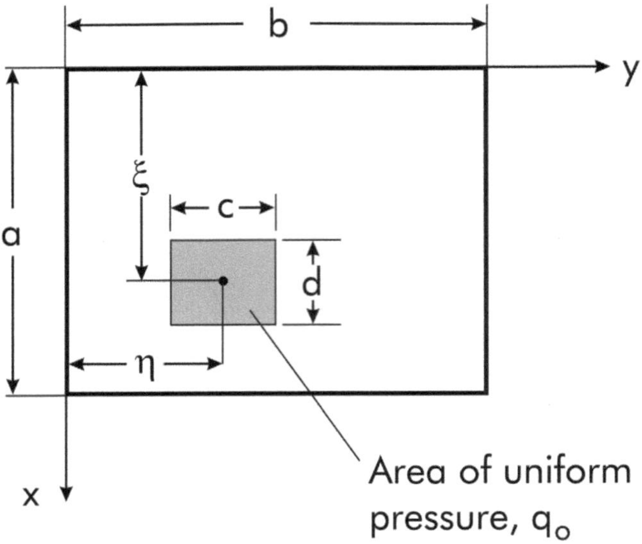

Figure 3.17 Rectangular panel loaded with uniform load over a rectangular region of the panel.

for a converged solution of the quantity of interest, i.e., deflection, strain or stress, must be examined for each case.

The analysis of a transversely loaded panels will here assume a simple one term case, i.e. $m = n = 1$, i.e. in Equation (3.70),

$$q(x, y) = q_0 \sin \frac{\pi x}{a} \sin \frac{\pi y}{b}. \tag{3.75}$$

Boundary conditions for rectangular panels are discussed in Section 3.2.3. In particular, the hinged-free tangential conditions defined in Equation (3.61) are assumed here. For the edges parallel to the y axis, $x = 0$ and a (Figure 3.16), the following conditions apply:

$$u_0 = 0, \tag{3.76a}$$

$$N_{xy} = A_{66} \left(\frac{\partial u_0}{\partial y} + \frac{\partial v_0}{\partial x} \right) = 0, \tag{3.76b}$$

$$w = 0, \tag{3.76c}$$

$$\psi_y = 0, \tag{3.76d}$$

$$M_x = D_{11} \frac{\partial \psi_x}{\partial x} + D_{12} \frac{\partial \psi_y}{\partial y} = 0. \tag{3.76e}$$

For the edges parallel to the x axis, $y = 0$ and b, the conditions are

$$v_0 = 0, \tag{3.77a}$$

$$N_{xy} = A_{66}\left(\frac{\partial u_0}{\partial y} + \frac{\partial v_0}{\partial x}\right) = 0, \tag{3.77b}$$

$$w = 0, \tag{3.77c}$$

$$\psi_x = 0, \tag{3.77d}$$

$$M_y = D_{12}\frac{\partial \psi_x}{\partial x} + D_{22}\frac{\partial \psi_y}{\partial y} = 0. \tag{3.77e}$$

A solution that satisfies the boundary conditions (3.76) and (3.77) consistent with the loading function, Equation (3.75), is given by

$$u_0 = A \sin\frac{\pi x}{a}\cos\frac{\pi y}{b}, \tag{3.78a}$$

$$v_0 = B \cos\frac{\pi x}{a}\sin\frac{\pi y}{b}, \tag{3.78b}$$

$$\psi_x = C \cos\frac{\pi x}{a}\sin\frac{\pi y}{b}, \tag{3.78c}$$

$$\psi_y = D \sin\frac{\pi x}{a}\cos\frac{\pi y}{b}, \tag{3.78d}$$

$$w = E \sin\frac{\pi x}{a}\sin\frac{\pi y}{b}. \tag{3.78e}$$

Substitution of Equations (3.78) into the governing equations (3.69) yields five equations conveniently expressed into the following matrix equation:

$$\begin{bmatrix} H_{11} & H_{12} & 0 & 0 & 0 \\ H_{12} & H_{22} & 0 & 0 & 0 \\ 0 & 0 & H_{33} & H_{34} & H_{35} \\ 0 & 0 & H_{34} & H_{44} & H_{45} \\ 0 & 0 & H_{35} & H_{45} & H_{55} \end{bmatrix} \begin{bmatrix} A \\ B \\ C \\ D \\ E \end{bmatrix} = \begin{bmatrix} 0 \\ 0 \\ 0 \\ 0 \\ q_0 \end{bmatrix}. \tag{3.79}$$

The elements H_{ij} of the symmetric 5×5 matrix are given by

$$H_{11} = \frac{A_{11}}{a^2} + \frac{A_{66}}{b^2}, \tag{3.80a}$$

$$H_{12} = \frac{A_{12} + A_{66}}{ab}, \tag{3.80b}$$

$$H_{22} = \frac{A_{22}}{b^2} + \frac{A_{66}}{a^2}, \tag{3.80c}$$

$$H_{33} = \frac{\pi^2 D_{11}}{a^2} + \frac{\pi^2 D_{66}}{b^2} + h_c G_{xz}, \tag{3.80d}$$

$$H_{34} = \frac{\pi (D_{12} + D_{66})}{ab}, \tag{3.80e}$$

$$H_{35} = \frac{\pi h_c G_{xz}}{a}, \tag{3.80f}$$

$$H_{44} = \frac{\pi^2 D_{22}}{b^2} + \frac{\pi^2 D_{66}}{a^2} + h_c G_{yz}, \tag{3.80g}$$

$$H_{45} = \frac{\pi h_c G_{yz}}{b}, \tag{3.80h}$$

$$H_{55} = \pi^2 h_c \left(\frac{G_{xz}}{a^2} + \frac{G_{yz}}{b^2} \right). \tag{3.80i}$$

Inversion of the H matrix, Equation (3.79), provides the constants A, B, C, D and E. The plate deflection is given by (Equation (3.78e))

$$w = E \sin \frac{\pi x}{a} \sin \frac{\pi y}{b}, \tag{3.81a}$$

with

$$E = h_{33} q_0, \tag{3.81b}$$

where

$$h_{33} = \frac{H_{33} H_{44} - H_{34}^2}{\det[H_{\text{sub}}]}. \tag{3.82}$$

$[H_{\text{sub}}]$ represents the following 3×3 submatrix defined by the last three rows and columns of the full matrix of Equation (3.79)

$$[H_{\text{sub}}] = \begin{bmatrix} H_{33} H_{34} H_{35} \\ H_{34} H_{44} H_{45} \\ H_{35} H_{45} H_{55} \end{bmatrix}. \tag{3.83}$$

The maximum deflection, $w_{\max}$, occurs at the panel center ($x = a/2$, $y = b/2$) and is given by

$$w_{\max} = E = h_{33} q_0. \tag{3.84}$$

A square ($a = b$) sandwich panel with unidirectional composite face sheets over an isotropic foam core is considered. The face sheets are assumed to consist of carbon/epoxy with the following mechanical properties:

$$E_1 = 125 \text{ GPa}, \quad E_2 = 5 \text{ GPa},$$

$$\nu_{12} = 0.25, \quad G_{12} = 2.5 \text{ GPa}.$$

The core considered is a H100 PVC foam with a shear modulus, $G = 60$ MPa. The thicknesses of the face sheets and core are $h_f = 2$ mm and $h_c = 16$ mm. Calculations of the deflection, $w_{\max}$, were conducted over a range of panel sizes. The side length, a, of the square panel was varied from 0.1 to 1.2 m. The bending stiffness elements D_{ij} were calculated from Equation (3.52c) which specifically applies to a symmetric sandwich with thin homogeneous face sheets.

Figure 3.18 displays the maximum deflection vs. the panel size, see the curve labeled SDPT (shear deformation plate theory). Also shown is the solution given by classical laminated plate theory (CLPT) (Whitney, 1987),

$$w_{\max} = \frac{q_0 a^4}{\pi^4 D(m = n = 1)}, \tag{3.85}$$

where $D(m = n = 1)$ is a bending stiffness term given by

$$D(m = n = 1) = D_{11} + 2(D_{12} + 2D_{66}) + D_{22}. \tag{3.86}$$

As discussed earlier, the CLPT formulation does not accommodate interlaminar shear deformation. Figure 3.18 shows that the CLPT results provide a lower bound to the deflection of the plate.

Figure 3.18 shows that the deflection of small plates with a thick core are quite substantially influenced by transverse shear deformation, while the deflection of larger panels is less affected by this mode of deformation and may be analyzed using CLPT.

3.4 Analysis of Sandwich Plate Twist Test

Tsai (1965) developed a plate twist test to determine the engineering elastic constants of orthotropic plates. The test utilized a square panel loaded at one corner and supported at the other corners. The panel response was analyzed using classical orthotropic plate theory (Lekhnitskii, 1968). By twisting a 0° panel and loading beams cut from the panel in bending, Tsai (1965) was able to determine all five elastic compliance constraints. Mure (1986) developed a two-point loading configuration of the plate twist test in order to determine the twisting stiffness (D_{66}) of corrugated core cardboard panels. Vinson

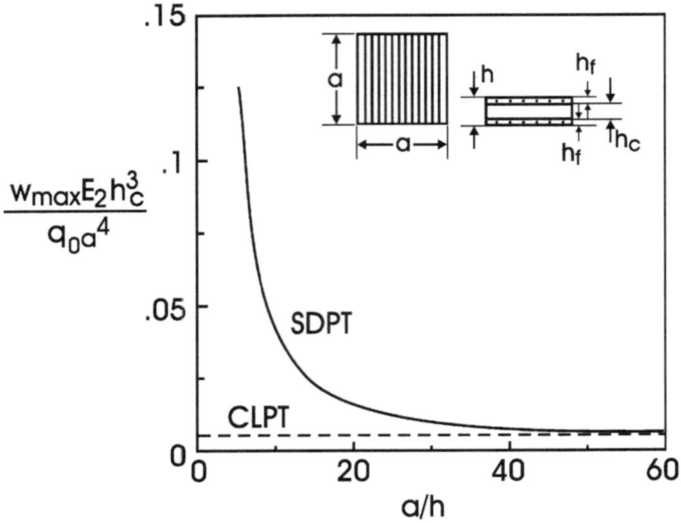

Figure 3.18 Maximum deflection of a square sandwich plate under transverse pressure load. CLPT and SDPT refer to classical laminated plate theory and shear deformation plate theory.

(1999) examined the test principle analytically in an effort to determine the in-plane shear strength of the face sheets, face/core adhesive or core, whatever constituent was the weakest. The plate twist test is also used to measure the in-plane shear modulus of plywood panels according to ASTM Standard D3044 (2000).

3.4.1 Classical Laminated Plate Theory Analysis

Figure 3.19 shows the two-point twist loading of a sandwich plate considered here. In this configuration, two diagonally opposite corners are loaded downwards by concentrated forces of magnitude $P/2$, while the other two corners are supported.

The sandwich plate is assumed to consist of identical isotropic or specially orthotropic faces. According to CLPT, the slopes of the middle surface are assumed to coincide (in magnitude) with the rotations of the cross-sections. Notice that this theory neglects transverse shear deformation. The following expressions for the plate curvatures defined in Equations (3.12) apply:

$$\kappa_x = -\frac{\partial^2 w}{\partial x^2}, \tag{3.87a}$$

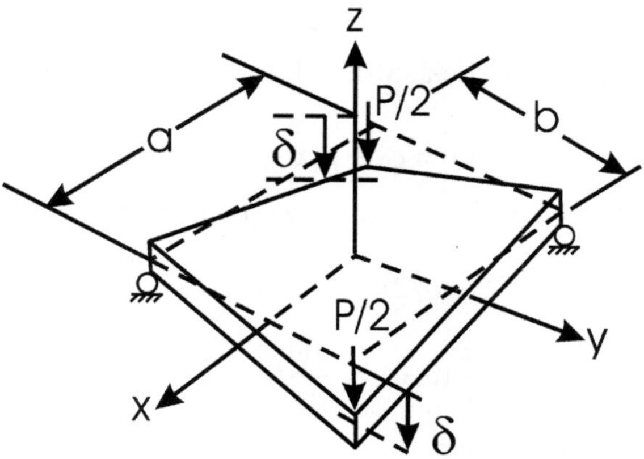

Figure 3.19 Illustration of loading and support conditions for the two-point sandwich plate twist test.

$$\kappa_y = -\frac{\partial^2 w}{\partial y^2}, \tag{3.87b}$$

$$\kappa_{xy} = -2\frac{\partial^2 w}{\partial y \partial x}. \tag{3.87c}$$

For a symmetric sandwich consisting of isotropic or specially orthotropic homogeneous face sheets, the pure twisting response is expressed by Equation (3.19)

$$M_{xy} = D_{66}\kappa_{xy}, \tag{3.88}$$

where the twisting stiffnesses, D_{66}, is given by Equation (3.20c) applied to a symmetric sandwich

$$D_{66} = (G_{12})_f h_f \left(\frac{2h_f^2}{3} + \frac{2h_c^2}{2} + h_f h_c \right) + \frac{(G_{12})_c h_c^3}{12}. \tag{3.89}$$

where $(G_{12})_f$ and $(G_{12})_c$ are the in-plane shear moduli of the face and core and h_f and h_c are the face and core thicknesses.

The twist loading shown in Figure 3.19 produces a twisting moment given by Timoshenko and Woinowsky-Krieger (1959),

$$M_{xy} = P/4. \tag{3.90}$$

A combination of Equations (3.88) and (3.90) yields

$$\kappa_{xy} = -2\frac{\partial^2 w}{\partial x \partial y} = \frac{P}{4D_{66}}. \tag{3.91}$$

The solution for the plate deflection representing a state of pure uniform twist curvature is, according to Lekhnitskii (1968), a second-order polynomial in x and y

$$w(x, y) = c_0 + c_1 x + c_2 y + c_3 xy. \tag{3.92}$$

The boundary conditions for the panel are (Figure 3.19),

$$w(a/2, -b/2) = w(-a/2, b/2) = 0, \tag{3.93a}$$

$$\frac{\partial w}{\partial x}(0, 0) = \frac{\partial w}{\partial y}(0, 0) = 0. \tag{3.93b}$$

Equations (3.91)–(3.93) yield

$$c_0 = \frac{ab}{4} c_5, \tag{3.94a}$$

$$c_1 = c_2 = 0, \tag{3.94b}$$

$$c_3 = \frac{-P}{8D_{66}}. \tag{3.94c}$$

Hence, the panel deflection becomes

$$w(x, y) = \frac{-P}{8D_{66}} \left(\frac{ab}{4} + xy \right). \tag{3.95}$$

The deflection of the loaded corners is given by $\delta = |w(a/2, b/2)|$. Equation (3.95) yields

$$\delta = \frac{Pab}{16D_{66}}. \tag{3.96}$$

The plate compliance is given by $C = \delta/P$. Hence,

$$C = \frac{ab}{16D_{66}}. \tag{3.97}$$

Measuring the compliance of the twisted plate should thus provide a means for determination of the twist stiffness D_{66}. It is noted that the only material parameters entering the expression for D_{66}, Equation (3.89), are the in-plane shear moduli for the faces and the core, $(G_{12})_f$ and $(G_{12})_c$. In most cases, the contribution to D_{66} from the core (the last term in Equation (3.89)) can be neglected because soft sandwich cores typically have $(G_{12})_c \ll (G_{12})_f$. For such a case, the twist test should provide a means to determine the in-plane shear modulus of the face sheets. As will be discussed in the next section, however, low modulus cores are susceptible to transverse shear deformation which will elevate the plate compliance.

3.4.2 Finite Element Analysis

It was pointed out that the compliance expression, Equation (3.97), does not include contributions due to transverse shear deformation. It is well known that shear deformation may influence the response of sandwich panels quite significantly, see Section 3.3. Furthermore, application of localized load to a sandwich panel is known to cause indentation deformation (Frostig et al., 1992; Thomsen, 1977). In order to further analyze, a sandwich panel under twist loading, finite element analysis was conducted by Aviles et al. (2009). A sandwich panel with a 25 mm thick foam core and 1 mm thick aluminum faces ($E = 70$ GPa, $\nu = 0.3$) was examined. The in-plane panel dimensions were 30.5 × 30.5 (cm). To reduce indentation deformation at load introduction and support points, square 10 × 10 (mm) areas were introduced at the contact regions at the corners, where the vertical displacement of the nodes was constrained to be uniform. Since sandwich panels are prone to transverse shear deformation in the core, a range of core shear moduli from 11.5 to 758 MPa was examined. The face sheets and core were modeled using the finite element code ANSYS (2006). All panels utilized the same mesh consisting of 3D eight-noded solid brick elements (SOLID 45).

Figure 3.20 presents the compliance of the sandwich panels calculated from FEA and CLPT vs. core-to-face shear modulus ratio (G_c/G_f). The compliance predicted by CLPT is fairly constant, since D_{66} is very little influenced by variations in G_c. The compliance predicted by FEA, on the other hand, decreases rapidly with increased core stiffness, until a plateau region is reached which coincides with the CLPT prediction. Convergence occurs for modulus ratios, $G_c/G_f > 14$. The large compliance for core materials with small core shear modulus is attributed to transverse shear deformation of the core (Aviles et al., 2009).

The results show that the compliance is extremely sensitive to changes in core shear modulus, G_c, when G_c is below about 100 MPa. Notice that several commercial PVC cores have shear modulus in this range (Table 1.3). Further, CLPT provides a very low estimate of the compliance unless the core is stiff. Thus, in general, the plate twist test is not a viable alternative to direct shear testing of the face sheets (Chapter 2). Further information on this test is provided in Aviles et al. (2009) and in Section 3.5.2.

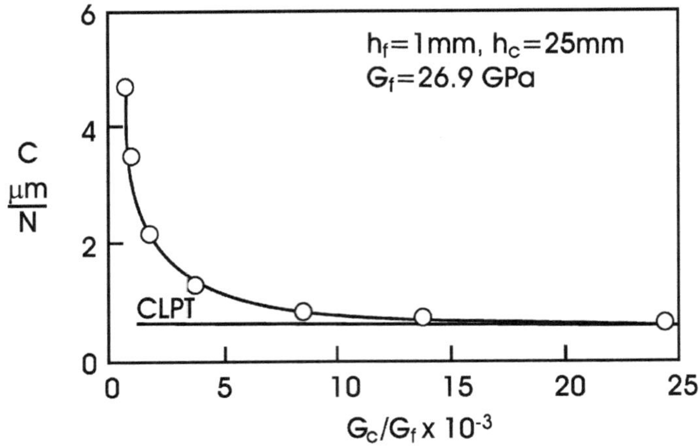

Figure 3.20 Compliance vs. core-to-face shear modulus ratio (G_c/G_f) for a 30.5 × 30.5 (cm) sandwich panel. $G_f = 26.9$ GPa.

3.5 Testing of Sandwich Panels

3.5.1 Pressure Loading of a Sandwich Plate

Testing of sandwich panels under distributed loading has been conducted by several investigators. The test is quite complex since it involves application of pressure over a quite large area in a controlled manner, while the edge conditions of the panel should be representative for those assumed in the analysis. Most reported test set-ups involve a pressurized water-filled rubber bladder to distribute the load over the panel surface in a uniform and controlled manner. The pressure in the bladder is controlled in order to achieve an accurate measure of the transverse load intensity, $q(x, y)$. Such an approach has been pursued by Rothschild et al. (1992), Bau-Madsen et al. (1992), Wennhage and Zenkert (1998), and Hayman et al. (1998).

Wennhage and Zenkert (1998) designed a test frame that was fitted in a large compression testing machine for testing of 0.85 × 0.85 (m) sandwich panels with a 25 mm thick H100 PVC foam core and 1 mm thick aluminum alloy faces ($E = 70$ GPa). The core mechanical properties are listed in Table 3.2. The faces were bonded to the core using a polyurethane adhesive. The panels were instrumented with four "shear plugs" consisting of 3 cm diameter cylindrical plugs of H100 core with a properly calibrated strain gage mounted at 45° to the cylinder axis (Moyer et al., 1992). The upper face of

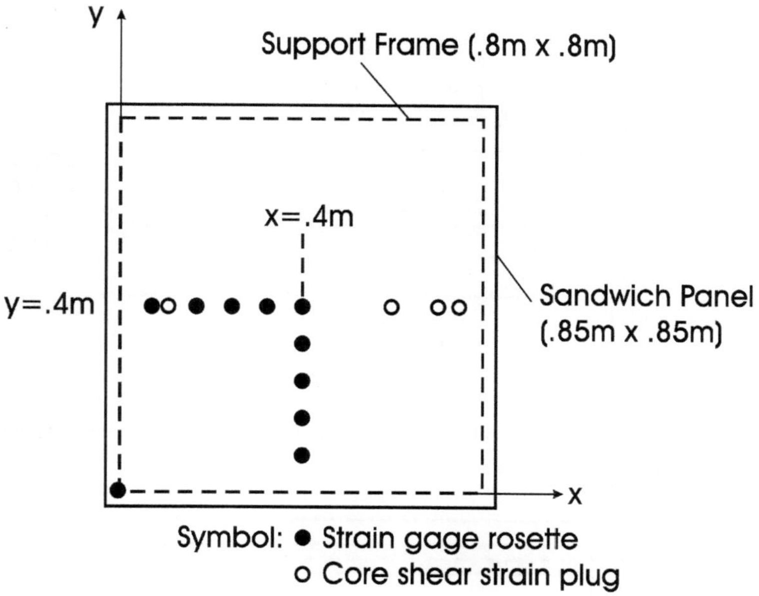

Figure 3.21 Location of surface mounted strain gages on sandwich panel (open circles) and core shear plugs (filled circles). The dotted line indicates line of support.

Table 3.1 Mechanical properties for H100 PVC foam core.

Shear modulus	Shear yield strength	Compressive strength
G	τ_y	σ_c
38 MPa	1.4 MPa	1.7 MPa

the sandwich panel was instrumented with ten 0°/90° strain gage rosettes, see the outline in Figure 3.21.

Notice that all strain gages, except for the one in the left corner, where the support lines intersect, were placed along the horizontal and vertical center lines representing symmetries of the panel. Figure 3.22 illustrates schematically the compression test procedure using the bladder system.

The testing rig consists of an upper frame made from rigid steel I beams with semi-circular rods of 20 mm diameter bolted to the lower surface of the I beams, Figure 3.22. The lower frame was made from rigid steel C beams. The primary purpose of the lower frame was to constrain the in-plane expansion of the rubber bladder. The rubber bladder was made from 2 mm thick EPDM rubber. It was filled with water and sealed prior to testing. The rounded steel bars are intended to provide simply supported edge boundary

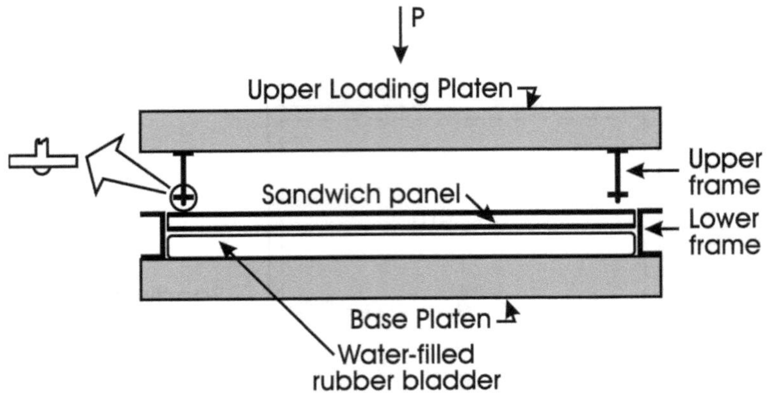

Figure 3.22 Principle of pressure loading of sandwich panel. Cross-sectional view in the x–z plane.

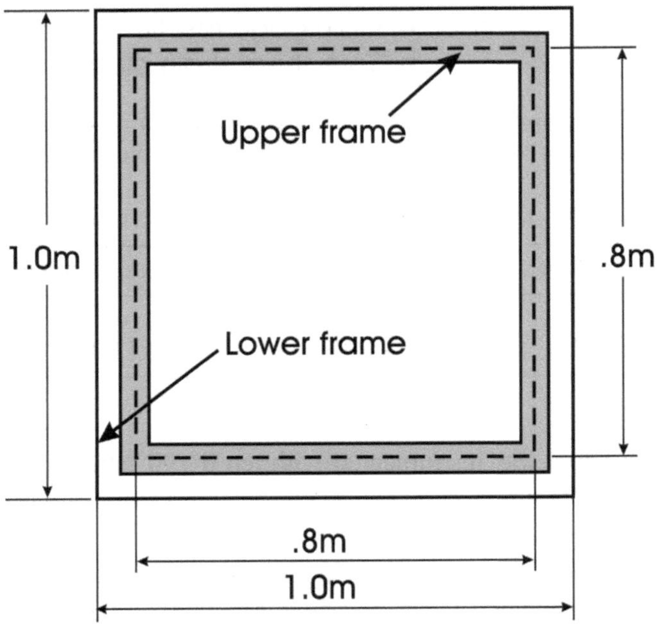

Figure 3.23 Top view of test panel and support structure.

conditions (Equation (3.59)) along the periphery of the panel, see the top view of the panel and the upper and lower frames shown in Figure 3.23.

In addition to the strain gage instrumentation (Figure 3.21), six displacement gages were attached to the upper loading platen of the test frame to measure the out-of-plane deflection of the panel along the symmetry lines

($x = 0.4$ and $y = 0.4$ m, Figure 3.21). The applied pressure loading was selected to maintain linear-elastic stress-strain behavior of all the constituents with a maximum stress less than or equal to 60% of the corresponding yield stress.

Test Results

The compressive load was increased to a maximum value of 50 kN, corresponding to a distributed pressure of 78 kPa over the 0.8×0.8 (m) test area. Such a pressure is not expected to cause compressive yielding of the H100 PVC foam core, see the material properties listed in Table 3.2.

It was found that the panel initially deviated somewhat from the ideal flat shape, i.e. it was slightly warped. All out-of-plane displacements readings were therefore corrected for deviation from flatness by subtracting the initial displacements from the actual readings. Upon further loading, the panels made full contact with the upper support fixture (Figure 3.22). Figure 3.24 shows experimental displacement data collected along symmetry lines ($x = 0.4$ and $y = 0.4$ m, Figure 3.21). Figure 3.24 also shows prediction of deflections from first-order sandwich plate analysis, see Section 3.5, using the loading function given by Equations (3.72) and simply supported boundary conditions (Equation (3.59)).

The experimental data reveal approximate symmetry with respect to the x and y directions, and there is good agreement between prediction and experimental data. Figure 3.25 shows the core shear stress results determined from the "shear plugs" embedded in the core. The shear stress is calculated from the measured shear strains using

$$\tau_{xz} = G_c \gamma_{xz}. \tag{3.98}$$

The predictions using sandwich plate theory analysis are overall in good agreement with the experimental data.

3.5.2 *Plate Twist Testing*

A test fixture for twist testing of sandwich panels was designed to represent the two-point configuration sketched in Figure 3.19. In this test configuration, two opposite corners are supported and two opposite corners are loaded. The fixture was designed for square panels with size up to about

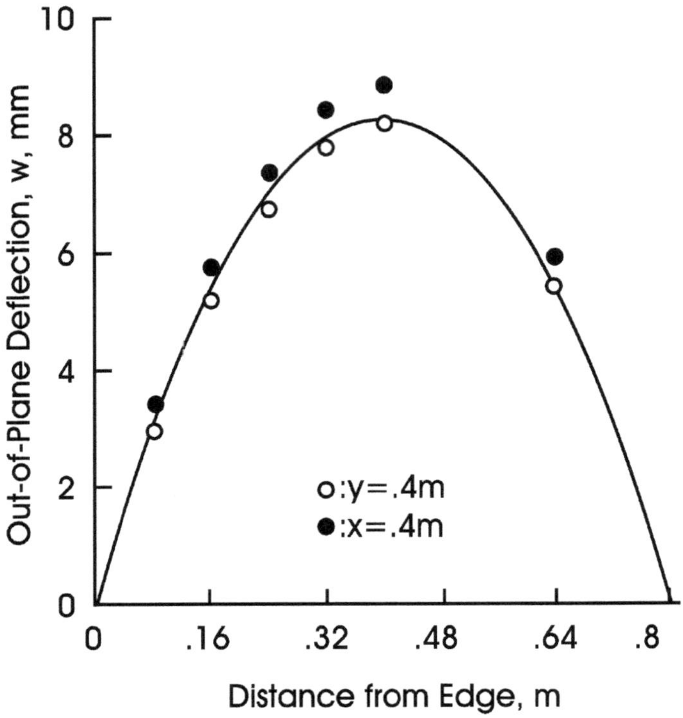

Figure 3.24 Measured and predicted plate deflections along symmetry lines ($x = 0.4$ m, $y = 0.4$ m).

30 × 30 (cm). Figure 3.26 shows a photograph of the test fixture with a slightly loaded sandwich panel. The fixture consists of a 30.5 × 30.5 × 1.28 (cm) aluminum plate bolted to the base of the test frame. To introduce load, a 46.5 cm long steel bar of cross-section 2.54 × 3.56 (cm) was attached to the load cell in the moving cross-head of the test machine. To allow testing of different size panels, multiple holes for attachment of support and loading pins were drilled in the bottom aluminum plate and in the steel bar, the outermost at a distance of 40.5 cm. The diagonally aligned holes in the aluminum base plate and longitudinal holes in the steel bar were spaced at increments of 1.27 cm. For loading and support of the panel at the desired points, steel pins of 9.5 mm diameter with hemispherical ends were inserted in the appropriate holes in the bottom plate and loading bar. The radius of the loading and support pin surfaces in contact with the panel was 4.75 mm.

A total of five square 30.5 × 30.5 (cm) (nominal dimensions) sandwich panels were prepared using aluminum face sheets and H80, H100, and H200 PVC foam cores. Panel dimensions and thicknesses of the constituents pan-

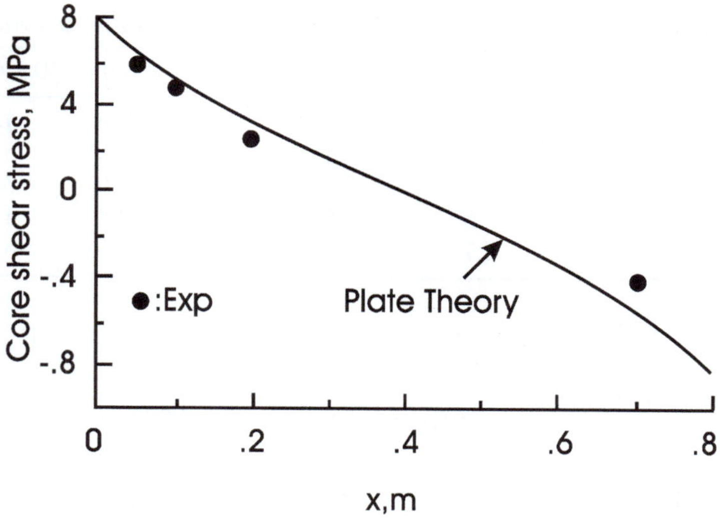

Figure 3.25 Core shear stress along symmetry section of the panel ($y = 0.4$ m).

Figure 3.26 Sandwich plate twist test fixture.

els are listed in Table 3.3. The face sheets were nominally 1.5 and 2.25 mm thick labeled "thin" and "thick". The nominal core thickness was 12.7 mm. Table 3.2 lists the twisting stiffness, D_{66}, for each panel calculated from CLPT (Equation (3.89)) based on nominal face and core thicknesses.

Table 3.2 Dimensions and twist stiffness (D_{66}) of sandwich panels.

Panel	Core	h_f (mm)	h_c (mm)	h (mm)	a (cm)	D_{66} (kNm)
H80/Thin	H80	1.50	12.5	16.4	30.6	4.09
H100/Thick1	H100	2.26	12.8	17.6	30.6	6.90
H100/Thick2	H100	2.25	12.5	15.6	30.6	7.00
H200/Thin	H200	1.50	12.5	17.1	30.2	4.10
H200/Thick	H200	2.24	12.4	17.4	30.6	6.83

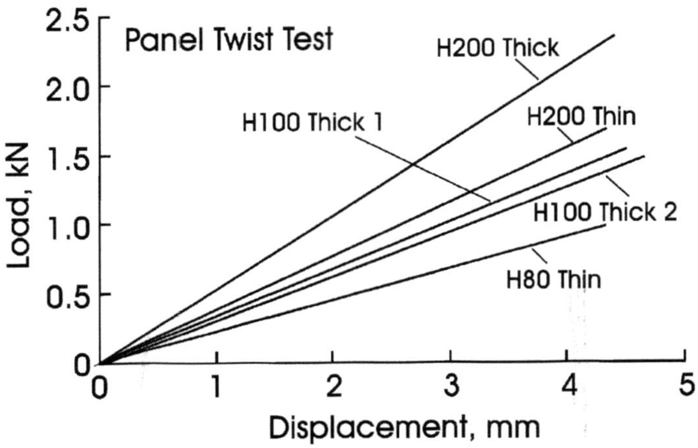

Figure 3.27 Load vs. deflection graphs for sandwich test panels.

Prior to testing of the sandwich panels, the machine/fixture compliance was determined by loading the fixture without a sandwich panel and measuring load and displacement. For testing a sandwich panel, the panel was inserted in the fixture, and adjusted so that the edges were aligned with the edges of the base plate, with the same amount of overhang at each edge. The deflection of the panel was measured using the cross-head displacement. The loading area was 27 × 27 (cm) in all tests. This corresponds to a nominal overhang length of about 1.75 cm. During testing of a sandwich panel, the first loading cycle revealed a stiffening nonlinear response at small loads. This was the result of local indentation deformation at load introduction and support locations, and slack in the fixture. Prior to the actual panel test, each panel was loaded to the maximum, and unloaded to about 500 N. This loading-unloading cycle was repeated a number of times until the loading and unloading load vs. displacement curves virtually coincided.

Figure 3.27 shows the experimental load vs. displacement curves for all the test panels (Table 3.3). Over the range of load levels from 0–2 kN, the

Table 3.3 Measured (Exp.), and CLPT and FEA predicted compliance values of sandwich panels.

Panel	C (μm/N)		
	Exp.	CLPT	FEA
H80/Thin	4.26	1.42	2.96
H100/Thick	3.01	0.85	1.91
H200/Thin	2.51	1.42	1.75
H200/Thick	1.82	0.85	1.27

response is linear and the loading and unloading curves virtually coincide. As expected, increased core density and face thickness result in a stiffer response.

After completion of the twist testing, the compliance of each specimen was determined from the load-displacement curves shown in Figure 3.27. Finite element analysis (FEA) of the actual test panels was conducted as explained in Section 3.4.2. The FEA was based on the nominal plate dimensions and an overhang of 15 mm. In all cases square 10×10 (mm) constrained areas were used at the load introduction and support regions.

Table 3.3 summarizes compliance values determined experimentally (Exp.) (corrected for machine compliance) and compliance values calculated from CLPT and FEA. The results in Table 3.3 reveal that CLPT substantially underestimates the compliance of the tested panels. The experimental results confirm the earlier assessment that CLPT is not accurate for analyzing the sandwich plate twist test, at least not for the sandwich panels considered here. The finite element predictions are much more close to the measured values, although the measured compliance values exceed the FEA predictions. The difference between FEA predictions and measured compliance value is attributed to more excessive indentation at load introduction and support points than accounted for in the finite element analysis.

Chapter 4
First-Order Shear Analysis of Sandwich Beams

In this chapter the first-order, two-dimensional shear deformation theory analysis presented in Chapter 3 is specialized to beams. First a general analysis of sandwich beams is developed which is subsequently applied to a three-point flexure loaded sandwich beam. Simplified beam analysis is developed by reducing the 3×3 plate stiffness matrices $[A]$, $[B]$, $[C]$ and $[D]$ to single stiffnesses A, B, C and D, and explicit expressions valid for symmetric beams with thin face sheets are derived. In the final section, three-point flexure testing of sandwich beams and analysis to determine the bending and shear stiffnesses from measured compliance data are outlined.

4.1 Analysis of Sandwich Beams

Although sandwich beams are rarely used as structural members, sandwich beams are convenient test specimens and such a test may constitute a simple way to determine the properties of a sandwich panel. Here we will consider beam specimens with thin faces. We will further assume the faces can be treated as homogeneous materials with the stiffnesses A_{ij}, B_{ij}, C_{ij} and D_{ij} given by Equation (3.51).

For the purpose of analysis of sandwich beams loaded in flexure, it is convenient to present Equations (3.41) and (3.42) in a compact form, Equation (3.56)

$$\begin{bmatrix} N \\ M \end{bmatrix} = \begin{bmatrix} A & B \\ C & D \end{bmatrix} \begin{bmatrix} \varepsilon^0 \\ \kappa \end{bmatrix}, \tag{4.1}$$

where the left-hand side is the 6×1 column vector formed by the six force and moment resultants, and the ABCD matrix is a 6×6 matrix formed by

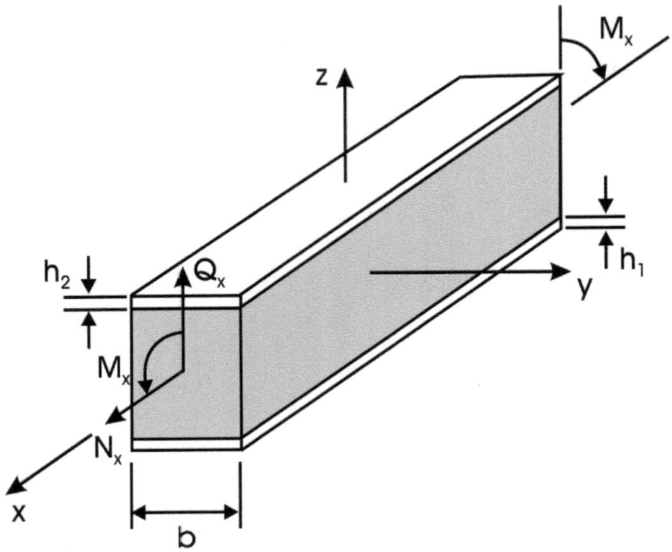

Figure 4.1 Bending of a sandwich beam.

the four 3 × 3 basic stiffness matrices in Equations (3.55), $[\varepsilon^0]$ and $[\kappa]$ represent the 6 × 1 column vector defined by the three mid-plane strains and three curvatures. To proceed, it is convenient to use the inverted form of Equations (4.1)

$$\begin{bmatrix} \varepsilon^0 \\ \kappa \end{bmatrix} = \begin{bmatrix} a & b \\ c & d \end{bmatrix} \begin{bmatrix} N \\ M \end{bmatrix}. \tag{4.2}$$

For bending of a beam by moments M, the components of the 6 × 1 column force and moment vector become

$$M_x = M/b, \tag{4.3a}$$

$$N_x = N_y = N_{xy} = M_y = M_{xy} = 0, \tag{4.3b}$$

where b is the width of the beam, see Figure 4.1.

The curvatures $[\kappa]$, are obtained from Equations (4.2) with $[N] = [0]$,

$$[\kappa] = [d][M]. \tag{4.4}$$

In particular, we will assume that the face laminates are free from bending/twisting coupling such that $d_{16} = 0$, or d_{16} is so small it can be neglected. Hence, application of bending moments M_x will not cause twisting of the beam, i.e., $\kappa_{xy} = 0$. The primary curvature of interest is κ_x which is given by

$$\kappa_x = d_{11} M_x, \tag{4.5}$$

where d_{11} is the flexural compliance of the sandwich beam.

The curvature, κ_x, is given by (Equation (3.36a))

$$\kappa_x = \frac{d\psi_x}{dx}. \tag{4.6}$$

The bending strain and transverse shear strain are given by Equations (3.35a) and (3.37a)

$$\varepsilon_x = \varepsilon_x^0 + z\kappa_x, \tag{4.7a}$$

$$\gamma_{xz} = \psi_x + \frac{dw}{dx}. \tag{4.7b}$$

Equation (3.54a) defines the constitutive equation for transverse shear

$$Q_x = h_c G_{xz}\left(\psi_x + \frac{dw}{dx}\right). \tag{4.8}$$

Substituting the moment, $M_x = \kappa_x/d_{11}$ (Equation (4.5)) with κ_x given by Equation (4.6) and the shear force given by Equation (4.8) substituted into the equilibrium Equation (3.60c) yields a differential equation for bending

$$\frac{d^2\psi_x}{dx^2} - \left(\psi_x + \frac{dw}{dx}\right) h_c G_{xz} d_{11} = 0 \tag{4.9}$$

Another differential equation governing bending of the beam is obtained by combining Equations (3.58c) and (4.8) which yields

$$\frac{dw}{dx} = -\psi_x + \frac{1}{h_c G_{xz}}\frac{dM_x}{dx}. \tag{4.10}$$

This equation is convenient for several beam-bending problems where the variation of moment along the beam, i.e., $M_x = M_x(x)$, is known from static equilibrium considerations.

4.2 Three-Point Flexure Loading of a Sandwich Beam

A symmetric sandwich beam under three-point flexure loading as shown in Figure 4.2 is considered. The faces are assumed to be thin symmetric and balanced composite laminates ($B_{ij} = A_{16} = A_{26} = 0$) or they could be made from an isotropic metal. Such laminates may contain non-zero D_{16} and D_{26}

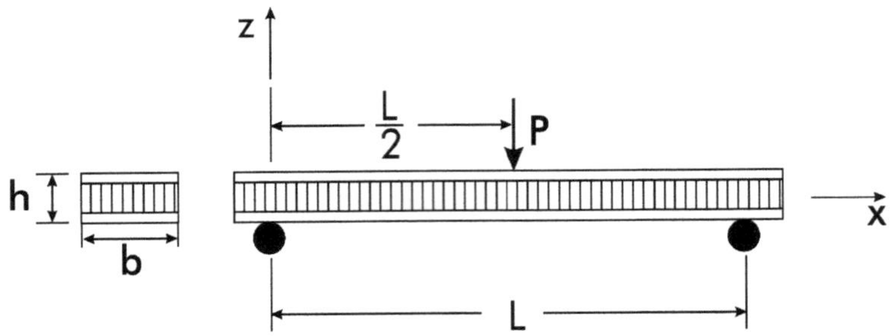

Figure 4.2 Sandwich beam under three-point flexural loading.

terms, but we will assume they are small and have negligible influence on the response.

With the xz coordinate system as shown in Figure 4.2 ($x = 0$ at the left support), the bending moment in the left half of the beam is

$$M_x = \frac{-Px}{2b}, \quad 0 \leq x \leq \frac{L}{2}. \tag{4.11}$$

This, combined with Equations (4.5) and (4.6), yields

$$\frac{d\psi_x}{dx} = \frac{-d_{11}Px}{2b} \tag{4.12}$$

which integrates to

$$\psi_x(x) = \frac{-d_{11}Px^2}{4b} + C_1. \tag{4.13}$$

To determine the constant, C_1, we consider the axial displacements of the beam (Equation (3.34a))

$$u(x) = u_0(x) + z\psi_x. \tag{4.14}$$

Symmetry of the beam displacements for the left and right halves of the beam, i.e., with respect to $x = L/2$, may be imposed by requiring vanishing of the axial displacements at the beam mid-span, i.e.,

$$u\left(\frac{L}{2}\right) = 0 \tag{4.15}$$

which, according to Equation (4.14), leads to

$$u_0\left(\frac{L}{2}\right) = \psi_x\left(\frac{L}{2}\right) = 0. \tag{4.16}$$

The condition $\psi_x\left(\frac{L}{2}\right) = 0$ in Equation (4.13) provides the constant of integration

$$C_1 = \frac{d_{11}PL^2}{16b}. \tag{4.17}$$

Substitution of Equation (4.13) with C_1 given by (4.17) into (4.10), yields

$$\frac{dw}{dx} = \frac{PL^2 d_{11}}{16b}\left(4\left(\frac{x}{L}\right)^2 - 1\right) - \frac{P}{2bh_c G_{xz}}. \tag{4.18}$$

Notice that the slope is non-zero at the center on the span, $x = L/2$

$$\frac{dw}{dx} = \frac{-P}{2bh_c G_{xz}}. \tag{4.19}$$

Analysis of the right half of the beam would yield a slope at $x = L/2$ given by

$$\frac{dw}{dx} = \frac{P}{2bh_c G_{xz}}. \tag{4.20}$$

Such a discontinuity of slope is consistent with the discontinuity of shear force (Q_x) across the beam center and the formulation of the first-order shear theory, although more refined theories would provide no such discontinuity.

To obtain the deflection of the beam, Equation (4.18) is integrated

$$w(x) = \frac{Pd_{11}x^3}{12b} - \frac{PL^2 d_{11}x}{16b} - \frac{Px}{2h_c bG_{xz}} + C_2, \tag{4.21}$$

where $C_2 = 0$ to guarantee $w(0) = 0$. Hence, the beam deflection is given by

$$w(x) = \frac{P}{2b}\left[\frac{d_{11}x^3}{6} - \frac{L^2 x d_{11}}{8} - \frac{x}{h_c G_{xz}}\right], \tag{4.22}$$

where $0 \leq x \leq L/2$.

In many cases, the magnitude of displacement, i.e., the deflection of the beam at the point of load application is of particular interest

$$\delta = |w(L/2)| = \frac{PL^3 d_{11}}{48b} + \frac{PL}{4h_c bG_{xz}}. \tag{4.23}$$

The bending compliance, d_{11}, is often replaced by the bending stiffness, D_x, in order to obtain a consistent formulation expressing the beam deflection in bending and shear stiffnesses rather than the mixed formulation in Equation (4.23)

$$D_x = \frac{1}{d_{11}}. \tag{4.24}$$

This yields

$$\delta = \frac{PL^3}{48bD_x} + \frac{PL}{4h_c bG_{xz}}. \tag{4.25}$$

This equation is often used as a basis for experimental determination of bending and shear stiffnesses of a sandwich beam (D_x and $h_c bG_{xz}$) as will be discussed later.

4.3 Simplified Analysis of Bending Stiffness of Sandwich Beams

Bending stiffness calculation in general requires calculation of the A, B, C, D stiffness matrices, compilation of those in the form of the 6×6 matrix in Equation (4.1), and inversion of the 6×6 matrix to arrive at the 6×6 compliance matrix in Equation (4.2) so that the bending compliance element, d_{11}, and bending stiffness, D_x (Equation (4.24)) can be determined. A set of matrix manipulations to accomplish this purpose are listed in Appendix B. Alternatively, the 6×6 $ABCD$ matrix may directly be inverted using a matrix solver. While this procedure is formally straightforward, it does not lend itself to "backside of the envelope" estimates of the bending stiffness, D_x.

A simplified analysis for calculations of D_x may be obtained by assuming a one-dimensional (plane) state of stress in each face sheet. Hence, the following simplified stress-strain relations apply:

$$\sigma_x(1) = E_1 \varepsilon_x(1), \tag{4.26a}$$

$$\sigma_x(2) = E_2 \varepsilon_x(2), \tag{4.26b}$$

where E_1 and E_2 are the effective Young's moduli of the bottom and top face sheets (1 and 2) (Section 3.1.5) respectively. The strain, ε_x, is given by Equation (4.7a)

$$\varepsilon_x = \varepsilon_x^0 + z\kappa_x, \tag{4.27}$$

where ε_x^0 is the mid-core strain, z is the thickness coordinate ($z = 0$ at the center of the core), Figure 4.3, and κ_x is the curvature at $z = 0$. Consequently,

$$\varepsilon_x(1) = \varepsilon_x^0 - \frac{(h_c - h_1)}{2}\kappa_x, \tag{4.28a}$$

$$\varepsilon_x(2) = \varepsilon_x^0 + \frac{(h_c + h_2)}{2}\kappa_x. \tag{4.28b}$$

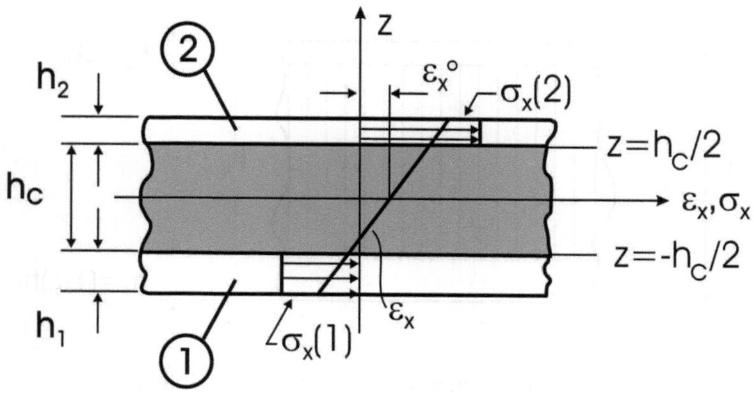

Figure 4.3 Variation of strain and stress in a sandwich beam under bending.

Integration of the stresses over the cross-section yields the axial force resultant, N_x, and bending moment, M_x, according to Equations (3.39a, b)

$$N_x = A\varepsilon_x^0 + B\kappa_x, \tag{4.29a}$$

$$M_x = C\varepsilon_x^0 + D\kappa_x, \tag{4.29b}$$

with the stiffness A, B, C, D given by

$$A = A_1 + A_2, \tag{4.30a}$$

$$B = \frac{1}{2}((h_c + h_2)A_2 - (h_c + h_1)A_1), \tag{4.30b}$$

$$C = B, \tag{4.30c}$$

$$D = \frac{1}{4}((h_c + h_2)^2 A_2 + (h_c + h_1)^2 A_1), \tag{4.30d}$$

where A_1 and A_2 are the extensional stiffnesses of the face sheets

$$A_1 = E_1 h_1, \tag{4.31a}$$

$$A_2 = E_2 h_2. \tag{4.31b}$$

Notice that Equation (4.29 constitutes the one-dimensional form of Equations (3.41) and (3.42). If the sandwich is symmetric, $h_1 = h_2 = h_f$, and $E_1 = E_2 = E_f$, Equations (4.30) and (4.31) give

$$A = 2E_f h_f, \tag{4.32a}$$

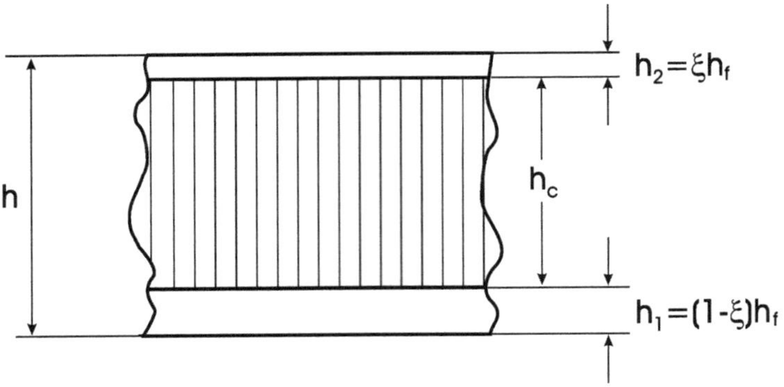

Figure 4.4 Geometrically unsymmetric sandwich.

$$B = C = 0, \tag{4.32b}$$

$$D = \frac{(h_c + h_f)^2}{2} E_f h_f. \tag{4.32c}$$

It is observed that the bending stiffness of a symmetric sandwich beam with thin faces is governed by the extensional stiffness of the face sheets, i.e. the product of face modulus and face thickness. Notice that Equation (4.32c) is identical to Equation (2.8) (with $d = h_c + h_f$) derived using the parallel axis theorem for a symmetric sandwich beam with thin faces.

4.3.1 Bending without Axial Force

Unsymmetric beams are characterized by non-zero B and C stiffnesses. For bending of such a beam without axial force, substitution of $N_x = 0$ in Equation (4.29a) gives

$$\varepsilon_x^0 = \frac{-B\kappa_x}{A}. \tag{4.33}$$

Substitution into Equation (4.29b) with $C = B$ yields

$$M_x = \left(D - \frac{B^2}{A}\right)\kappa_x, \tag{4.34}$$

which enables calculation of the bending stiffness, D_x (ratio of M_x/κ_x)

$$D_x = D - \frac{B^2}{A}. \tag{4.35}$$

Since $B \neq 0$ for unsymmetric beams, this equation shows that such beams will have reduced bending stiffness compared to a symmetric beam. Substitution of Equations (4.30) into (4.35) yields

$$D_x = \frac{(2h_c + h_1 + h_2)^2 A_1 A_2}{4(A_1 + A_2)}. \tag{4.36}$$

To illustrate the influence of asymmetry on the bending stiffness of a sandwich beam, consider the asymmetric sandwich beam shown in Figure 4.4. The faces are assumed to be of the same material, i.e., $E_1 = E_2 = E_f$, but of different thicknesses, while the sum of the face sheet thicknesses, $h_1 + h_2 = h_f$, is constant,

$$h_1 = (1 - \xi)h_f, \tag{4.37a}$$

$$h_2 = \xi h_f. \tag{4.37b}$$

The parameter, $0 = \xi = 1$, partitions the thicknesses of the faces. Notice that the total thickness, h, of the sandwich is constant, irrespective of the value of ξ.

It is recognized that $\xi = 0.5$ corresponds to a symmetric sandwich, while any other value corresponds to a (geometrically) unsymmetric panel. It should be pointed out that asymmetry could also be introduced by using different face materials ($E_1 \neq E_2$).

For the sandwich in Figure 4.4, it may be readily shown that the bending stiffness is given by

$$D_x = \frac{(2h_c + h_f)^2 E_c h_f \xi(1 - \xi)}{4}. \tag{4.38}$$

D_x may be normalized with its (maximum) value for a symmetric sandwich beam ($\xi = 0.5$)

$$D_x^s = \frac{(2h_c + h_f)^2 E h_f}{16}. \tag{4.39}$$

Figure 4.5 shows D_x plotted vs. the thickness parameter ξ. It is observed that the asymmetry reduces the bending stiffness slightly when ξ is close to 0.5. The curve is symmetric around $\xi = 0.5$ verifying invariance of bending stiffness if the thicker face is on top or bottom. For highly asymmetric beams, where one face is much stiffer than the other, the reduction in bending stiffness is quite severe.

The results here demonstrate that the optimal bending stiffness is achieved by making sandwich beams symmetric. Other design considerations, however, may demand an asymmetric sandwich (see, e.g., Satapathy and Vinson, 2000).

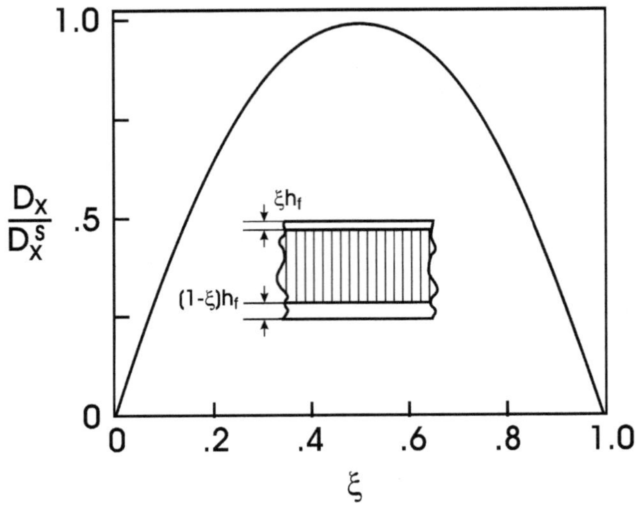

Figure 4.5 Influence of asymmetry on bending stiffness of sandwich beam.

4.4 Three-Point Flexure Testing of Sandwich Beams

Testing and data reduction for a sandwich beam loaded in three-point flexure (Figure 4.2) will be discussed. Testing of sandwich beams may be conducted for a variety of reasons. For testing in the linear-elastic regime at small loads, the purpose is most commonly determination of the shear and bending stiffnesses of the beam. Testing the beam to failure may be done to determine the tensile or compressive strengths of the faces or the shear strength of the core. These failure modes may be achieved by specific design of the sandwich beam and testing geometry.

4.4.1 Evaluation of Flexural and Shear Stiffnesses

Three-point flexure loading of sandwich beams with a soft core or web core is complicated by local out-of-plane deformations of the face sheets at support and load introduction regions, as schematically illustrated in Figure 4.6. For web-core sandwich beams, indentation is accentuated if the loading and support locations coincide with unsupported regions of the web core. If the core is very weak in the out-of-plane direction, the sandwich beam could be reinforced by removing a section of the core at load introduction and support

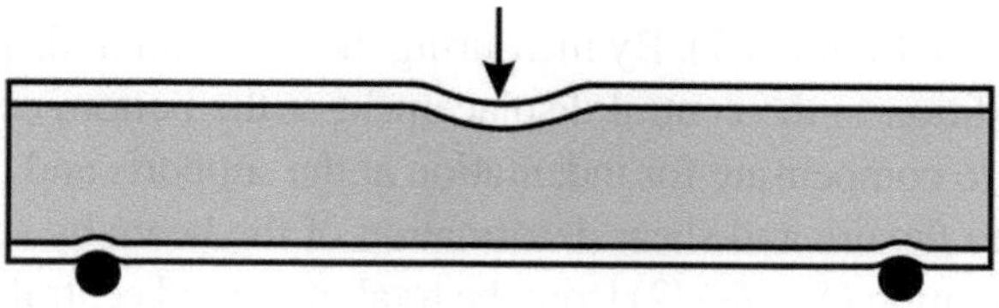

Figure 4.6 Local deformation of a face sheet in a flexure-loaded sandwich beam.

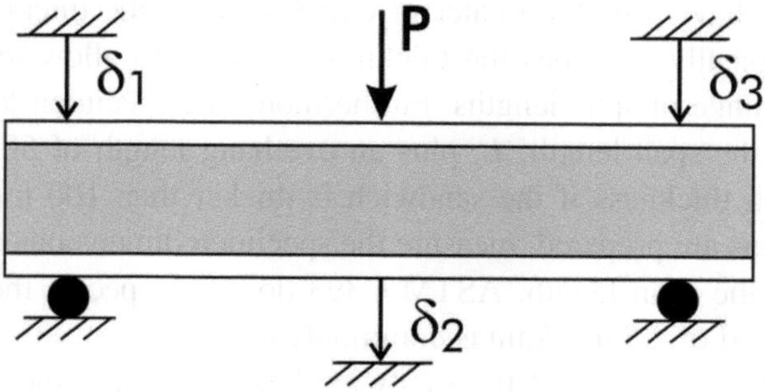

Figure 4.7 Method to compensate for the indentation at supports in three-point flexure testing.

regions and replace it with a stronger (high density) core. Such reinforcements, however, tend also to increase the shear stiffness of the beam.

Local deformation under the central load introduction line is expected to be more severe than at the supports because this load is twice as large as the support loads. If local deformation is expected or detected in trial tests, the sandwich beam flexure test standard ASTM C393 (2000) recommends placing narrow strips between load introduction and support points and the sandwich in order to distribute the load over a larger area. Rubber pads may also be used. ASTM C393 also provides some general instructions for the selection of testing machines, fixturing, and instrumentation. It is common procedure to use circular steel rods of 25.4 mm diameter at the load introduction and support regions. For measuring deflection, δ, of the beam under the central loading line, a displacement gage with an accuracy within ± 0.025 mm should be used. A linear variable differential transformer (LVDT) (see Adams et al., 2003), is often a good choice.

A method to compensate for the possible influence of local indentation at the support regions in a three-point flexure test (Figure 4.3), was implemented by Alfredsson et al. (2007) by placing two additional LVDTs above the supports (see Figure 4.7). By measuring the indentation displacements δ_1 and δ_3, in addition to the central displacement at the bottom of the beam δ_2, it is possible to compensate for indentation at the supports and isolate the deflection due to flexure and shear deformation of the beam by subtracting the average indentation $(\delta_1+\delta_3)/2)$ from the total measured central displacement δ_2. This method also compensates for load-cell and fixture deformations.

The ASTM C293 standard stipulates that the sandwich beam test specimen should have a width greater or equal to twice the thickness, and the length is typically 25 times the thickness in order to allow testing over a reasonable range of span lengths. Furthermore, the specimen length should be equal to the span length, L, plus an overhang length of 50 mm or half the sandwich thickness if the sandwich is thicker than 100 mm. After the test specimens are prepared, measure the specimen dimensions at several locations and the span length. ASTM C393 does not specify the cross-head speed. A speed of 2.5 mm/min is commonly used.

The bending stiffness of the sandwich beam, D_x, and the shear modulus, G_{xz}, of the core may be determined using the three-point flexure test. Consider the deflection formula (Equation (4.25))

$$\delta = \frac{PL^3}{48bD_x} + \frac{PL}{4h_c bG_{xz}}. \tag{4.40}$$

The compliance, $C = \delta/P$, of the three-point flexure specimen becomes

$$C = \frac{L^3}{48bD_x} + \frac{L}{4h_c bG_{xz}}. \tag{4.41}$$

Allen (1969) proposed data reduction for the three-point flexure test based on Equation (4.41) expressed in two ways

$$\frac{C}{L} = \frac{L^2}{48bD_x} + \frac{1}{4bh_c G_{xz}}, \tag{4.42a}$$

$$\frac{C}{L^3} = \frac{1}{4bh_c G_{xz}} \frac{1}{L^2} + \frac{1}{48bD_x}. \tag{4.42b}$$

Consequently, if the compliance of a sandwich beam loaded in three-point flexure is determined over a range of span lengths and the results are plotted in linear graphs, i.e., C/L vs. L^2, and C/L^3 vs. $1/L^2$, see Figure 4.8, it is possible to determine the flexural and shear stiffnesses from the slopes m_1 and m_2 of the graphs, according to Equation (4.42)

$$m_1 = \frac{1}{48bD_x}, \tag{4.43a}$$

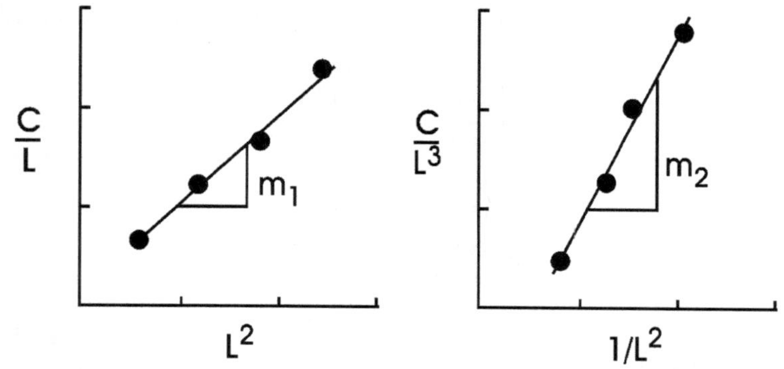

Figure 4.8 Graphical evaluation of bending and shear stiffnesses of a sandwich beam.

$$m_2 = \frac{1}{4bh_c G_{xz}}. \tag{4.43b}$$

The slopes are best evaluated by fitting a linear equation to the data. Once the slopes are established, the required quantities, D_x and G_{xz}, are easily solved.

Example

To demonstrate the procedure for determining bending and shear stiffness, actual test data for a sandwich beam determined in a laboratory course at Florida Atlantic are considered. The sandwich beam consisted of 3.02 mm thick 6061-T6 aluminum face sheets over a 26 mm thick H80 PVC foam core. The total length of the beam was 610 mm and the width (b) 52 mm. Notice that the width is twice the beam thickness, and the length about 24 times the beam thickness satisfying the ASTM C293 requirements. The beam was tested in three-point flexure at span lengths, L, of 25.4, 35.6, 45.7, and 55.9 cm. The deflection, δ, under the central load was measured with a LVDT. The load levels were adjusted to obtain a reasonable part of the initial linear response region without causing yielding of the face sheets and core. Span lengths and measured compliance values are listed in Table 4.1.

First, it is verified that the face sheets may be considered "thin". Equation (1.7) stipulates that a sandwich with a core-to-face thickness ratio greater than 5.35 may be considered as "thin faced". For the specific panel considered, the core-to-face thickness ratio is $26/3.02 = 8.61$. Hence, we can apply the theory outlined above. Figure 4.9 shows the compliance data listed in Table 4.1 formatted and graphed according to Figure 4.8.

Table 4.1 Span lengths and compliance values for a three-point flexure loaded sandwich beam.

L (cm)	C (μm/N)
25.4	1.59
35.6	2.38
45.7	3.18
55.9	4.13

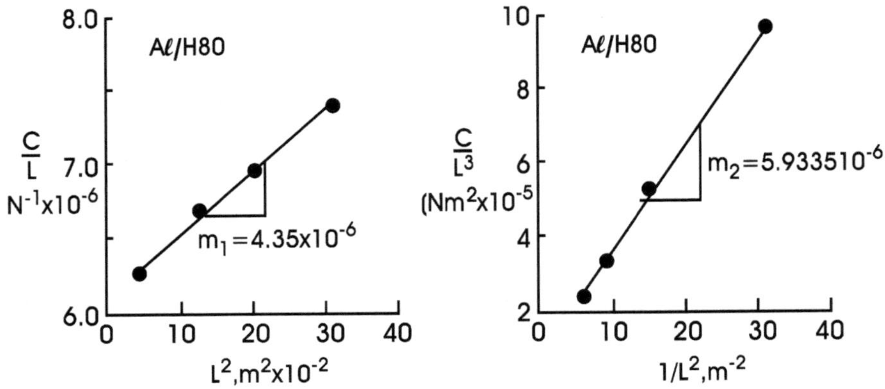

Figure 4.9 Determination of bending and shear stiffnesses for sandwich panel consisting of aluminum faces and H80 PVC foam core.

Fitting straight lines to the data allowed determination of the slopes, m_1 and m_2,

$$m_1 = 4.35 \times 10^{-6}\ (\mathrm{Nm}^2)^{-1},$$

$$m_2 = 5.933 \times 10^{-6}\ \mathrm{N}^{-1}.$$

Solving for D_x and G_{xy}, using the slopes and Equations (4.43) yields $D_x = 9.21 \times 10^4$ Nm, and $G_{xz} = 31.2$ MPa. Calculation of the face modulus from the value of D_x using Equation (4.32c) yields $E_f = 72.5$ GPa which is in close agreement with the value listed for aluminum in Table 1.1. The core shear modulus determined here (31.2 MPa) is very close to that provided for H80 PVC foam by DIAB, $G_{xz} = 31$ MPa (www.diabgroup.com).

4.4.2 Evaluation of Core Shear Strength and Face Compression Strength

In addition to measurement of the elastic flexural and shear stiffnesses, the three-point flexure test may be used to determine the shear strength of the core and compression strength of the face sheets. Such testing requires specific design of the test since both failure modes will not occur simultaneously. To achieve the desired failure mode of the beam, the stresses in the face sheets and core needs to be determined. We will here examine a symmetric sandwich beam, i.e. with identical face sheets, each of thickness h_f and elastic modulus E_f bonded to a low modulus core.

The tensile and compressive stresses in the face sheets are obtained from Equations (4.26), (4.28), (4.33) and (4.34).

$$\sigma_x(1) = \frac{-E_f(h_c + h_f)M_x}{2D}, \tag{4.44a}$$

$$\sigma_x(2) = \frac{E_f(h_c + h_f)M_x}{2D}, \tag{4.44b}$$

where 1 and 2 refer to the bottom and top face sheets, respectively, and h_f and h_c are the face and core thicknesses. M_x is the bending moment per unit width and D is bending stiffness. Substitution of the expression for the bending stiffness Equation (4.32c), yields

$$\sigma_x(1) = \frac{-M_x}{(h_c + h_f)h_f}, \tag{4.45a}$$

$$\sigma_x(2) = \frac{M_x}{(h_c + h_f)h_f}. \tag{4.45b}$$

Consequently, the magnitudes of bending stress are the same in both face sheets.

To determine the variation of the magnitude of bending stress ($|\sigma_x(1)| = |\sigma_x(2)| = \sigma_x$) along the span ($x$), the bending moment given by Equation (4.11) is substituted into (4.45)

$$\sigma_x(x) = \frac{Px}{2b(h_c + h_f)h_f}, \quad 0 \le x \le \frac{L}{2}, \tag{4.46}$$

where $\sigma_x(x)$ is the magnitude of the stress. Notice that the stress in the lower face sheet (#1) is tensile (> 0) and the stress in the top face (#2) is compressive. The highest stress $\sigma_x(x)$ occurs at the center span, $x = L/2$,

$$\sigma_x(\max) = \frac{PL}{4b(h_f + h_c)h_f}. \tag{4.47}$$

Another possible failure mode of the sandwich is core shear failure if the core shear stress, τ_{xz}, becomes excessive. Hence, analysis of the core shear stress is required. The transverse shear force, Q_x, is determined from the variation of bending moment, i.e., using the equilibrium equation (3.58c), applied to (4.11),

$$Q_x = \frac{dM_x}{dx} = \frac{-P}{2b}. \tag{4.48}$$

The core shear stress τ_{xz} is determined from the assumption that the core carries the entire shear force

$$\tau_{xz} = \frac{-P}{2bh_c}. \tag{4.49}$$

Consequently, the shear stress is uniform though the thickness of the core and along the length of the beam (although a sign reversal occurs at $x = L/2$).

The ratio of maximum bending stress in the face sheet and shear stress in the core (Equations (4.47) and (4.49)) becomes

$$\frac{\sigma_x(\max)}{\tau_{xz}} = \frac{L}{2h_f(1 + h_f/h_c)}. \tag{4.50}$$

This expression can be utilized for determining the geometry where a transition from core shear failure and face sheet failure occurs. Consider, for example, a sandwich beam with 1 mm thick aluminum face sheets over a 30 mm thick PVC H100 foam core. Face and core material properties are listed in Tables 1.1 and 1.3. In Table 1.1, it is noted that the tensile and compressive strengths of aluminum are the same ($X_L = X_C = 414$ MPa). The core shear strength is 1.4 MPa. Substitution of the face and core strengths and thicknesses into Equation (4.50) yields a transition span length, $L = 0.61$ m. Hence, if such a beam is loaded in flexure at a span length $L < 0.61$ m, the beam would fail by core shear. Beams with span lengths greater than 0.61 m would fail by face failure. Equation (4.50) may be also used for sizing the thickness of the face sheets so that the beam fails in the desired failure mode. Tensile failure rarely occurs in symmetric sandwich beams since structural materials used as face sheets typically are less strong in compression than in tension, see Table 1.1. If the material is stronger in tension than in compression, and tensile failure of such a beam is desired, a thicker top face must be used. This corresponds to an unsymmetric sandwich, however, and strength determination requires analysis of this specific case with $C = B \neq 0$ in Section 4.3.

A sandwich beam could also fail by face sheet wrinkling on the compression side, as discussed in Chapter 8. The face wrinkling stress, σ_w, determined using Equation (8.13) is $\sigma_f = 614$ MPa ($E_f = 73$ GPa, $E_z^c = 105$ MPa, $G_{xz}^c = 40$ MPa). Since σ_w exceeds the strength of aluminum (414 MPa), this particular sandwich beam is expected to fail on the compression side by face yielding failure.

To measure the core shear strength using the three-point flexure test, it is necessary to set the span length, L, below the transitional value provided by Equation (4.50). If a nominal value of the core shear strength is not available, it is necessary to reduce the span length until the specimen actually fails by core shear. For determination of the strength of the face sheet, it is necessary to use a long span length L or thin face h_f according to Equation (4.50). If the strength ratio is not known, it is necessary to test specimens at increasing span lengths until face failure is observed. The possibility for competing failure modes for the compression loaded face sheet, such as local buckling (face wrinkling or face dimpling) (Chapter 8) should be examined.

Chapter 5
Elasticity Solutions for Sandwich Structures

This chapter presents the theory of elasticity solutions for sandwich plates or shells. Elasticity solutions are significant because they provide a benchmark for assessing the performance of the various plate or shell theories or various numerical methods such as the finite element method. Most of these solutions are an extension of the corresponding solutions for monolithic anisotropic bodies which have been developed primarily by Lekhnitskii (1963). This chapter does not cover all problems of the theory of elasticity for sandwich bodies, but presents only some of the most studied ones in an attempt to collect the accumulated recent progress in this field. Section 5.1 on sandwich rectangular plates is adapted from Pagano (1970a), which was extended to the case of positive discriminant materials by Kardomateas (2008a) and Section 5.2 on sandwich shells from Kardomateas (2001).

5.1 A Rectangular Sandwich Plate with Orthotropic Face Sheets and Core

We consider a sandwich plate consisting of orthotropic face sheets of thickness $h_1 = f_1$ and $h_2 = f_2$ and an orthotropic core of thickness $h_c = 2c$, such that the various axes of elastic symmetry are parallel to the plate axes x, y, and z (Figure 5.1). The plate is simply supported. A normal traction $\sigma_z = q_0(x, y)$ is applied on the upper surface but the lower surface is traction-free.

Let us denote each layer by i where $i = f_1$ for the upper face-sheet, $i = c$ for the core and $i = f_2$ for the lower face-sheet. Then, for each layer, the orthotropic strain-stress relations are

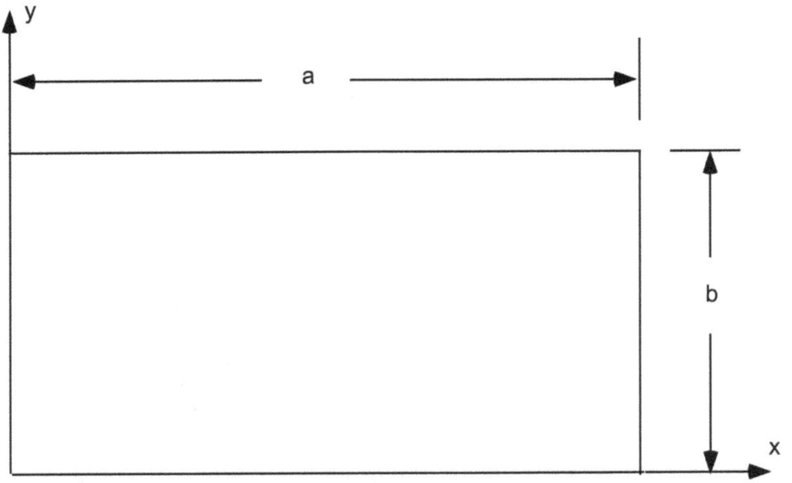

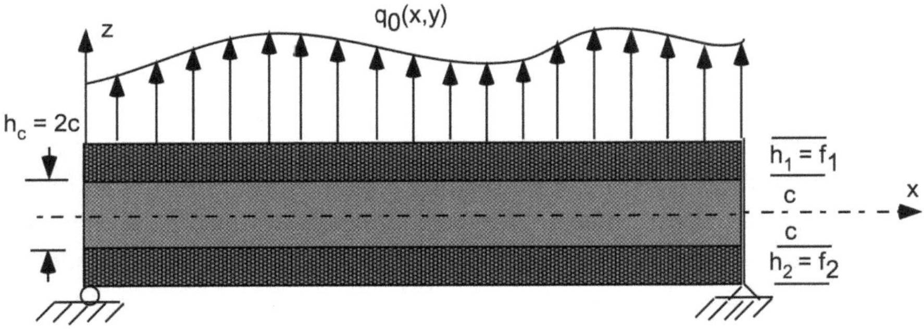

Figure 5.1 Definition of geometry and loading for the sandwich plate.

$$\begin{bmatrix} \sigma_{xx}^{(i)} \\ \sigma_{yy}^{(i)} \\ \sigma_{zz}^{(i)} \\ \tau_{yz}^{(i)} \\ \tau_{xz}^{(i)} \\ \tau_{xy}^{(i)} \end{bmatrix} = \begin{bmatrix} c_{11}^i & c_{12}^i & c_{13}^i & 0 & 0 & 0 \\ c_{12}^i & c_{22}^i & c_{23}^i & 0 & 0 & 0 \\ c_{13}^i & c_{23}^i & c_{33}^i & 0 & 0 & 0 \\ 0 & 0 & 0 & c_{44}^i & 0 & 0 \\ 0 & 0 & 0 & 0 & c_{55}^i & 0 \\ 0 & 0 & 0 & 0 & 0 & c_{66}^i \end{bmatrix} \begin{bmatrix} \varepsilon_{xx}^{(i)} \\ \varepsilon_{yy}^{(i)} \\ \varepsilon_{zz}^{(i)} \\ \gamma_{yz}^{(i)} \\ \gamma_{xz}^{(i)} \\ \gamma_{xy}^{(i)} \end{bmatrix}, \quad (i = f_1, c, f_2), \tag{5.1}$$

where c_{ij}^i are the stiffness constants (we have used the notation $1 \equiv x$, $2 \equiv y$, $3 \equiv z$).

Using the strain-displacement relations

$$\varepsilon_{xx} = u_{,x}; \quad \varepsilon_{yy} = v_{,y}; \quad \varepsilon_{zz} = w_{,z}, \tag{5.2a}$$

$$\gamma_{yz} = w_{,y} + v_{,z}; \quad \gamma_{xz} = u_{,z} + w_{,x}; \quad \gamma_{xy} = u_{,y} + v_{,x}, \tag{5.2b}$$

and the equilibrium relations

$$\sigma_{xx,x} + \tau_{xy,y} + \tau_{xz,z} = 0, \tag{5.3a}$$

$$\tau_{xy,x} + \sigma_{yy,y} + \tau_{yz,z} = 0, \tag{5.3b}$$

$$\tau_{xz,x} + \tau_{yz,y} + \sigma_{zz,z} = 0, \tag{5.3c}$$

leads to the following governing field equations in terms of the displacements for each layer:

$$c_{11}^i u_{,xx} + c_{66}^i u_{,yy} + c_{55}^i u_{,zz} + (c_{12}^i + c_{66}^i) v_{,xy} + (c_{13}^i + c_{55}^i) w_{,xz} = 0, \tag{5.4a}$$

$$(c_{12}^i + c_{66}^i) u_{,xy} + c_{66}^i v_{,xx} + c_{22}^i v_{,yy} + c_{44}^i v_{,zz} + (c_{23}^i + c_{44}^i) w_{,yz} = 0, \tag{5.4b}$$

$$(c_{13}^i + c_{55}^i) u_{,xz} + (c_{23}^i + c_{44}^i) v_{,yz} + c_{55}^i w_{,xx} + c_{44}^i w_{,yy} + c_{33}^i w_{,zz} = 0. \tag{5.4c}$$

In the following, we shall drop the superscript i that refers to the layers (core or face sheets) on the understanding that the derived relations will hold for each layer.

For a simply supported plate, an appropriate solution for the displacements would be in the form

$$u = U(z) \cos px \sin qy, \tag{5.5a}$$

$$v = V(z) \sin px \cos qy, \tag{5.5b}$$

$$w = W(z) \sin px \sin qy, \tag{5.5c}$$

where

$$p = n\pi/a; \quad q = m\pi/b \quad (n, m = 1, 2, 3, \ldots). \tag{5.5d}$$

These displacements, in conjunction with the corresponding strains and stresses from (5.2) and (5.1), would satisfy the simple support edge conditions:

$$\text{at} \quad x = 0, a: \quad w = v = \sigma_{xx} = 0. \tag{5.5e}$$

$$\text{at} \quad y = 0, b: \quad w = u = \sigma_{yy} = 0. \tag{5.5f}$$

Assuming that

$$[U(z), V(z), W(z)] = [U_0, V_0, W_0]e^{sz}, \tag{5.5g}$$

where U_0, V_0 and W_0 are constants, and substituting (5.5) into (5.4) results in the following system of algebraic equations:

$$(c_{11}p^2+c_{66}q^2-c_{55}s^2)U_0+(c_{12}+c_{66})pq\,V_0-(c_{13}+c_{55})ps\,W_0=0, \quad (5.6a)$$

$$(c_{12}+c_{66})pq\,U_0+(c_{22}q^2+c_{66}p^2-c_{44}s^2)V_0-(c_{23}+c_{44})qs\,W_0=0, \quad (5.6b)$$

$$(c_{13}+c_{55})ps\,U_0+(c_{23}+c_{44})qs\,V_0+(c_{55}p^2+c_{44}q^2-c_{33}s^2)W_0=0. \quad (5.6c)$$

Non-trivial solutions of this system exist only if the determinant of the coefficients vanishes, which leads to

$$A_0s^6+A_1s^4+A_2s^2+A_3=0, \tag{5.7}$$

where

$$A_0=-c_{33}c_{44}c_{55}, \tag{5.8a}$$

$$A_1=p^2\left[c_{44}(c_{11}c_{33}-c_{13}^2)+c_{55}(c_{33}c_{66}-2c_{13}c_{44})\right]+ \\ +q^2\left[c_{55}(c_{22}c_{33}-c_{23}^2)+c_{44}(c_{33}c_{66}-2c_{23}c_{55})\right], \tag{5.8b}$$

$$\begin{aligned}A_2=&-p^4\left[c_{66}(c_{11}c_{33}-c_{13}^2)+c_{55}(c_{11}c_{44}-2c_{13}c_{66})\right]\\&+p^2q^2\big[-c_{11}(c_{22}c_{33}-c_{23}^2)-2(c_{12}+c_{66})(c_{13}+c_{55})(c_{23}+c_{44})\\&-2c_{44}c_{55}c_{66}+2c_{11}c_{23}c_{44}+c_{12}c_{33}(c_{12}+2c_{66})+c_{13}c_{22}(c_{13}+2c_{55})\big]\\&-q^4\left[c_{66}(c_{22}c_{33}-c_{23}^2)+c_{44}(c_{22}c_{55}-2c_{23}c_{66})\right],\end{aligned} \tag{5.8c}$$

$$\begin{aligned}A_3=&\,p^6c_{11}c_{55}c_{66}+p^4q^2\left[c_{55}(c_{11}c_{22}-c_{12}^2)+c_{66}(c_{11}c_{44}-2c_{12}c_{55})\right]\\&+p^2q^4\left[c_{44}(c_{11}c_{22}-c_{12}^2)+c_{66}(c_{22}c_{55}-2c_{12}c_{44})\right]+q^6c_{22}c_{44}c_{66}.\end{aligned} \tag{5.8d}$$

With the substitution

$$\beta=s^2, \tag{5.9}$$

Equation (5.7), which defines the parameter s, can be written in the form of a cubic equation as

$$\beta^3+a_1\beta^2+a_2\beta+a_3=0, \quad a_i=A_i/A_0 \quad (i=1,2,3). \tag{5.10}$$

Let

$$Q=\frac{3a_2-a_1^2}{9}; \quad R=\frac{9a_1a_2-27a_3-2a_1^3}{54}; \quad D=Q^3+R^2. \tag{5.11a}$$

The last quantity, D, is the discriminant, which determines the nature of the solution.

5.1.1 Negative Discriminant D

If $D < 0$, then all roots are real and different as follows:

$$\beta_1 = 2\sqrt{-Q}\cos\left(\frac{\theta}{3}\right) - \frac{a_1}{3}, \tag{5.11b}$$

$$\beta_2 = 2\sqrt{-Q}\cos\left(\frac{\theta + 2\pi}{3}\right) - \frac{a_1}{3}, \tag{5.11c}$$

$$\beta_3 = 2\sqrt{-Q}\cos\left(\frac{\theta + 4\pi}{3}\right) - \frac{a_1}{3}, \tag{5.11d}$$

where

$$\cos\theta = R/\sqrt{-Q^3}\,. \tag{5.11e}$$

Corresponding to the three roots are the displacements functions defined in Equations (5.5a–c)

$$U(z) = \sum_{j=1}^{3} U_j(z); \quad V(z) = \sum_{j=1}^{3} V_j(z); \quad W(z) = \sum_{j=1}^{3} W_j(z). \tag{5.12}$$

If $\beta_j < 0$ then $s_j = \pm i\sqrt{|\beta_j|}$ and if we set

$$m_j = \sqrt{|\beta_j|}\,, \tag{5.13a}$$

then $s_j = \pm i m_j$. From (5.5g) for each pair of roots s_j, we can write

$$U_j(z) = U_{cj}\cos m_j z + U_{sj}\sin m_j z, \tag{5.13b}$$

$$V_j(z) = V_{cj}\cos m_j z + V_{sj}\sin m_j z, \tag{5.13c}$$

$$W_j(z) = W_{cj}\sin m_j z + W_{Sj}\cos m_j z, \tag{5.13d}$$

Substituting directly into (5.5a–c) and then into the equilibrium equations (5.4b, c), leads to the following two equations for V_{cj} and W_{sj}:

- from (5.4b) collecting terms of $\cos m_j z$:

$$V_{cj}\left(c_{66}p^2 + c_{22}q^2 + c_{44}m_j^2\right) - W_{sj}(c_{23}+c_{44})qm_j = -(c_{12}+c_{66})pqU_{cj}, \tag{5.13e}$$

- from (5.4c) collecting terms of $\sin m_j z$:

$$V_{cj}(c_{23}+c_{44})qm_j + W_{sj}\left(c_{55}p^2 + c_{44}q^2 + c_{33}m_j^2\right) = (c_{13}+c_{55})pm_jU_{cj}, \tag{5.13f}$$

These two equations can be solved for V_{cj} and W_{sj} in terms of U_{cj}. Similar equations can be derived for V_{sj} and W_{sj} by collecting terms of $\sin m_j z$ in (5.4b) and of $\cos m_j z$ in (5.4c). In the end, we obtain the following expressions:

$$U_j(z) = U_{cj} \cos m_j z + U_{sj} \sin m_j z, \tag{5.13g}$$

$$V_j(z) = \frac{B_j}{\Delta_j} U_{cj} \cos m_j z + \frac{B_j}{\Delta_j} U_{sj} \sin m_j z, \tag{5.13h}$$

$$W_j(z) = -\frac{C_j}{\Delta_j} U_{sj} \cos m_j z + \frac{C_j}{\Delta_j} U_{cj} \sin m_j z, \tag{5.13i}$$

where

$$\Delta_j = \left(c_{66}p^2 + c_{22}q^2 + c_{44}m_j^2\right)\left(c_{55}p^2 + c_{44}q^2 + c_{33}m_j^2\right) + (c_{23} + c_{44})^2 q^2 m_j^2, \tag{5.13j}$$

$$B_j = pq\big[-(c_{12} + c_{66})\left(c_{55}p^2 + c_{44}q^2 + c_{33}m_j^2\right) + (c_{13} + c_{55})(c_{23} + c_{44})m_j^2\big], \tag{5.13k}$$

$$C_j = pm_j\big[\left(c_{66}p^2 + c_{22}q^2 + c_{44}m_j^2\right)(c_{13} + c_{55}) + (c_{12} + c_{66})(c_{23} + c_{44})q^2\big]. \tag{5.13l}$$

If $\beta_j > 0$, we set

$$m_j = \sqrt{\beta_j}\,. \tag{5.14a}$$

By following an analogous procedure, we can write

$$U_j(z) = U_{cj} \cosh m_j z + U_{sj} \sinh m_j z, \tag{5.14b}$$

$$V_j(z) = \frac{B_j}{\Delta_j} U_{cj} \cosh m_j z + \frac{B_j}{\Delta_j} U_{sj} \sinh m_j z, \tag{5.14c}$$

$$W_j(z) = \frac{C_j}{\Delta_j} U_{cj} \sinh m_j z + \frac{C_j}{\Delta_j} U_{sj} \cosh m_j z, \tag{5.14d}$$

where

$$\Delta_j = \left(c_{66}p^2 + c_{22}q^2 - c_{44}m_j^2\right)\left(c_{55}p^2 + c_{44}q^2 - c_{33}m_j^2\right) + (c_{23} + c_{44})^2 q^2 m_j^2, \tag{5.14e}$$

$$B_j = -pq\big[(c_{12}+c_{66})\left(c_{55}p^2+c_{44}q^2-c_{33}m_j^2\right) + (c_{13}+c_{55})(c_{23}+c_{44})m_j^2\big], \tag{5.14f}$$

$$C_j = pm_j\big[-\left(c_{66}p^2+c_{22}q^2-c_{44}m_j^2\right)(c_{13}+c_{55}) + (c_{12}+c_{66})(c_{23}+c_{44})q^2\big]. \tag{5.14g}$$

Hence, the independent parameters are the six constants U_{c1}, U_{c2}, U_{c3}, U_{s1}, U_{s2}, U_{s3} which, for convenience, we rename g_1, g_2, g_3, g_4, g_5 and g_6, respectively. Then the displacements are as follows:

$$U(z) = d_{u1}g_1 + d_{u2}g_2 + d_{u3}g_3 + d_{u4}g_4 + d_{u5}g_5 + d_{u6}g_6, \tag{5.15a}$$

with the z-dependent coefficients defined for $j = 1, 2, 3$,

$$d_{uj} = \begin{cases} \cos m_j z, & \text{if } \beta_j < 0, \\ \cosh m_j z, & \text{if } \beta_j > 0, \end{cases} \tag{5.15b}$$

$$d_{u(j+3)} = \begin{cases} \sin m_j z, & \text{if } \beta_j < 0, \\ \sinh m_j z, & \text{if } \beta_j > 0. \end{cases} \tag{5.15c}$$

In the following expressions (5.16–5.20), Δ_j, B_j and C_j refer to (5.13j–l) if $\beta_j < 0$, and to (5.14e–g) if $\beta_j > 0$. With this remark we can set $V(z)$ in the form

$$V(z) = d_{v1}g_1 + d_{v2}g_2 + d_{v3}g_3 + d_{v4}g_4 + d_{v5}g_5 + d_{v6}g_6, \tag{5.16a}$$

where, again, for $j = 1, 2, 3$,

$$d_{vj} = \begin{cases} \dfrac{B_j}{\Delta_j}\cos m_j z, & \text{if } \beta_j < 0 \\ \dfrac{B_j}{\Delta_j}\cosh m_j z, & \text{if } \beta_j > 0, \end{cases} \tag{5.16b}$$

$$d_{v(j+3)} = \begin{cases} \dfrac{B_j}{\Delta_j}\sin m_j z, & \text{if } \beta_j < 0 \\ \dfrac{B_j}{\Delta_j}\sinh m_j z, & \text{if } \beta_j > 0, \end{cases} \tag{5.16c}$$

and

$$W(z) = d_{w1}g_1 + d_{w2}g_2 + d_{w3}g_3 + d_{w4}g_4 + d_{w5}g_5 + d_{w6}g_6, \tag{5.17a}$$

where the z-dependent coefficients again are defined for $j = 1, 2, 3$,

$$d_{wj} = \begin{cases} \dfrac{C_j}{\Delta_j} \sin m_j z, & \text{if} \quad \beta_j < 0, \\ \dfrac{C_j}{\Delta_j} \sinh m_j z, & \text{if} \quad \beta_j > 0, \end{cases} \tag{5.17b}$$

$$d_{w(j+3)} = \begin{cases} -\dfrac{C_j}{\Delta_j} \cos m_j z, & \text{if} \quad \beta_j < 0, \\ \dfrac{C_j}{\Delta_j} \cosh m_j z, & \text{if} \quad \beta_j > 0. \end{cases} \tag{5.17c}$$

The corresponding stresses are derived by substituting the above displacement expressions into (5.5), (5.1) and (5.2). We present the explicit expressions for the stresses σ_{zz}, τ_{yz} and τ_{xz} because these enter into the interface conditions. The stress σ_{zz} can be written in the form

$$\sigma_{zz} = (b_{zz1}g_1 + b_{zz2}g_2 + b_{zz3}g_3 + b_{zz4}g_4 + b_{zz5}g_5 + b_{zz6}g_6) \sin px \sin qy, \tag{5.18a}$$

The z-dependent coefficients are defined for $j = 1, 2, 3$ as

$$b_{zzj} = \begin{cases} -\left(c_{13}p + c_{23}q\dfrac{B_j}{\Delta_j} - c_{33}\dfrac{C_j}{\Delta_j}m_j\right) \cos m_j z, & \text{if} \quad \beta_j < 0 \\ -\left(c_{13}p + c_{23}q\dfrac{B_j}{\Delta_j} - c_{33}\dfrac{C_j}{\Delta_j}m_j\right) \cosh m_j z, & \text{if} \quad \beta_j > 0 \end{cases} \tag{5.18b}$$

$$b_{zz(j+3)} = \begin{cases} -\left(c_{13}p + c_{23}q\dfrac{B_j}{\Delta_j} - c_{33}\dfrac{C_j}{\Delta_j}m_j\right) \sin m_j z, & \text{if} \quad \beta_j < 0 \\ -\left(c_{13}p + c_{23}q\dfrac{B_j}{\Delta_j} - c_{33}\dfrac{C_j}{\Delta_j}m_j\right) \sinh m_j z, & \text{if} \quad \beta_j > 0 \end{cases} \tag{5.18c}$$

Next,

$$\tau_{yz} = \left(b_{yz1}g_1 + b_{yz2}g_2 + b_{yz3}g_3 + b_{yz4}g_4 + b_{yz5}g_5 + b_{yz6}g_6\right) \sin px \cos qy, \tag{5.19a}$$

with the z-dependent coefficients defined for $j = 1, 2, 3$ as

$$b_{yzj} = \begin{cases} c_{44}\left(q\dfrac{C_j}{\Delta_j} - m_j\dfrac{B_j}{\Delta_j}\right) \sin m_j z, & \text{if} \quad \beta_j < 0 \\ c_{44}\left(q\dfrac{C_j}{\Delta_j} + m_j\dfrac{B_j}{\Delta_j}\right) \sinh m_j z, & \text{if} \quad \beta_j > 0 \end{cases} \tag{5.19b}$$

$$b_{yz(j+3)} = \begin{cases} -c_{44}\left(q\dfrac{C_j}{\Delta_j} - m_j\dfrac{B_j}{\Delta_j}\right)\cos m_j z, & \text{if } \beta_j < 0 \\ c_{44}\left(q\dfrac{C_j}{\Delta_j} + m_j\dfrac{B_j}{\Delta_j}\right)\cosh m_j z, & \text{if } \beta_j > 0 \end{cases} \tag{5.19c}$$

Finally,

$$\tau_{xz} = (b_{xz1}g_1 + b_{xz2}g_2 + b_{xz3}g_3 + b_{xz4}g_4 + b_{xz5}g_5 + b_{xz6}g_6)\cos px \sin qy, \tag{5.20a}$$

with the z-dependent coefficients defined for $j = 1, 2, 3$ as

$$b_{xzj} = \begin{cases} c_{55}\left(p\dfrac{C_j}{\Delta_j} - m_j\right)\sin m_j z, & \text{if } \beta_j < 0 \\ c_{55}\left(p\dfrac{C_j}{\Delta_j} + m_j\right)\sinh m_j z, & \text{if } \beta_j > 0 \end{cases} \tag{5.20b}$$

$$b_{xz(j+3)} = \begin{cases} -c_{55}\left(p\dfrac{C_j}{\Delta_j} - m_j\right)\cos m_j z, & \text{if } \beta_j < 0 \\ c_{55}\left(p\dfrac{C_j}{\Delta_j} + m_j\right)\cosh m_j z, & \text{if } \beta_j > 0 \end{cases} \tag{5.20c}$$

5.1.2 Positive Discriminant D

If $D > 0$, where the discriminant D is defined in (5.11a), then the cubic equation (5.10) has one real root and two complex conjugates.

With R and D defined in (5.11a), we further define

$$S = \sqrt[3]{R + \sqrt{D}}\,; \quad T = \sqrt[3]{R - \sqrt{D}}. \tag{5.21a}$$

Then with

$$\mu_R = -\frac{1}{2}(S+T) - \frac{a_1}{3}\,; \quad \mu_I = \frac{1}{2}\sqrt{3}(S-T), \tag{5.21b}$$

the two complex conjugate roots are

$$\beta_1 = \mu_R + i\mu_I; \quad \beta_2 = \mu_R - i\mu_I. \tag{5.21c}$$

The real root is

$$\beta_3 = S + T - \frac{a_1}{3}\,. \tag{5.21d}$$

The real root is dealt with in the same manner as for the case of a negative discriminant (Section 5.1.1).

Next we shall consider how to deal with the complex conjugate roots. In terms of the modulus r and amplitude θ of these complex numbers,

$$r = \sqrt{\mu_R^2 + \mu_I^2}\,; \quad \theta = \tan^{-1}\left(\frac{\mu_I}{\mu_R}\right) \tag{5.21e}$$

these roots can be set in the form

$$\beta_1 = r(\cos\theta + i\,\sin\theta); \quad \beta_2 = r(\cos\theta - i\,\sin\theta). \tag{5.21f}$$

From (5.9), we now seek the square roots of β_1 and β_2. Thus, in terms of

$$\gamma_1 = \sqrt{r}\cos\frac{\theta}{2}\,; \quad \gamma_2 = \sqrt{r}\sin\frac{\theta}{2}, \tag{5.21g}$$

the corresponding roots of the sixth-order equation (5.7), s_i, are

$$s_{1,2} = \pm(\gamma_1 + i\gamma_2); \quad s_{3,4} = \pm(\gamma_1 - i\gamma_2). \tag{5.21h}$$

Corresponding to these four roots, the displacement functions take the form

$$\begin{aligned} U_\eta(z) &= a_{1\eta}e^{\gamma_1 z}\cos\gamma_2 z + a_{2\eta}e^{\gamma_1 z}\sin\gamma_2 z \\ &\quad + a_{3\eta}e^{-\gamma_1 z}\cos\gamma_2 z + a_{4\eta}e^{-\gamma_1 z}\sin\gamma_2 z, \end{aligned} \tag{5.22}$$

where $\eta = u, v, w$ corresponds to the U, V, W displacements and the $a_{i\eta}$ are constants. Of the 12 constants appearing in (5.22) only four are independent. The eight relations that exist among these constants are found by substituting the displacements along with (5.5) into the equilibrium equations (5.4).

For convenience, let us set

$$r_1 = c_{44}(\gamma_1^2 + \gamma_2^2) + c_{66}p^2 + c_{22}q^2, \tag{5.23a}$$

$$r_2 = c_{44}(\gamma_1^2 + \gamma_2^2) - c_{66}p^2 - c_{22}q^2, \tag{5.23b}$$

$$r_3 = c_{55}(\gamma_1^2 + \gamma_2^2) + c_{11}p^2 + c_{66}q^2, \tag{5.23c}$$

$$r_4 = c_{55}(\gamma_1^2 + \gamma_2^2) - c_{11}p^2 - c_{66}q^2, \tag{5.23d}$$

and

$$e_1 = r_1(c_{13} + c_{55}) - q^2(c_{12} + c_{66})(c_{23} + c_{44}), \tag{5.23e}$$

$$e_2 = r_2(c_{13} + c_{55}) + q^2(c_{12} + c_{66})(c_{23} + c_{44}), \tag{5.23f}$$

$$e_3 = r_3(c_{23} + c_{44}) - p^2(c_{12} + c_{66})(c_{13} + c_{55}), \tag{5.23g}$$

$$e_4 = r_4(c_{23} + c_{44}) + p^2(c_{12} + c_{66})(c_{13} + c_{55}), \tag{5.23h}$$

In this way, we obtain the following relations for the coefficients in the displacement expression for $V(z)$, Equation (5.22), in terms of the coefficients in the expression for $U(z)$:

$$a_{1v} = \xi_{11}a_{1u} + \xi_{12}a_{2u}, \quad a_{2v} = \xi_{21}a_{1u} + \xi_{22}a_{2u}, \tag{5.24a}$$

$$a_{3v} = \xi_{33}a_{3u} + \xi_{34}a_{4u}, \quad a_{4v} = \xi_{43}a_{3u} + \xi_{44}a_{4u}, \tag{5.24b}$$

where

$$\xi_{11} = \xi_{22} = \xi_{33} = \xi_{44} = \frac{q(e_1e_3\gamma_2^2 + e_2e_4\gamma_1^2)}{p(\gamma_2^2e_1^2 + \gamma_1^2e_2^2)}. \tag{5.24c}$$

$$\xi_{12} = -\xi_{21} = -\xi_{34} = \xi_{43} = \frac{q\gamma_1\gamma_2(e_2e_3 - e_1e_4)}{p(\gamma_2^2e_1^2 + \gamma_1^2e_2^2)}. \tag{5.24d}$$

Also, the following relations for the coefficients in the expression for $W(z)$, Equation (5.22), in terms of the coefficients in the expression for $U(z)$:

$$a_{1w} = f_{11}a_{1u} + f_{12}a_{2u}, \quad a_{2w} = f_{21}a_{1u} + f_{22}a_{2u}, \tag{5.25a}$$

$$a_{3w} = f_{33}a_{3u} + f_{34}a_{4u}, \quad a_{4w} = f_{43}a_{3u} + f_{44}a_{4u}, \tag{5.25b}$$

where

$$\begin{aligned} f_{11} &= f_{22} = -f_{33} = -f_{44} \\ &= \frac{(c_{12} + c_{66})pq\gamma_1 - r_2\gamma_1\xi_{11} - r_1\gamma_2\xi_{21}}{q(c_{23} + c_{44})(\gamma_1^2 + \gamma_2^2)}, \end{aligned} \tag{5.25c}$$

$$\begin{aligned} f_{12} &= -f_{21} = f_{34} = -f_{43} \\ &= -\frac{(c_{12} + c_{66})pq\gamma_2 + r_2\gamma_1\xi_{12} + r_1\gamma_2\xi_{22}}{q(c_{23} + c_{44})(\gamma_1^2 + \gamma_2^2)}. \end{aligned} \tag{5.25d}$$

Now, coming to the real root β_3, we set

$$m_3 = \sqrt{|\beta_3|}, \tag{5.26}$$

then if $\beta_3 < 0$ and following (5.13g–l) we can write

$$U_3(z) = a_{5u} \cos m_3z + a_{6u} \sin m_3z, \tag{5.27a}$$

$$V_3(z) = \frac{B_3}{\Delta_3} a_{5u} \cos m_3 z + \frac{B_3}{\Delta_3} a_{6u} \sin m_3 z, \tag{5.27b}$$

$$W_3(z) = -\frac{C_3}{\Delta_3} a_{6u} \cos m_3 z + \frac{C_3}{\Delta_3} a_{5u} \sin m_3 z, \tag{5.27c}$$

where

$$\Delta_3 = \left(c_{66}p^2 + c_{22}q^2 + c_{44}m_3^2\right)\left(c_{55}p^2 + c_{44}q^2 + c_{33}m_3^2\right) + (c_{23}+c_{44})^2 q^2 m_3^2, \tag{5.27d}$$

$$B_3 = pq\big[-(c_{12}+c_{66})\left(c_{55}p^2 + c_{44}q^2 + c_{33}m_3^2\right) + (c_{13}+c_{55})(c_{23}+c_{44})m_3^2\big], \tag{5.27e}$$

$$C_3 = pm_3\big[\left(c_{66}p^2 + c_{22}q^2 + c_{44}m_3^2\right)(c_{13}+c_{55}) + (c_{12}+c_{66})(c_{23}+c_{44})q^2\big]. \tag{5.27f}$$

If $\beta_3 > 0$ then, following (5.14b–g)

$$U_3(z) = a_{5u} \cosh m_3 z + a_{6u} \sinh m_3 z, \tag{5.28a}$$

$$V_3(z) = \frac{B_3}{\Delta_3} a_{5u} \cosh m_3 z + \frac{B_3}{\Delta_3} a_{6u} \sinh m_3 z, \tag{5.28b}$$

$$W_3(z) = \frac{C_3}{\Delta_3} a_{5u} \sinh m_3 z + \frac{C_3}{\Delta_3} a_{6u} \cosh m_3 z, \tag{5.28c}$$

where

$$\Delta_3 = \left(c_{66}p^2 + c_{22}q^2 - c_{44}m_3^2\right)\left(c_{55}p^2 + c_{44}q^2 - c_{33}m_3^2\right) + (c_{23}+c_{44})^2 q^2 m_3^2, \tag{5.28d}$$

$$B_3 = -pq\big[(c_{12}+c_{66})\left(c_{55}p^2 + c_{44}q^2 - c_{33}m_3^2\right) + (c_{13}+c_{55})(c_{23}+c_{44})m_3^2\big], \tag{5.28e}$$

$$C_3 = pm_3\big[-\left(c_{66}p^2 + c_{22}q^2 - c_{44}m_3^2\right)(c_{13}+c_{55}) + (c_{12}+c_{66})(c_{23}+c_{44})q^2\big]. \tag{5.28f}$$

Hence, if we consider the constants a_{1u}, a_{2u}, a_{3u}, a_{4u}, a_{5u}, a_{6u} as independent, which for convenience we rename again as g_1, g_2, g_3, g_4, g_5, g_6, respectively, the displacement $U(z)$ is of the form (5.15a) with the z-dependent coefficients defined as

$$d_{u1} = e^{\gamma_1 z} \cos \gamma_2 z; \quad d_{u2} = e^{\gamma_1 z} \sin \gamma_2 z, \tag{5.29a}$$

$$d_{u3} = e^{-\gamma_1 z} \cos \gamma_2 z; \quad d_{u4} = e^{-\gamma_1 z} \sin \gamma_2 z, \tag{5.29b}$$

$$d_{u5} = \begin{cases} \cos m_3 z, & \text{if} \quad \beta_3 < 0, \\ \cosh m_3 z, & \text{if} \quad \beta_3 > 0, \end{cases} \tag{5.29c}$$

$$d_{u6} = \begin{cases} \sin m_3 z, & \text{if} \quad \beta_3 < 0, \\ \sinh m_3 z, & \text{if} \quad \beta_3 > 0. \end{cases} \tag{5.29d}$$

In the following expressions (5.30–5.34), Δ_3, B_3 and C_3 are from (5.27d–f) if $\beta_3 < 0$, and from (5.28d–f) if $\beta_3 > 0$. With this observation, the displacement $V(z)$ is of the form (5.16a), where

$$d_{v1} = (\xi_{11} \cos \gamma_2 z + \xi_{21} \sin \gamma_2 z)e^{\gamma_1 z}\,; \quad d_{v2} = (\xi_{12} \cos \gamma_2 z + \xi_{22} \sin \gamma_2 z)e^{\gamma_1 z}, \tag{5.30a}$$

$$d_{v3} = (\xi_{33} \cos \gamma_2 z + \xi_{43} \sin \gamma_2 z)e^{-\gamma_1 z}\,; \quad d_{v4} = (\xi_{34} \cos \gamma_2 z + \xi_{44} \sin \gamma_2 z)e^{-\gamma_1 z}, \tag{5.30b}$$

$$d_{v5} = \begin{cases} \dfrac{B_3}{\Delta_3} \cos m_3 z, & \text{if} \quad \beta_3 < 0 \\ \dfrac{B_3}{\Delta_3} \cosh m_3 z, & \text{if} \quad \beta_3 > 0 \end{cases} \tag{5.30c}$$

$$d_{v6} = \begin{cases} \dfrac{B_3}{\Delta_3} \sin m_3 z, & \text{if} \quad \beta_3 < 0 \\ \dfrac{B_3}{\Delta_3} \sinh m_3 z, & \text{if} \quad \beta_3 > 0. \end{cases} \tag{5.30d}$$

Similarly, the displacement $W(z)$ is of the form (5.17a) with the z-dependent coefficients:

$$d_{w1} = (f_{11} \cos \gamma_2 z + f_{21} \sin \gamma_2 z)e^{\gamma_1 z}; \quad d_{w2} = (f_{12} \cos \gamma_2 z + f_{22} \sin \gamma_2 z)e^{\gamma_1 z}, \tag{5.31a}$$

$$d_{w3} = (f_{33} \cos \gamma_2 z + f_{43} \sin \gamma_2 z)e^{-\gamma_1 z}; \quad d_{w4} = (f_{34} \cos \gamma_2 z + f_{44} \sin \gamma_2 z)e^{-\gamma_1 z}, \tag{5.31b}$$

$$d_{w5} = \begin{cases} \dfrac{C_3}{\Delta_3} \sin m_3 z, & \text{if} \quad \beta_3 < 0, \\ \dfrac{C_3}{\Delta_3} \sinh m_3 z, & \text{if} \quad \beta_j > 0, \end{cases} \tag{5.31c}$$

$$d_{w6} = \begin{cases} -\dfrac{C_3}{\Delta_3} \cos m_3 z, & \text{if} \quad \beta_3 < 0, \\ \dfrac{C_3}{\Delta_3} \cosh m_3 z, & \text{if} \quad \beta_3 > 0. \end{cases} \tag{5.31d}$$

The corresponding stresses are derived by substituting the above displacement expressions into (5.5), (5.1), and (5.2). We present the explicit expressions for σ_{zz}, τ_{yz} and τ_{xz}, which enter into the interface conditions. σ_{zz} is of the form (5.18a) with the z-dependent coefficients defined as

$$\begin{aligned} b_{zz1} &= [c_{33}(f_{11}\gamma_1 + f_{21}\gamma_2) - c_{13}p - c_{23}q\xi_{11}]e^{\gamma_1 z}\cos\gamma_2 z \\ &\quad + [c_{33}(f_{21}\gamma_1 - f_{11}\gamma_2) - c_{23}q\xi_{21}]e^{\gamma_1 z}\sin\gamma_2 z, \end{aligned} \tag{5.32a}$$

$$\begin{aligned} b_{zz2} &= [c_{33}(f_{12}\gamma_1 + f_{22}\gamma_2) - c_{23}q\xi_{12}]e^{\gamma_1 z}\cos\gamma_2 z \\ &\quad + [c_{33}(f_{22}\gamma_1 - f_{12}\gamma_2) - c_{13}p - c_{23}q\xi_{22}]e^{\gamma_1 z}\sin\gamma_2 z, \end{aligned} \tag{5.32b}$$

$$\begin{aligned} b_{zz3} &= -[c_{33}(f_{33}\gamma_1 - f_{43}\gamma_2) + c_{13}p + c_{23}q\xi_{33}]e^{-\gamma_1 z}\cos\gamma_2 z \\ &\quad - [c_{33}(f_{43}\gamma_1 + f_{33}\gamma_2) + c_{23}q\xi_{43}]e^{-\gamma_1 z}\sin\gamma_2 z, \end{aligned} \tag{5.32c}$$

$$\begin{aligned} b_{zz4} &= -[c_{33}(f_{34}\gamma_1 - f_{44}\gamma_2) + c_{23}q\xi_{34}]e^{-\gamma_1 z}\cos\gamma_2 z \\ &\quad - [c_{33}(f_{44}\gamma_1 + f_{34}\gamma_2) + c_{13}p + c_{23}q\xi_{44}]e^{-\gamma_1 z}\sin\gamma_2 z. \end{aligned} \tag{5.32d}$$

$$b_{zz5} = \begin{cases} -\left(c_{13}p + c_{23}q\dfrac{B_3}{\Delta_3} - c_{33}\dfrac{C_3}{\Delta_3}m_3\right)\cos m_3 z, & \text{if } \beta_3 < 0 \\ -\left(c_{13}p + c_{23}q\dfrac{B_3}{\Delta_3} - c_{33}\dfrac{C_3}{\Delta_3}m_3\right)\cosh m_3 z, & \text{if } \beta_3 > 0 \end{cases} \tag{5.32e}$$

$$b_{zz6} = \begin{cases} -\left(c_{13}p + c_{23}q\dfrac{B_3}{\Delta_3} - c_{33}\dfrac{C_3}{\Delta_3}m_3\right)\sin m_3 z, & \text{if } \beta_3 < 0 \\ -\left(c_{13}p + c_{23}q\dfrac{B_j}{\Delta_3} - c_{33}\dfrac{C_3}{\Delta_3}m_3\right)\sinh m_3 z, & \text{if } \beta_3 > 0 \end{cases} \tag{5.32f}$$

τ_{yz} is of the form (5.19a) with the z-dependent coefficients defined as

$$\begin{aligned} b_{yz1} &= c_{44}e^{\gamma_1 z}[(\xi_{11}\gamma_1 + \xi_{21}\gamma_2 + qf_{11})\cos\gamma_2 z \\ &\quad + (\xi_{21}\gamma_1 - \xi_{11}\gamma_2 + qf_{21})\sin\gamma_2 z], \end{aligned} \tag{5.33a}$$

$$\begin{aligned} b_{yz2} &= c_{44}e^{\gamma_1 z}[(\xi_{12}\gamma_1 + \xi_{22}\gamma_2 + qf_{12})\cos\gamma_2 z \\ &\quad + (\xi_{22}\gamma_1 - \xi_{12}\gamma_2 + qf_{22})\sin\gamma_2 z], \end{aligned} \tag{5.33b}$$

$$\begin{aligned} b_{yz3} &= c_{44}e^{-\gamma_1 z}[(qf_{33} + \xi_{43}\gamma_2 - \xi_{33}\gamma_1)\cos\gamma_2 z \\ &\quad + (qf_{43} - \xi_{33}\gamma_2 - \xi_{43}\gamma_1)\sin\gamma_2 z], \end{aligned} \tag{5.33c}$$

$$\begin{aligned} b_{yz4} &= c_{44}e^{-\gamma_1 z}[(qf_{34} + \xi_{44}\gamma_2 - \xi_{34}\gamma_1)\cos\gamma_2 z \\ &\quad + (qf_{44} - \xi_{34}\gamma_2 - \xi_{44}\gamma_1)\sin\gamma_2 z], \end{aligned} \tag{5.33d}$$

$$b_{yz5} = \begin{cases} c_{44}\left(q\dfrac{C_3}{\Delta_3} - m_3\dfrac{B_3}{\Delta_3}\right)\sin m_3 z, & \text{if } \beta_3 < 0, \\ c_{44}\left(q\dfrac{C_3}{\Delta_3} + m_3\dfrac{B_3}{\Delta_3}\right)\sinh m_3 z, & \text{if } \beta_3 > 0, \end{cases} \tag{5.33e}$$

$$b_{yz6} = \begin{cases} -c_{44}\left(q\dfrac{C_3}{\Delta_3} - m_3\dfrac{B_3}{\Delta_3}\right)\cos m_3 z, & \text{if } \beta_3 < 0, \\ c_{44}\left(q\dfrac{C_3}{\Delta_3} + m_3\dfrac{B_3}{\Delta_3}\right)\cosh m_3 z, & \text{if } \beta_3 > 0. \end{cases} \tag{5.33f}$$

Finally, τ_{xz} is of the form (5.20a) with the z-dependent coefficients defined as

$$b_{xz1} = c_{55}e^{\gamma_1 z}[(\gamma_1 + pf_{11})\cos\gamma_2 z + (pf_{21} - \gamma_2)\sin\gamma_2 z], \tag{5.34a}$$

$$b_{xz2} = c_{55}e^{\gamma_1 z}[(\gamma_2 + pf_{12})\cos\gamma_2 z + (pf_{22} + \gamma_1)\sin\gamma_2 z], \tag{5.34b}$$

$$b_{xz3} = c_{55}e^{-\gamma_1 z}[(pf_{33} - \gamma_1)\cos\gamma_2 z + (pf_{43} - \gamma_2)\sin\gamma_2 z], \tag{5.34c}$$

$$b_{xz4} = c_{55}e^{-\gamma_1 z}[(pf_{34} + \gamma_2)\cos\gamma_2 z + (pf_{44} - \gamma_1)\sin\gamma_2 z], \tag{5.34d}$$

$$b_{xz5} = \begin{cases} c_{55}\left(p\dfrac{C_3}{\Delta_3} - m_3\right)\sin m_3 z, & \text{if } \beta_3 < 0\,, \\ c_{55}\left(p\dfrac{C_3}{\Delta_3} + m_3\right)\sinh m_3 z, & \text{if } \beta_3 > 0, \end{cases} \tag{5.34e}$$

$$b_{xz6} = \begin{cases} -c_{55}\left(p\dfrac{C_3}{\Delta_3} - m_3\right)\cos m_3 z, & \text{if } \beta_3 < 0, \\ c_{55}\left(p\dfrac{C_3}{\Delta_3} + m_3\right)\cosh m_3 z, & \text{if } \beta_3 > 0. \end{cases} \tag{5.34f}$$

5.1.3 Isotropic Layers

In the event that one of the layers in the sandwich panel is isotropic (this is more common for the core) with extensional modulus E and Poisson's ratio ν, then the following relationships for the material constants hold:

$$c_{11} = c_{22} = c_{33} = E\frac{1-\nu}{(1-2\nu)(1+\nu)}, \tag{5.35a}$$

$$c_{12} = c_{13} = c_{23} = c_{11}\frac{\nu}{1-\nu}; \quad c_{66} = c_{55} = c_{44} = c_{11}\frac{1-2\nu}{2(1-\nu)}. \tag{5.35b}$$

In this case we find that D vanishes and the solution to Equation (5.10) consists of three equal roots, $\beta_i = p^2 + q^2$. Therefore, the solutions to (5.7) occur in the form of three repeated pairs of roots, $s_i = \pm\lambda$, where

$$\lambda = (p^2 + q^2)^{1/2}. \tag{5.36}$$

In this case, the displacement functions take the form

$$U_\eta(z) = (a_{1\eta} + a_{3\eta}z + a_{5\eta}z^2)e^{\lambda z} + (a_{2\eta} + a_{4\eta}z + a_{6\eta}z^2)e^{-\lambda z}, \tag{5.37}$$

where $\eta = u, v, w$ corresponds to the U, V, W displacements and the $a_{i\eta}$ are constants. Of the 18 constants appearing in (5.37), only six are independent. The various relations that exist among these constants are found by substituting (5.37) and (5.5) into (5.4), in which the relations (5.35) for the isotropic material constants are used. In this way we deduce the following 12 relations:

$$a_{5\eta} = a_{6\eta} = 0; \quad \eta = u, v, w, \tag{5.38a}$$

$$qa_{3u} = pa_{3v}; \quad \lambda a_{3u} = pa_{3w}, \tag{5.38b}$$

$$qa_{4u} = pa_{4v}; \quad \lambda a_{4u} = -pa_{4w}, \tag{5.38c}$$

$$pa_{1u} + qa_{1v} - \lambda a_{1w} = -\frac{\lambda}{p}(4\nu - 3)a_{3u}, \tag{5.38d}$$

$$pa_{2u} + qa_{2v} + \lambda a_{2w} = \frac{\lambda}{p}(4\nu - 3)a_{4u}. \tag{5.38e}$$

Hence, if we consider the constants a_{1u}, a_{2u}, a_{3u}, a_{4u}, a_{1v}, and a_{2v} as independent, which for convenience we rename g_1, g_2, g_3, g_4, g_5, g_6, respectively, the displacement $U(z)$ is of the form (5.15a) with the z-dependent coefficients defined as

$$d_{u1} = e^{\lambda z}; \quad d_{u2} = e^{-\lambda z}; \quad d_{u3} = ze^{\lambda z}; \quad d_{u4} = ze^{-\lambda z}; \quad d_{u5} = d_{u6} = 0. \tag{5.39}$$

The displacement $V(z)$ is of the form (5.16a) where

$$d_{v1} = d_{v2} = 0; \; d_{v3} = \frac{q}{p}ze^{\lambda z}; \; d_{v4} = \frac{q}{p}ze^{-\lambda z}; \; d_{v5} = e^{\lambda z}; \; d_{v6} = e^{-\lambda z}, \tag{5.40}$$

and the displacement $W(z)$ is of the form (5.17a) where,

$$d_{w1} = \frac{p}{\lambda}e^{\lambda z}; \quad d_{w2} = -\frac{p}{\lambda}e^{-\lambda z}; \quad d_{w3} = \left(\frac{4\nu - 3}{p} + \frac{\lambda}{p}z\right)e^{\lambda z}, \tag{5.41a}$$

$$d_{w4} = \left(\frac{4\nu - 3}{p} - \frac{\lambda}{p} z \right) e^{-\lambda z} \, ; \quad d_{w5} = \frac{q}{\lambda} e^{\lambda z} \, ; \quad d_{w6} = -\frac{q}{\lambda} e^{-\lambda z}. \tag{5.41b}$$

The corresponding stresses are derived by substituting the above displacement expressions into (5.5), (5.1), and (5.2). We present again the explicit expressions for σ_{zz}, τ_{yz}, and τ_{xz}, which come into the interface conditions. σ_{zz} is of the form (5.18a) with the z-dependent coefficients defined as

$$b_{zz1} = c_{11} p \frac{1 - 2\nu}{1 - \nu} e^{\lambda z} \, ; \quad b_{zz2} = c_{11} p \frac{1 - 2\nu}{1 - \nu} e^{-\lambda z}, \tag{5.42a}$$

$$b_{zz3} = c_{11} \frac{\lambda (1 - 2\nu)}{p (1 - \nu)} e^{\lambda z} \left[\lambda z - 2(1 - \nu) \right], \tag{5.42b}$$

$$b_{zz4} = c_{11} \frac{\lambda (1 - 2\nu)}{p (1 - \nu)} e^{-\lambda z} \left[\lambda z + 2(1 - \nu) \right], \tag{5.42c}$$

$$b_{zz5} = c_{11} q \frac{1 - 2\nu}{1 - \nu} e^{\lambda z} \, ; \quad b_{zz6} = c_{11} q \frac{1 - 2\nu}{1 - \nu} e^{-\lambda z}. \tag{5.42d}$$

τ_{yz} is of the form (5.19a) with the z-dependent coefficients defined as

$$b_{yz1} = \frac{c_{11}(1 - 2\nu)}{2(1 - \nu)} \frac{pq}{\lambda} e^{\lambda z} \, ; \quad b_{yz2} = -\frac{c_{11}(1 - 2\nu)}{2(1 - \nu)} \frac{pq}{\lambda} e^{-\lambda z}, \tag{5.43a}$$

$$b_{yz3} = \frac{c_{11}(1 - 2\nu)}{(1 - \nu)} (2\nu - 1 + \lambda z) \frac{q}{p} e^{\lambda z}, \tag{5.43b}$$

$$b_{yz4} = \frac{c_{11}(1 - 2\nu)}{(1 - \nu)} (2\nu - 1 - \lambda z) \frac{q}{p} e^{-\lambda z}, \tag{5.43c}$$

$$b_{yz5} = \frac{c_{11}(1 - 2\nu)}{2(1 - \nu)} \left(\frac{q^2}{\lambda} + \lambda \right) e^{\lambda z}; \quad b_{yz6} = -\frac{c_{11}(1 - 2\nu)}{2(1 - \nu)} \left(\frac{q^2}{\lambda} + \lambda \right) e^{-\lambda z}. \tag{5.43d}$$

τ_{xz} is of the form (5.20a) with the z-dependent coefficients defined as

$$b_{xz1} = \frac{c_{11}(1 - 2\nu)}{2(1 - \nu)} \left(\frac{p^2}{\lambda} + \lambda \right) e^{\lambda z}; \quad b_{xz2} = -\frac{c_{11}(1 - 2\nu)}{2(1 - \nu)} \left(\frac{p^2}{\lambda} + \lambda \right) e^{-\lambda z}, \tag{5.44a}$$

$$b_{xz3} = \frac{c_{11}(1 - 2\nu)}{(1 - \nu)} (2\nu - 1 + \lambda z) \, e^{\lambda z}, \tag{5.44b}$$

$$b_{xz4} = \frac{c_{11}(1 - 2\nu)}{(1 - \nu)} (2\nu - 1 - \lambda z) \, e^{-\lambda z}, \tag{5.44c}$$

$$b_{xz5} = \frac{c_{11}(1 - 2\nu)}{2(1 - \nu)} \frac{pq}{\lambda} e^{\lambda z}; \quad b_{xz6} = -\frac{c_{11}(1 - 2\nu)}{2(1 - \nu)} \frac{pq}{\lambda} e^{-\lambda z}. \tag{5.44d}$$

From this analysis, we can see that the stresses in each layer (i), where $i = f_1, c, f_2$, are described by six constants: $g_j^{(i)}$, $g_{j+3}^{(i)}$, $j = 1, 2, 3$. Therefore, for the sandwich panel, a total of 18 constants are to be determined.

There are three traction conditions at each of the two core/face-sheet interfaces, giving a total of six conditions. In a similar fashion, there are three displacement continuity conditions at each of the two core/face-sheet interfaces, giving another six conditions. Finally, there are three traction boundary conditions on each of the two plate outer surfaces, giving an another six conditions, i.e. a total of 18 equations.

Finally, for completeness, we also give the detailed expressions for the in-plane stresses σ_{xx}, σ_{yy} and τ_{xy}. σ_{xx} can be written in the form

$$\sigma_{xx} = (b_{xx1}g_1 + b_{xx2}g_2 + b_{xx3}g_3 + b_{xx4}g_4 + b_{xx5}g_5 + b_{xx6}g_6) \sin px \sin qy, \tag{5.45}$$

where the z-dependent coefficients b_{xxj} are found from the b_{zzj} expressions (5.18b–c) and (5.32a–f) by replacing c_{33} with c_{13}, c_{13} with c_{11} and c_{23} with c_{12}. In the same manner, σ_{yy} is given by

$$\sigma_{yy} = \left(b_{yy1}g_1 + b_{yy2}g_2 + b_{yy3}g_3 + b_{yy4}g_4 + b_{yy5}g_5 + b_{yy6}g_6\right) \sin px \sin qy, \tag{5.46}$$

where the z-dependent coefficients b_{yyj} are again found from the b_{zzj} expressions (5.18b–c) and (5.32a–f) by now replacing c_{33} with c_{23}, c_{13} with c_{12} and c_{23} with c_{22}. Finally, the shear stress, τ_{xy}, is

$$\tau_{xy} = \left(b_{xy1}g_1 + b_{xy2}g_2 + b_{xy3}g_3 + b_{xy4}g_4 + b_{xy5}g_5 + b_{xy6}g_6\right) \cos px \sin qy. \tag{5.47}$$

For orthotropic layers with $D < 0$, the z-dependent coefficients are defined for $j = 1, 2, 3$ as

$$b_{xyj} = \begin{cases} c_{66}\left(q + p\dfrac{B_j}{\Delta_j}\right) \cos m_j z, & \text{if } \beta_j < 0, \\ c_{66}\left(q + p\dfrac{B_j}{\Delta_j}\right) \cosh m_j z, & \text{if } \beta_j > 0, \end{cases} \tag{5.48a}$$

$$b_{xy(j+3)} = \begin{cases} c_{66}\left(q + p\dfrac{B_j}{\Delta_j}\right) \sin m_j z, & \text{if } \beta_j < 0, \\ c_{66}\left(q + p\dfrac{B_j}{\Delta_j}\right) \sinh m_j z, & \text{if } \beta_j > 0. \end{cases} \tag{5.48b}$$

In the expressions (5.48) and (5.49), Δ_j and B_3 refer to (5.13j–k) if $\beta_j < 0$, and to (5.14e–f) if $\beta_j > 0$. Further, Δ_3 and B_3 refer to (5.27d–e) if $\beta_3 < 0$ and to (5.28d–e) if $\beta_3 > 0$. With this note, for orthotropic layers with $D > 0$, the z-dependent coefficients are

$$b_{xy1} = c_{66} e^{\gamma_1 z}[(q + p\xi_{11}) \cos \gamma_2 z + p\xi_{21} \sin \gamma_2 z], \tag{5.49a}$$

$$b_{xy2} = c_{66} e^{\gamma_1 z}[p\xi_{12} \cos \gamma_2 z + (q + p\xi_{22}) \sin \gamma_2 z], \tag{5.49b}$$

$$b_{xy3} = c_{66} e^{-\gamma_1 z}[(q + p\xi_{33}) \cos \gamma_2 z + p\xi_{43} \sin \gamma_2 z], \tag{5.49c}$$

$$b_{xy4} = c_{66} e^{-\gamma_1 z}[p\xi_{34} \cos \gamma_2 z + (q + p\xi_{44}) \sin \gamma_2 z], \tag{5.49d}$$

$$b_{xy5} = \begin{cases} c_{66}\left(q + p\dfrac{B_3}{\Delta_3}\right) \cos m_3 z, & \text{if} \quad \beta_3 < 0 \\ c_{66}\left(q + p\dfrac{B_3}{\Delta_3}\right) \cosh m_3 z, & \text{if} \quad \beta_3 > 0 \end{cases} \tag{5.49e}$$

$$b_{xy6} = \begin{cases} c_{66}\left(q + p\dfrac{B_3}{\Delta_3}\right) \sin m_3 z, & \text{if} \quad \beta_3 < 0 \\ c_{66}\left(q + p\dfrac{B_3}{\Delta_3}\right) \sinh m_3 z, & \text{if} \quad \beta_3 > 0 \end{cases} \tag{5.49f}$$

For isotropic materials, the z-dependent coefficients are:

$$b_{xy1} = \frac{c_{11}(1 - 2\nu)}{2(1 - \nu)} q e^{\lambda z}; \quad b_{xy2} = \frac{c_{11}(1 - 2\nu)}{2(1 - \nu)} q e^{-\lambda z}, \tag{5.50a}$$

$$b_{xy3} = \frac{c_{11}(1 - 2\nu)}{1 - \nu} q z e^{\lambda z}, \tag{5.50b}$$

$$b_{xy4} = \frac{c_{11}(1 - 2\nu)}{1 - \nu} q z e^{-\lambda z}, \tag{5.50c}$$

$$b_{xy5} = \frac{c_{11}(1 - 2\nu)}{2(1 - \nu)} p e^{\lambda z}; \quad b_{xy6} = \frac{c_{11}(1 - 2\nu)}{2(1 - \nu)} p e^{-\lambda z}. \tag{5.50d}$$

5.1.4 Examples

As an illustration of the above, let us consider a symmetric sandwich plate with unidirectional carbon/epoxy faces and a hexagonal glass/phenolic honeycomb core. This material combination is very common in the aerospace/rotorcraft industry (although the faces would be multidirectional for most applications). The orthotropic face moduli are (in GPa):

$E_1^f = 181$, $E_2^f = E_3^f = 10.3$, $G_{23}^f = 5.96$, $G_{12}^f = G_{31}^f = 7.17$; and the face Poisson's ratios: $\nu_{12}^f = \nu_{13}^f = 0.277$, $\nu_{32}^f = 0.400$. The orthotropic honeycomb core moduli are (in GPa): $E_1^c = E_2^c = 0.032$, $E_3^c = 0.300$, $G_{23}^c = G_{31}^c = 0.048$, $G_{12}^c = 0.013$; and the core Poisson's ratios: $\nu_{12}^c = \nu_{32}^c = \nu_{31}^c = 0.25$. The thickness of each face sheet is $f_1 = f_2 = 2$ mm and the core thickness $2c = 16$ mm. The plate is square with $a = b = 10h$, where h is the total thickness of the sandwich plate.

We further assume that a transverse load is applied at the top face sheet of the form

$$q_0(x, y) = Q_0 \sin px \sin qy, \tag{5.51}$$

and in the definition of p and q in (5.5d), we further assume $m = n = 1$. Note that a general load can be expanded in a series of terms of the type (5.51).

For each layer, the compliance constants are given by

$$a_{11} = \frac{1}{E_1}; \quad a_{12} = \frac{-\nu_{21}}{E_2}; \quad a_{13} = \frac{-\nu_{31}}{E_3}, \tag{5.52a}$$

$$a_{22} = \frac{1}{E_2}; \quad a_{23} = \frac{-\nu_{32}}{E_3}; \quad a_{33} = \frac{1}{E_3}, \tag{5.52b}$$

$$a_{44} = \frac{1}{G_{23}}; \quad a_{55} = \frac{1}{G_{13}}; \quad a_{66} = \frac{1}{G_{12}}. \tag{5.52c}$$

The stiffness matrix is the inverse of the compliance matrix. The inversion leads to the following formulas for the c_{ij}:

$$c_{11} = E_1\frac{(1-\nu_{23}\nu_{32})}{C_0}; \quad c_{12} = E_2\frac{(\nu_{12}+\nu_{13}\nu_{32})}{C_0}; \quad c_{13} = E_3\frac{(\nu_{13}+\nu_{12}\nu_{23})}{C_0}, \tag{5.52d}$$

$$c_{22} = E_2\frac{(1-\nu_{13}\nu_{31})}{C_0}; \quad c_{23} = E_3\frac{(\nu_{23}+\nu_{21}\nu_{13})}{C_0}; \quad c_{33} = E_3\frac{(1-\nu_{12}\nu_{21})}{C_0}, \tag{5.52e}$$

$$c_{44} = G_{23}; \quad c_{55} = G_{13}; \quad c_{66} = G_{12}, \tag{5.52f}$$

where

$$C_0 = 1 - (\nu_{12}\nu_{21} + \nu_{23}\nu_{32} + \nu_{13}\nu_{31}) - (\nu_{12}\nu_{23}\nu_{31} + \nu_{21}\nu_{13}\nu_{32}). \tag{5.52g}$$

Substituting the corresponding constants leads to the following β's:

- *Face sheets*, $D > 0$, therefore two complex conjugate roots and one real root:

$$\beta_1^f = 342.5 + i316.3; \quad \beta_2^f = 342.5 - i316.3; \quad \beta_3^f = 6150.2.$$

- *Core*, $D > 0$, therefore again two complex conjugate roots and one real root:

$$\beta_1^c = 158.9 + i49.2; \quad \beta_2^c = 158.9 - i49.2\,; \quad \beta_3^c = 131.6.$$

Since we have a positive discriminant for both the face sheet and the core, the corresponding positive discriminant formulas for the coefficients in the expressions of the displacements and stresses are applicable. The solution is obtained by imposing the following.

There are three traction conditions at the lower face-sheet/core interface, $z = -c$:

(a) $\sigma_{zz}^{(c)} = \sigma_{zz}^{(f_2)}|_{z=-c}$, which gives

$$\sum_{j=1}^{6} b_{zzj}^{(c)}|_{z=-c} g_j^{(c)} = \sum_{j=1}^{6} b_{zzj}^{(f_2)}|_{z=-c} g_j^{(f_2)}, \tag{5.53a}$$

(b) $\tau_{yz}^{(c)} = \tau_{yz}^{(f_2)}|_{z=-c}$, which gives

$$\sum_{j=1}^{6} b_{yzj}^{(c)}|_{z=-c} g_j^{(c)} = \sum_{j=1}^{6} b_{yzj}^{(f_2)}|_{z=-c} g_j^{(f_2)}, \tag{5.53b}$$

and

(c) $\tau_{xz}^{(c)} = \tau_{xz}^{(f_2)}|_{z=-c}$, which gives

$$\sum_{j=1}^{6} b_{xzj}^{(c)}|_{z=-c} g_j^{(c)} = \sum_{j=1}^{6} b_{xzj}^{(f_2)}|_{z=-c} g_j^{(f_2)}. \tag{5.53c}$$

There are also three displacement continuity conditions at the lower core/face-sheet interfaces:

(a) $U^{(c)} = U^{(f_2)}$ at $z = -c$, which results in

$$\sum_{j=1}^{6} d_{uj}^{(c)}|_{z=-c} g_j^{(c)} = \sum_{j=1}^{6} d_{uj}^{(f_2)}|_{z=-c} g_j^{(f_2)}, \tag{5.53d}$$

(b) $V^{(c)} = V^{(f_2)}$ at $z = -c$, which gives

$$\sum_{j=1}^{6} d_{vj}^{(c)}|_{z=-c} g_j^{(c)} = \sum_{j=1}^{6} d_{vj}^{(f_2)}|_{z=-c} g_j^{(f_2)}, \tag{5.53e}$$

and

(c) $W^{(c)} = W^{(f_2)}$ at $z = -c$, which gives

$$\sum_{j=1}^{6} d_{wj}^{(c)}|_{z=-c} g_j^{(c)} = \sum_{j=1}^{6} d_{wj}^{(f_2)}|_{z=-c} g_j^{(f_2)}. \tag{5.53f}$$

Next, there are three traction conditions at the upper face-sheet/core interface, $z = c$:

(a) $\sigma_{zz}^{(f_1)} = \sigma_{zz}^{(c)}|_{z=c}$, which gives

$$\sum_{j=1}^{6} b_{zzj}^{(c)}|_{z=c} g_j^{(c)} = \sum_{j=1}^{6} b_{zzj}^{(f_1)}|_{z=c} g_j^{(f_1)}, \tag{5.54a}$$

(b) $\tau_{yz}^{(f_1)} = \tau_{yz}^{(c)}|_{z=c}$, which gives

$$\sum_{j=1}^{6} b_{yzj}^{(c)}|_{z=c} g_j^{(c)} = \sum_{j=1}^{6} b_{yzj}^{(f_1)}|_{z=c} g_j^{(f_1)}, \tag{5.54b}$$

and

(c) $\tau_{xz}^{(f_1)} = \tau_{xz}^{(c)}|_{z=c}$, which gives

$$\sum_{j=1}^{6} b_{xzj}^{(c)}|_{z=c} g_j^{(c)} = \sum_{j=1}^{6} b_{xzj}^{(f_1)}|_{z=c} g_j^{(f_1)}. \tag{5.54c}$$

The corresponding displacement continuity conditions at the upper face-sheet/core interface, $z = c$ are

(a) $U^{(f_1)} = U^{(c)}$ at $z = c$, which gives

$$\sum_{j=1}^{6} d_{uj}^{(c)}|_{z=c} g_j^{(c)} = \sum_{j=1}^{6} d_{uj}^{(f_1)}|_{z=c} g_j^{(f_1)}, \tag{5.54d}$$

(b) $V^{(f_1)} = V^{(c)}$ at $z = c$, which gives

$$\sum_{j=1}^{6} d_{vj}^{(c)}|_{z=c} g_j^{(c)} = \sum_{j=1}^{6} d_{vj}^{(f_1)}|_{z=c} g_j^{(f_1)}, \tag{5.54e}$$

and

(c) $W^{(f_1)} = W^{(c)}$ at $z = c$, which gives

$$\sum_{j=1}^{6} d_{wj}^{(c)}|_{z=c} g_j^{(c)} = \sum_{j=1}^{6} d_{wj}^{(f_1)}|_{z=c} g_j^{(f_1)}. \tag{5.54f}$$

Finally, three traction conditions exist on each of the two outer surfaces. The traction free conditions at the lower outer surface, $z = -(c + f_2)$, can be written as follows:

(a) $\sigma_{zz}|_{z=-(c+f_2)} = 0$, which gives

$$\sum_{j=1}^{6} b_{zzj}^{(f_2)}|_{z=-(c+f_2)} g_j^{(f_2)} = 0, \tag{5.55a}$$

(b) $\tau_{yz}|_{z=-(c+f_2)} = 0$, which gives

$$\sum_{j=1}^{6} b_{yzj}^{(f_2)}|_{z=-(c+f_2)} g_j^{(f_2)} = 0, \tag{5.55b}$$

and

(c) $\tau_{xz}|_{z=-(c+f_2)}$, which gives

$$\sum_{j=1}^{6} b_{xzj}^{(f_2)}|_{z=-(c+f_2)} g_j^{(f_2)} = 0. \tag{5.55c}$$

For the upper surface, where the transverse pressure q_0 is applied:

(a) $\sigma_{zz}|_{z=(c+f_1)} = q_0$, which gives

$$\sum_{j=1}^{6} b_{zzj}^{(f_1)}|_{z=(c+f_1)} g_j^{(f_1)} = Q_0, \tag{5.55d}$$

(b) $\tau_{yz}|_{z=(c+f_1)} = 0$, which gives

$$\sum_{j=1}^{6} b_{yzj}^{(f_1)}|_{z=(c+f_1)} g_j^{(f_1)} = 0, \tag{5.55e}$$

and

(c) $\tau_{xz}|_{z=(c+f_1)} = 0$, which gives

$$\sum_{j=1}^{6} b_{xzj}^{(f_1)}|_{z=(c+f_1)} g_j^{(f_1)} = 0. \tag{5.55f}$$

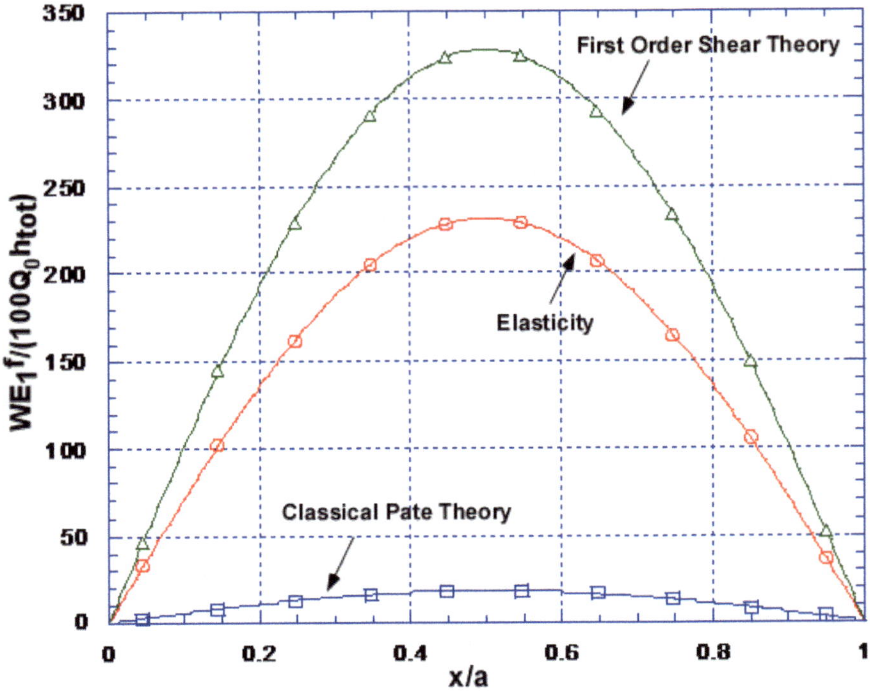

Figure 5.2 Transverse displacement, W, at the top face sheet and at $y = b/2$, as a function of x for $a = b = 10h_{\text{tot}}$ (carbon/epoxy faces and glass/phenolic honeycomb core).

Therefore, we have a system of 18 linear algebraic equations with 18 unknowns, $g_j^{(f_2)}$, $g_j^{(c)}$ and $g_j^{(f_1)}$, $j = 1, 6$.

A square sandwich panel with size $a = b = 10h_{\text{tot}}$, where h_{tot} $(= h)$ is the thickness of the panel, was considered first. The resulting transverse displacement profile w at the top surface, $z = c + f_1$, and at $y = b/2$, is shown in Figure 5.2. The displacement is normalized with $100hQ_0/E_1^f$. In this figure, we also show the predictions from the classical plate theory which does not include transverse shear. Furthermore, the displacement profile obtained from the first-order core shear theory is also shown. The classical and first-order shear theories are outlined in detail in Chapter 3. It can be seen that the classical plate theory is unconservative and quite inaccurate. Furthermore, the first-order shear is too conservative and also quite inaccurate (although considerably better than the classical plate theory).

To illustrate the effect of plate size, Figure 5.3 shows the displacement profiles for a plate five times larger, i.e., with $a = b = 50h$. For this case,

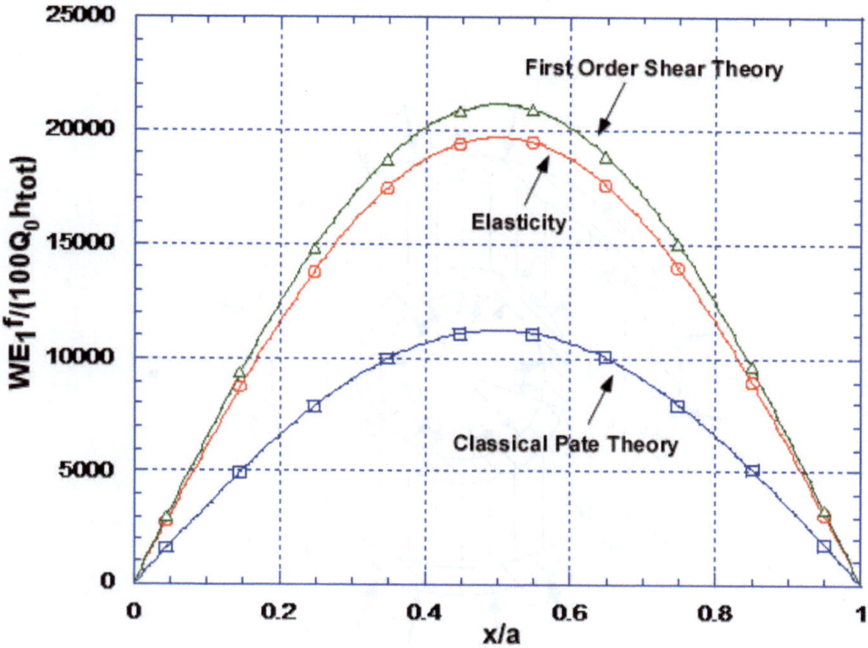

Figure 5.3 Transverse displacement, W, at the top face sheet and at $y = b/2$, as a function of x for $a = b = 50h_{\text{tot}}$.

the first-order shear theory is close to the elasticity, as expected. The classical plate theory is still quite inaccurate. These figures demonstrate clearly the large effect of transverse shear, which is an important feature of sandwich structures.

5.2 A Cylindrical Sandwich Shell with Orthotropic Layers

We consider next the elastic equilibrium of a body in the form of a hollow round cylinder (a tube) of sandwich construction which consists of two face-sheets and a core (Figure 5.4). All three layers are made from a material with cylindrical orthotropy. The body is under the influence of stresses distributed along the lateral surface and on the ends. Let us assume that (1) the axis of orthotropy coincides with the geometric axis of the body; (2) there are planes of elastic symmetry normal to the axis of the cylinder; (3) the stresses acting on the outer and inner surfaces are normal and distributed uniformly, and (4) the stresses which act on the end surfaces reduce to forces

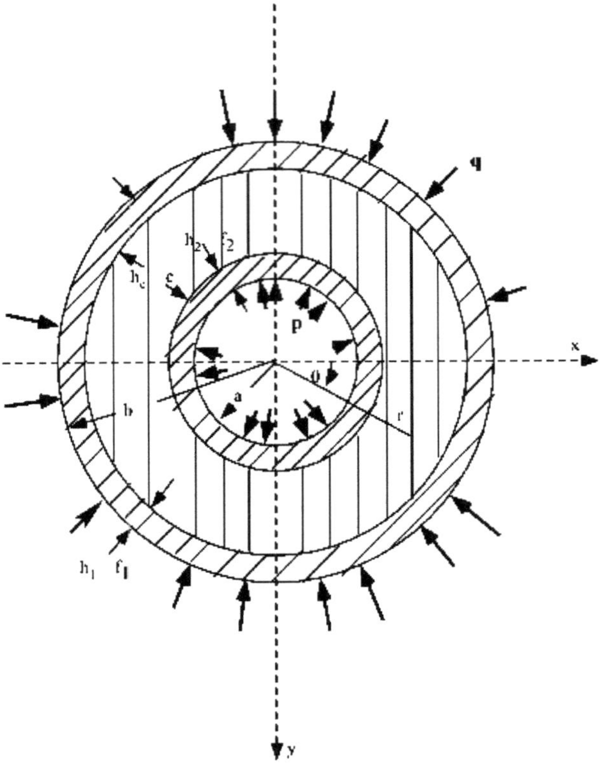

Figure 5.4 Cross-section of a cylindrical sandwich shell under internal pressure (p), external pressure (q), and axial loading (P, normal to the plane of the figure).

which are directed along the axis and to twisting moments. We denote the thickness of the outer face-sheet by $h_1 = f_1$, that of the inner face-sheet by $h_2 = f_2$, and that of the core by $h_c = c$. The inner radius is a and the outer b, where, of course, $b = a + f_2 + c + f_1$. The shell thickness is $h = b - a$.

Let us denote each layer by i where $i = f_1$ for the outer face-sheet, $i = c$ for the core and $i = f_2$ for the inner face-sheet. Then, for each layer, the orthotropic strain-stress relations are

$$\begin{bmatrix} \varepsilon_{rr}^{(i)} \\ \varepsilon_{\theta\theta}^{(i)} \\ \varepsilon_{zz}^{(i)} \\ \gamma_{\theta z}^{(i)} \\ \gamma_{rz}^{(i)} \\ \gamma_{r\theta}^{(i)} \end{bmatrix} = \begin{bmatrix} a_{11}^i & a_{12}^i & a_{13}^i & 0 & 0 & 0 \\ a_{12}^i & a_{22}^i & a_{23}^i & 0 & 0 & 0 \\ a_{13}^i & a_{23}^i & a_{33}^i & 0 & 0 & 0 \\ 0 & 0 & 0 & a_{44}^i & 0 & 0 \\ 0 & 0 & 0 & 0 & a_{55}^i & 0 \\ 0 & 0 & 0 & 0 & 0 & a_{66}^i \end{bmatrix} \begin{bmatrix} \sigma_{rr}^{(i)} \\ \sigma_{\theta\theta}^{(i)} \\ \sigma_{zz}^{(i)} \\ \tau_{\theta z}^{(i)} \\ \tau_{rz}^{(i)} \\ \tau_{r\theta}^{(i)} \end{bmatrix}, \quad (i = f_1, c, f_2) \tag{5.56}$$

where a_{ij}^i are the compliance constants (we have used the notation $1 \equiv r$, $2 \equiv \theta, 3 \equiv z$).

We have taken the axis of the body as the z axis of the cylindrical coordinate system, and the polar x axis is arbitrary in the plane of one of the end sections. The following notations are introduced: p and q are the internal and external pressures, respectively; P is the axial force; M is the twisting moment. Let us introduce the following notation for certain constants which enter into the stress formulas and depend on the elastic constants:

$$\beta_{11}^i = a_{11}^i - \frac{a_{13}^{i\,2}}{a_{33}^i}; \quad \beta_{22}^i = a_{22}^i - \frac{a_{23}^{i\,2}}{a_{33}^i}, \tag{5.57a}$$

$$\beta_{12}^i = a_{12}^i - \frac{a_{13}^i a_{23}^i}{a_{33}^i}, \tag{5.57b}$$

and

$$k_i = \sqrt{\frac{\beta_{11}^i}{\beta_{22}^i}}; \quad \mu_i = \frac{1}{a_{44}^i}; \quad \xi_i = \frac{a_{13}^i - a_{23}^i}{\beta_{22}^i - \beta_{11}^i}, \tag{5.58}$$

where $i = f_1, c, f_2$.

Remark. In the case of isotropy ($a_{13}^i = a_{23}^i$ and $\beta_{22}^i = \beta_{11}^i$), ξ equals zero and all the formulas in this section will still be valid.

Now, if we assume that the applied external stresses are the same at all the cross-sections (do not vary with z) and, in addition, that the stresses depend only on the distance r from the axis, then the stresses in each of the orthotropic layers can be written in terms of two stress functions, $F^{(i)}(r)$ and $\Psi^{(i)}(r)$, $(i = f_1, c, f_2)$ so that

$$\sigma_{rr}^{(i)}(r) = \frac{F^{(i)\prime}(r)}{r}; \quad \sigma_{\theta\theta}^{(i)}(r) = F^{(i)\prime\prime}(r), \tag{5.59a}$$

$$\tau_{r\theta}^{(i)} = 0; \quad \tau_{rz}^{(i)} = 0; \quad \tau_{\theta z}^{(i)} = -\Psi^{(i)\prime}(r), \tag{5.59b}$$

$$\sigma_{zz}^{(i)} = C^{(i)} - \frac{1}{a_{33}}\left[a_{13}^{i}\sigma_{rr}^{(i)} + a_{23}^{i}\sigma_{\theta\theta}^{(i)}\right], \tag{5.59c}$$

where $i = f_1, c, f_2$.

Under the aforementioned assumptions, the equations of equilibrium and the condition that the displacements are single-valued functions of the coordinates, will be satisfied if

$$F^{(i)}(r) = \frac{C^{(i)}}{2}\xi_i r^2 + \frac{C_2^{(i)}}{1+k_i}r^{1+k_i} + \frac{C_3^{(i)}}{1-k_i}r^{1-k_i}, \tag{5.60a}$$

$$\Psi^{(i)}(r) = -\frac{\bar{\theta}^{(i)}\mu_i}{2}r^2. \tag{5.60b}$$

where $i = f_1, c, f_2$.

The constants $C^{(i)}$, $C_2^{(i)}$, $C_3^{(i)}$, $\bar{\theta}^{(i)}$ are found from the conditions on the cylindrical lateral surfaces (e.g. applied uniform internal and/or external pressure) and the conditions on the ends (e.g. applied axial load or axial strain or twisting moment).

Therefore, from Equations (5.59), the stresses are

$$\sigma_{rr}^{(i)}(r) = C^{(i)}\xi_i + C_2^{(i)}r^{k_i-1} + C_3^{(i)}r^{-k_i-1}, \tag{5.61a}$$

$$\sigma_{\theta\theta}^{(i)}(r) = C^{(i)}\xi_i + C_2^{(i)}k_i r^{k_i-1} - C_3^{(i)}k_i r^{-k_i-1}, \tag{5.61b}$$

$$\tau_{\theta z}^{(i)}(r) = \bar{\theta}^{(i)}\mu_i r, \tag{5.61c}$$

$$\begin{aligned}\sigma_{zz}^{(i)}(r) = C^{(i)}\left[1 - \frac{(a_{13}^{i} + a_{23}^{i})}{a_{33}^{i}}\xi_i\right] - C_2^{(i)}\frac{(a_{13}^{i} + a_{23}^{i}k_i)}{a_{33}^{i}}r^{k_i-1}\\ - C_3^{(i)}\frac{(a_{13}^{i} - a_{23}^{i}k_i)}{a_{33}^{i}}r^{-k_i-1}.\end{aligned} \tag{5.61d}$$

where $i = f_1, c, f_2$.

Denoting by $u_r^{(i)}$, $u_\theta^{(i)}$ and $w^{(i)}$ the displacements in the radial, circumferential and axial direction, respectively, the displacement field for this case, excluding rigid body translation and rotation, is given as

$$\begin{aligned}&u_r^{(i)}(r,z) = U^{(i)}(r)\,; \quad u_\theta^{(i)}(r,z) = \bar{\theta}^{(i)}zr + V^{(i)}(r);\\ &w^{(i)}(r,z) = C^{(i)}a_{33}^{i}z + W^{(i)}(r),\end{aligned} \tag{5.62}$$

where $U^{(i)}$, $V^{(i)}$ and $W^{(i)}$ are found from the strain-displacement relations and the stress field (5.59) from the following:

$$\frac{\partial U^{(i)}}{\partial r} = \beta_{11}^i \sigma_{rr}^{(i)} + \beta_{12}^i \sigma_{\theta\theta}^{(i)} + a_{13}^i C^{(i)}, \tag{5.63a}$$

$$\frac{1}{r}\frac{\partial V^{(i)}}{\partial \theta} + \frac{U^{(i)}}{r} = \beta_{12}^i \sigma_{rr}^{(i)} + \beta_{22}^i \sigma_{\theta\theta}^{(i)} + a_{23}^i C^{(i)}, \tag{5.63b}$$

$$\frac{1}{r}\frac{\partial U^{(i)}}{\partial \theta} + \frac{\partial V^{(i)}}{\partial r} - \frac{V^{(i)}}{r} = 0; \quad \frac{\partial W^{(i)}}{\partial r} = 0; \quad \frac{1}{r}\frac{\partial W^{(i)}}{\partial \theta} = 0. \tag{5.63c}$$

Therefore, with the definitions (5.57) for k_i and ξ_i, the displacement field which satisfies these equations and would result in strains, is found by integrating (5.63), as

$$U^{(i)}(r) = C^{(i)}\left[a_{13}^i + \xi_i(\beta_{11}^i + \beta_{12}^i)\right] r + C_2^{(i)} \frac{(\beta_{11}^i + k_i \beta_{12}^i)}{k_i} r^{k_i}$$

$$- C_3^{(i)} \frac{(\beta_{11}^i - k_i \beta_{12}^i)}{k_i} r^{-k_i}, \tag{5.64a}$$

$$V^{(i)}(r) = 0\,; \quad W^{(i)}(r) = 0. \tag{5.64b}$$

5.2.1 *Generalized Plane Deformation of an Orthotropic Sandwich Tube Subjected to Internal and/or External Pressures*

Let us assume that the sandwich cylinder considered in the previous section is subject to pressures p and q distributed uniformly on the inner and outer surfaces, respectively, and has infinite length (generalized plane deformation assumption). Then, not only the stresses, but also the displacements do not depend on z. Alternatively, this is the assumption we would make if the cylinder were securely fixed at the ends ($\varepsilon_z = 0$). Consequently, we can assume

$$C^{(i)} = \bar{\theta}^{(i)} = 0. \tag{5.65}$$

The traction conditions at the core/face-sheet interfaces give

$$\sigma_{rr}^{(f_2)}|_{r=a+f_2} = \sigma_{rr}^{(c)}|_{r=a+f_2}; \quad \sigma_{rr}^{(c)}|_{r=b-f_1} = \sigma_{rr}^{(f_1)}|_{r=b-f_1}. \tag{5.66}$$

Applying (5.61) and (5.65), this gives

$$C_2^{(f_2)}(a+f_2)^{k_{f_2}-1} + C_3^{(f_2)}(a+f_2)^{-k_{f_2}-1}$$

$$= C_2^{(c)}(a+f_2)^{k_c-1} + C_3^{(c)}(a+f_2)^{-k_c-1}, \tag{5.67a}$$

$$C_2^{(c)}(b-f_1)^{k_c-1} + C_3^{(c)}(b-f_1)^{-k_c-1}$$

$$= C_2^{(f_1)}(b-f_1)^{k_{f_1}-1} + C_3^{(f_1)}(b-f_1)^{-k_{f_1}-1}. \tag{5.67b}$$

The displacement continuity at the core/face-sheet interfaces is, in turn,

$$U^{(f_2)}|_{r=a+f_2} = U^{(c)}|_{r=a+f_2}; \quad U^{(c)}|_{r=b-f_1} = U^{(f_1)}|_{r=b-f_1}. \tag{5.68}$$

which, by use of (5.64a) and (5.65), gives

$$C_2^{(f_2)} \frac{(\beta_{11}^{f_2} + k_{f_2}\beta_{12}^{f_2})}{k_{f_2}} (a+f_2)^{k_{f_2}} - C_3^{(f_2)} \frac{(\beta_{11}^{f_2} - k_{f_2}\beta_{12}^{f_2})}{k_{f_2}} (a+f_2)^{-k_{f_2}}$$
$$= C_2^{(c)} \frac{(\beta_{11}^{c} + k_c\beta_{12}^{c})}{k_c} (a+f_2)^{k_c} - C_3^{(c)} \frac{(\beta_{11}^{c} - k_c\beta_{12}^{c})}{k_c} (a+f_2)^{-k_c}, \tag{5.69a}$$

$$C_2^{(c)} \frac{(\beta_{11}^{c} + k_c\beta_{12}^{c})}{k_c} (b-f_1)^{k_c} - C_3^{(c)} \frac{(\beta_{11}^{c} - k_c\beta_{12}^{c})}{k_c} (b-f_1)^{-k_c}$$
$$= C_2^{(f_1)} \frac{(\beta_{11}^{f_1} + k_{f_1}\beta_{12}^{f_1})}{k_{f_1}} (b-f_1)^{k_{f_1}} - C_3^{(f_1)} \frac{(\beta_{11}^{f_1} - k_{f_1}\beta_{12}^{f_1})}{k_{f_1}} (b-f_1)^{-k_{f_1}}. \tag{5.69b}$$

The conditions of applied internal and external pressures on the inner and outer surfaces ($r = a, b$) are

$$\sigma_{rr}^{(f_2)}|_{r=a} = -p\,; \quad \sigma_{rr}^{(f_1)}|_{r=b} = -q, \tag{5.70}$$

which gives

$$C_2^{(f_2)} a^{k_{f_2}-1} + C_3^{(f_2)} a^{-k_{f_2}-1} = -p; \quad C_2^{(f_1)} b^{k_{f_1}-1} + C_3^{(f_1)} b^{-k_{f_1}-1} = -q, \tag{5.71}$$

The six unknowns $C_2^{(i)}, C_3^{(i)}$ ($i = f_1, c, f_2$) are solved in terms of p and q using a system of six linear equations formed by Equations (5.67a, b), (5.69a, b) and (5.71). Then, the stresses are found by Equations (5.61).

Since there is no stress $\tau_{\theta z}$, there is no resultant twisting moment. The stresses σ_{zz} on the ends and in any cross-section reduce to an axial force P which can be found from

$$\frac{P}{2\pi} = \int_a^b \sigma_{zz} r\,dr = \int_a^{a+f_2} \sigma_{zz}^{(f_2)} r\,dr + \int_{a+f_2}^{b-f_1} \sigma_{zz}^{(c)} r\,dr + \int_{b-f_1}^{b} \sigma_{zz}^{(f_1)} r\,dr. \tag{5.72}$$

Using (5.61d), this becomes

$$\frac{P}{2\pi} = -(D_2 + D_3). \tag{5.73a}$$

where

$$D_2 = C_2^{(f_1)} \frac{(a_{13}^{f_1} + a_{23}^{f_1} k_{f_1})}{a_{33}^{f_1} (k_{f_1} + 1)} \left[b^{(k_{f_1}+1)} - (b - f_1)^{(k_{f_1}+1)} \right]$$
$$+ C_2^{(c)} \frac{(a_{13}^c + a_{23}^c k_c)}{a_{33}^c (k_c + 1)} \left[(b - f_1)^{(k_c+1)} - (a + f_2)^{(k_c+1)} \right]$$
$$+ C_2^{(f_2)} \frac{(a_{13}^{f_2} + a_{23}^{f_2} k_{f_2})}{a_{33}^{f_2} (k_{f_2} + 1)} \left[(a + f_2)^{(k_{f_2}+1)} - a^{(k_{f_2}+1)} \right], \quad (5.73b)$$

$$D_3 = C_3^{(f_1)} \frac{(a_{13}^{f_1} - a_{23}^{f_1} k_{f_1})}{a_{33}^{f_1} (-k_{f_1} + 1)} \left[b^{(-k_{f_1}+1)} - (b - f_1)^{(-k_{f_1}+1)} \right]$$
$$+ C_3^{(c)} \frac{(a_{13}^c - a_{23}^c k_c)}{a_{33}^c (-k_c + 1)} \left[(b - f_1)^{(-k_c+1)} - (a + f_2)^{(-k_c+1)} \right]$$
$$+ C_3^{(f_2)} \frac{(a_{13}^{f_2} - a_{23}^{f_2} k_{f_2})}{a_{33}^{f_2} (-k_{f_2} + 1)} \left[(a + f_2)^{(-k_{f_2}+1)} - a^{(-k_{f_2}+1)} \right]. \quad (5.73c)$$

5.2.2 *An Orthotropic Hollow Sandwich Cylinder Loaded by an Axial Force*

We now assume that the shell is loaded by stresses distributed on the ends and which reduce to a tensile force P. The stresses at the ends are applied so that a uniformly distributed constant axial strain, ε_0, exists throughout the section. Note also that no resultant twisting moment is assumed to exist and $\bar{\theta}^i = 0$.

From (5.62) the axial strain is $C^{(i)} a_{33}^i$, and the first condition is

$$C^{(f_2)} a_{33}^{f_2} = C^{(c)} a_{33}^c = C^{(f_1)} a_{33}^{f_1} = \varepsilon_0. \quad (5.74)$$

i.e., the constants $C^{(i)}$ are non-zero.

Next, the traction conditions (5.66) at the face-sheet/core interfaces give by use of (5.61a) and (5.74):

$$\varepsilon_0 \frac{\xi_{f_2}}{a_{33}^{f_2}} + C_2^{(f_2)} (a + f_2)^{k_{f_2}-1} + C_3^{(f_2)} (a + f_2)^{-k_{f_2}-1}$$
$$= \varepsilon_0 \frac{\xi_c}{a_{33}^c} + C_2^{(c)} (a + f_2)^{k_c-1} + C_3^{(c)} (a + f_2)^{-k_c-1}, \quad (5.75a)$$

$$\varepsilon_0 \frac{\xi_c}{a_{33}^c} + C_2^{(c)}(b - f_1)^{k_c - 1} + C_3^{(c)}(b - f_1)^{-k_c - 1}$$
$$= \varepsilon_0 \frac{\xi_{f_1}}{a_{33}^{f_1}} + C_2^{(f_1)}(b - f_1)^{k_{f_1} - 1} + C_3^{(f_1)}(b - f_1)^{-k_{f_1} - 1}. \quad (5.75\text{b})$$

The displacement continuity at the face-sheet/core interfaces, (5.68), by use of (5.64a) and (5.74) becomes

$$\varepsilon_0 \frac{(a_{13}^{f_2} + \xi_{f_2}(\beta_{11}^{f_2} + \beta_{12}^{f_2}))}{a_{33}^{f_2}}(a + f_2) + C_2^{(f_2)} \frac{(\beta_{11}^{f_2} + k_{f_2}\beta_{12}^{f_2})}{k_{f_2}}(a + f_2)^{k_{f_2}}$$
$$- C_3^{(f_2)} \frac{(\beta_{11}^{f_2} - k_{f_2}\beta_{12}^{f_2})}{k_{f_2}}(a + f_2)^{-k_{f_2}} = \varepsilon_0 \frac{(a_{13}^c + \xi_c(\beta_{11}^c + \beta_{12}^c))}{a_{33}^c}(a + f_2)$$
$$+ C_2^{(c)} \frac{(\beta_{11}^c + k_c\beta_{12}^c)}{k_c}(a + f_2)^{k_c} - C_3^{(c)} \frac{(\beta_{11}^c - k_c\beta_{12}^c)}{k_c}(a + f_2)^{-k_c}, \quad (5.76\text{a})$$

$$\varepsilon_0 \frac{(a_{13}^c + \xi_c(\beta_{11}^c + \beta_{12}^c))}{a_{33}^c}(b - f_1) + C_2^{(c)} \frac{(\beta_{11}^c + k_c\beta_{12}^c)}{k_c}(b - f_1)^{k_c}$$
$$- C_3^{(c)} \frac{(\beta_{11}^c - k_c\beta_{12}^c)}{k_c}(b - f_1)^{-k_c} = \varepsilon_0 \frac{(a_{13}^{f_1} + \xi_{f_1}(\beta_{11}^{f_1} + \beta_{12}^{f_1}))}{a_{33}^{f_1}}(b - f_1)$$
$$+ C_2^{(f_1)} \frac{(\beta_{11}^{f_1} + k_{f_1}\beta_{12}^{f_1})}{k_{f_1}}(b - f_1)^{k_{f_1}} - C_3^{(f_1)} \frac{(\beta_{11}^{f_1} - k_{f_1}\beta_{12}^{f_1})}{k_{f_1}}(b - f_1)^{-k_{f_1}}. \quad (5.76\text{b})$$

Next, the condition of traction-free lateral surfaces is expressed by

$$\sigma_{rr}^{(f_2)}|_{r=a} = 0; \quad \sigma_{rr}^{(f_1)}|_{r=b} = 0, \quad (5.77)$$

which gives

$$\varepsilon_0 \frac{\xi_{f_2}}{a_{33}^{f_2}} + C_2^{(f_2)} a^{k_{f_2} - 1} + C_3^{(f_2)} a^{-k_{f_2} - 1} = 0, \quad (5.78\text{a})$$

$$\varepsilon_0 \frac{\xi_{f_1}}{a_{33}^{f_1}} + C_2^{(f_1)} b^{k_{f_1} - 1} + C_3^{(f_1)} b^{-k_{f_1} - 1} = 0. \quad (5.78\text{b})$$

Again, the solution is found by solving for the six constants $C_2^{(i)}$, $C_3^{(i)}$, $(i = f_1, c, f_2)$ in terms of ε_0, from the six linear equations (5.75a, b), (5.76a, b) and (5.78a, b).

An expression for the resultant applied force P in terms of ε_0 can be found by integrating σ_{zz} as in (5.72), and this now gives, by using (5.61d),

$$\frac{P}{2\pi} = -(D_1 + D_2 + D_3), \tag{5.79a}$$

where D_2 and D_3 are given by (5.73b, c) and

$$\begin{aligned} D_1/\varepsilon_0 = {} & \left[1 - \frac{(a_{13}^{f_1} + a_{23}^{f_1})}{a_{33}^{f_1}}\xi_{f_1}\right] \frac{\left[b^2 - (b - f_1)^2\right]}{2a_{33}^{f_1}} \\ & + \left[1 - \frac{(a_{13}^{c} + a_{23}^{c})}{a_{33}^{c}}\xi_{c}\right] \frac{\left[(b - f_1)^2 - (a + f_2)^2\right]}{2a_{33}^{c}} \\ & + \left[1 - \frac{(a_{13}^{f_2} + a_{23}^{f_2})}{a_{33}^{f_2}}\xi_{f_2}\right] \frac{\left[(a + f_2)^2 - a^2\right]}{2a_{33}^{f_2}}. \end{aligned} \tag{5.79b}$$

Of course, the axial stress σ_{zz} is non-uniformly distributed over the cross-section as opposed to the axial strain, ε_0, assumed to be uniform.

5.2.3 Sandwich Shell Theory Expressions

We refer to a cylindrical coordinate system z, θ and r, in which z and θ are in the axial and circumferential directions and r is in the (radial) direction. The corresponding displacements at any point are denoted by w, v and u.

In addition to Equation (5.56) which is in terms of the compliance constants, we shall use the stress-strain relations in terms of the stiffness constants, as follows:

$$\begin{bmatrix} \sigma_{rr}^{(i)} \\ \sigma_{\theta\theta}^{(i)} \\ \sigma_{zz}^{(i)} \\ \tau_{\theta z}^{(i)} \\ \tau_{rz}^{(i)} \\ \tau_{r\theta} \end{bmatrix} = \begin{bmatrix} c_{11}^i & c_{12}^i & c_{13}^i & 0 & 0 & 0 \\ c_{12}^i & c_{22}^i & c_{23}^i & 0 & 0 & 0 \\ c_{13}^i & c_{23}^i & c_{33}^i & 0 & 0 & 0 \\ 0 & 0 & 0 & c_{44}^i & 0 & 0 \\ 0 & 0 & 0 & 0 & c_{55}^i & 0 \\ 0 & 0 & 0 & 0 & 0 & c_{66}^i \end{bmatrix} \begin{bmatrix} \varepsilon_{rr}^{(i)} \\ \varepsilon_{\theta\theta}^{(i)} \\ \varepsilon_{zz}^{(i)} \\ \gamma_{\theta z}^{(i)} \\ \gamma_{rz}^{(i)} \\ \gamma_{r\theta}^{(i)} \end{bmatrix}, \quad (i = f_1, c, f_2), \tag{5.80}$$

where we have again used the notation $1 \equiv r, 2 \equiv \theta, 3 \equiv z$.

The sandwich shell theory employed is a version of Love's (1927) shell theory extended to shear deformable structures (but note the absence of shear in this case of orthotropy). The core is assumed to carry only shear stresses and the face sheets carry the normal stresses, therefore the extensional and bending stiffnesses of the shell are based exclusively on the face-sheet stiffnesses. On the contrary, transverse shear stress resultants (should they exist) are based exclusively on the shear stiffnesses of the core.

Taking into account the displacement distribution through the thickness assumed in the shell theory, we can easily see that in the generalized plane deformation problems under consideration, the displacement field throughout the shell is

$$u(r,\theta,z) = u_0\,; \quad v(r,\theta,z) = 0 \quad w(r,\theta,z) = \varepsilon_0 z, \tag{5.81a}$$

where u_0 is a constant and ε_0 is the uniform axial strain.

The relationships for the strains throughout the shell, corresponding to Love's (1927) shell theory are

$$\varepsilon_{rr} = 0\,; \quad \varepsilon_{\theta\theta} = \frac{u_0}{R}\,; \quad \varepsilon_{zz} = \varepsilon_0, \tag{5.81b}$$

where R is the mid-surface radius. The shear strains are all zero. Notice that in these simplified, axisymmetric, generalized plane deformation problems, there is no difference between first-order shear deformation and classical solutions.

The stress resultants of interest are

$$N_\theta = C_{22}\varepsilon^0_{\theta\theta} + C_{23}\varepsilon^0_{zz}; \quad N_z = C_{23}\varepsilon^0_{\theta\theta} + C_{33}\varepsilon^0_{zz}; \quad N_{z\theta} = 0, \tag{5.81c}$$

where ε^0_{ij} are the mid-surface strains, identical to the ones in (5.81b). Moreover, the C_{ij} are the shell stiffness constants, determined by the face-sheets (in the context of sandwich shell formulation) by

$$C_{ij} = f_1 c^{f_1}_{ij} + f_2 c^{f_2}_{ij}, \quad (i,j = 2,3). \tag{5.81d}$$

For *external pressure*, the equilibrium equations in terms of the stress resultants are satisfied if

$$N_\theta = -qR. \tag{5.82a}$$

Furthermore, based on the assumptions of the problem for the external pressure case, $\varepsilon_0 = 0$. Then (5.81c) and (5.81b) give

$$u_0 = -qR^2/C_{22}\,; \quad \varepsilon_{\theta\theta} = -qR/C_{22}. \tag{5.82b}$$

Subsequently by using (5.80) the stresses are:

$$\sigma_{rr} = -q\frac{c^i_{12}R}{C_{22}}; \quad \sigma_{\theta\theta} = -q\frac{c^i_{22}R}{C_{22}}; \quad \sigma_{zz} = -q\frac{c^i_{23}R}{C_{22}} \quad (i = f_2, c, f_1), \tag{5.82c}$$

For *axial loading* with a uniform axial strain ε_0, the equilibrium equations are satisfied if $N_\theta = 0$, which, by using (5.81c) and (5.81b), gives

$$u_0 = -\varepsilon_0 RC_{23}/C_{22}\,; \quad \varepsilon_{\theta\theta} = -\varepsilon_0 C_{23}/C_{22}. \tag{5.83a}$$

Subsequently, N_z can be obtained from (5.81c) as

$$N_z = \varepsilon_0\left(C_{33} - \frac{C^2_{23}}{C_{22}}\right). \tag{5.83b}$$

Then the stresses are found by using (5.80):

$$\sigma_{rr} = \varepsilon_0\left(c^i_{13} - c^i_{12}\frac{C_{23}}{C_{22}}\right); \quad \sigma_{\theta\theta} = \varepsilon_0\left(c^i_{23} - c^i_{22}\frac{C_{23}}{C_{22}}\right);$$

$$\sigma_{zz} = \varepsilon_0\left(c^i_{33} - c^i_{23}\varepsilon_0\frac{C_{23}}{C_{22}}\right), \tag{5.83c}$$

where $i = f_1, c, f_2$.

As an illustrative example, the stress and displacement distribution was determined for a sandwich composite circular cylindrical shell of outer radius $b = 1$ m, a ratio of outside over inside radii, $b/a = 1.20$, ratios of face-sheet thicknesses over shell thickness, $f_2/h = f_1/h = 0.10$.

The face sheets are made from unidirectional E-glass/polyester with the fiber direction along the circumference, with moduli in GPa: $E_2^{(f_1,f_2)} = 40$, $E_1^{(f_1,f_2)} = E_3^{(f_1,f_2)} = 10$, $G_{13}^{(f_1,f_2)} = 3.5$, $G_{12}^{(f_1,f_2)} = G_{23}^{(f_1,f_2)} = 4.5$, and Poisson's ratios $\nu_{31}^{(f_1,f_2)} = 0.40$, $\nu_{21}^{(f_1,f_2)} = \nu_{23}^{(f_1,f_2)} = 0.26$. Note that 1 is the radial (r), 2 is the circumferential (θ), and 3 the axial (z) direction. The core modulus and Poisson's ratio are assumed to be $E^c = 75$ MPa and $\nu^c = 0.30$. Notice that the compliance constants for the orthotropic face sheets are given by (5.32a–c).

For the case of pure *external pressure*, q, Figure 5.5 shows the radial displacement $U(r)$, normalized with qR^2/C_{22} (C_{22} is defined in (5.81d)) plotted vs. r/R. The elasticity solution (Section 5.2.1) predicts a non-uniform displacement as opposed to the shell theory.

For the case of pure *axial loading* by a uniform applied axial strain ε_0, Figure 5.6 shows the displacement, $U(r)$, normalized with $\varepsilon_0 RC_{23}/C_{22}$. Again, the elasticity solution (Section 5.2.2) predicts a non-uniform displacement distribution as opposed to the shell theory.

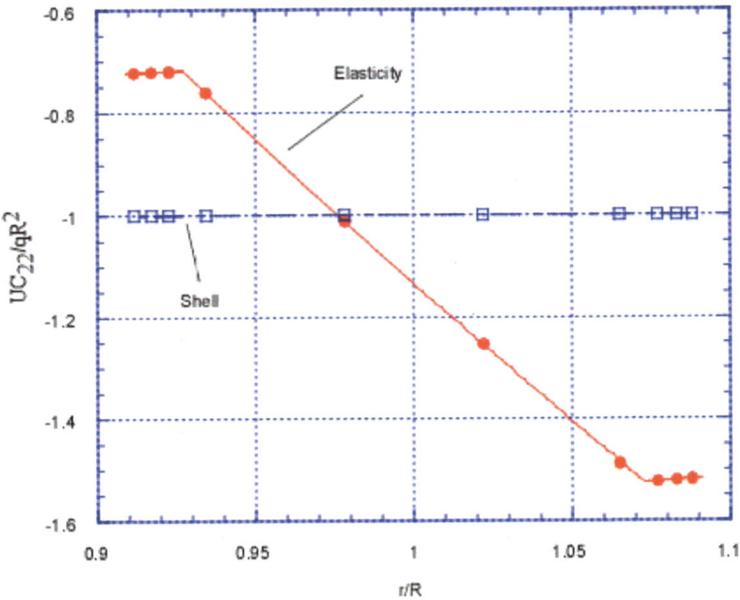

Figure 5.5 Radial displacement, $U(r)$ vs. normalized radius (r/R), for a cylindrical sandwich shell of mean radius, R, under uniform external pressure, q.

Also note that because of the orthotropy and the axisymmetric geometry, there are no shear stresses generated from internal/external pressure and axial loading. Therefore even a first-order shear deformation theory would not result in improved shell theory predictions.

Finally, it should be pointed out that the concept of sandwich construction may not be ideal for the loading and structure analyzed. This is because in the case considered there is no shear in the core and to really take advantage of the sandwich concept, the core should carry the shear and the face sheet should support the normal stresses. If the cylinder is loaded in compression, however, and buckling occurs, then the core would support the shear, and the solution presented can be used as the exact pre-buckling state of stress and displacement in the formulation of the buckling problem.

5.2.4 Torsion of a Sandwich Shaft

Let us consider the more general case of off-axis orientation of the material, but with one plane of elastic symmetry normal to the cylinder axis. Hence, a^i_{45}, a^i_{16}, a^i_{26} and a^i_{36} are non-zero, and the strain-stress relations becomes

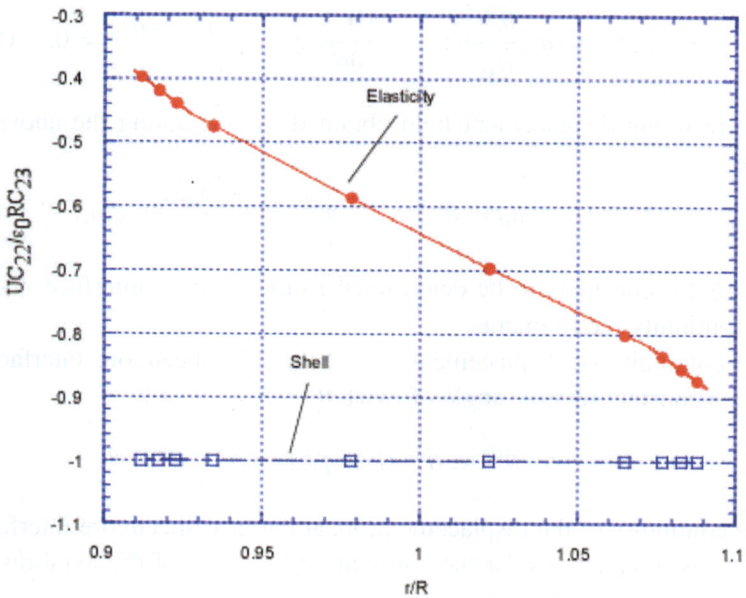

Figure 5.6 Radial displacement, $U(r)$ vs. normalized radius (r/R), for a cylindrical sandwich shell of mean radius, R, under uniformly applied axial strain, ε_0.

$$\begin{bmatrix} \varepsilon_{rr}^{(i)} \\ \varepsilon_{\theta\theta}^{(i)} \\ \varepsilon_{zz}^{(i)} \\ \gamma_{\theta z}^{(i)} \\ \gamma_{rz}^{(i)} \\ \gamma_{r\theta}^{(i)} \end{bmatrix} = \begin{bmatrix} a_{11}^i & a_{12}^i & a_{13}^i & 0 & 0 & a_{16}^i \\ a_{12}^i & a_{22}^i & a_{23}^i & 0 & 0 & a_{26}^i \\ a_{13}^i & a_{23}^i & a_{33}^i & 0 & 0 & a_{36}^i \\ 0 & 0 & 0 & a_{44}^i & a_{45}^i & 0 \\ 0 & 0 & 0 & a_{45}^i & a_{55}^i & 0 \\ a_{16}^i & a_{26}^i & a_{36}^i & 0 & 0 & a_{66}^i \end{bmatrix} \begin{bmatrix} \sigma_{rr}^{(i)} \\ \sigma_{\theta\theta}^{(i)} \\ \sigma_{zz}^{(i)} \\ \tau_{\theta z}^{(i)} \\ \tau_{rz}^{(i)} \\ \tau_{r\theta}^{(i)} \end{bmatrix}, \quad (i = f_1, c, f_2). \tag{5.84a}$$

The equations of equilibrium are satisfied for

$$\sigma_{rr}^{(i)} = \sigma_{\theta\theta}^{(i)} = \sigma_{zz}^{(i)} = \tau_{r\theta}^{(i)} = \tau_{rz}^{(i)} = 0\,; \quad \tau_{\theta z}^{(i)} = \frac{\bar{\theta}^{(i)}}{a_{44}^i} r, \tag{5.84b}$$

and the displacement field (excluding rigid body rotation and translation) that results from these stresses can be found from the strain-displacement relations and the strain-stress relations, which in this case become

$$\frac{\partial U^{(i)}}{\partial r} = 0\,; \quad \frac{1}{r}\frac{\partial V^{(i)}}{\partial \theta} + \frac{U^{(i)}}{r} = 0\,; \quad \frac{1}{r}\frac{\partial U^{(i)}}{\partial \theta} + \frac{\partial V^{(i)}}{\partial r} - \frac{V^{(i)}}{r} = 0, \tag{5.85a}$$

where $i = f_1, c, f_2$.

$$\frac{\partial W^{(i)}}{\partial r} = a^i_{45}\tau^{(i)}_{\theta z} = a^i_{45}\frac{\bar{\theta}^{(i)}}{a^i_{44}}r\,; \quad \frac{1}{r}\frac{\partial W^{(i)}}{\partial \theta} = a^i_{44}\tau^{(i)}_{\theta z} - \bar{\theta}^{(i)}r = 0. \quad (5.85b)$$

The resulting displacement field obtained by integrating the above relations is

$$u^{(i)}_r = 0\,; \quad u_\theta = \bar{\theta}^{(i)}rz\,; \quad w^{(i)} = \bar{\theta}^{(i)}\frac{a^{(i)}_{45}}{a^{(i)}_{45}}\frac{r^2}{2} + d_i. \qquad (5.86)$$

where d_i are constants to be determined from face/core interface displacement continuity requirements.

The continuity of displacement, u_θ, at the face-sheet/core interfaces results in a constant relative angle of twist, $\bar{\theta}^{(i)}$:

$$\bar{\theta}^{(i)} = \bar{\theta} \quad (i = f_1, c, f_2). \qquad (5.87)$$

The continuity of the displacement, w, at the face-sheet/core interfaces in turn results in equations for the constants d_i in terms of the axial displacement, w, expressions (5.86).

The resultant twisting moment, M, is then found from

$$\frac{M}{2\pi} = \int_a^b \tau_{\theta z}r^2dr = \int_a^{a+f_2} \tau^{(f_2)}_{\theta z}r^2dr + \int_{a+f_2}^{b-f_1} \tau^{(c)}_{\theta z}r^2dr + \int_{b-f_1}^{b} \tau^{(f_1)}_{\theta z}r^2dr. \qquad (5.88)$$

Using (5.84b) and (5.87) results in the following expression:

$$\frac{M}{2\pi} = \frac{\bar{\theta}}{4}\left\{\frac{[(a+f_2)^4 - a^4]}{a^{(f_2)}_{44}} + \frac{[(b-f_1)^4 - (a+f_2)^4]}{a^{(c)}_{44}} + \frac{[b^4 - (b-f_1)^4]}{a^{(f_1)}_{44}}\right\}. \qquad (5.89)$$

If $a^i_{45} = 0$ for all three layers, then $w^{(i)} = 0$, and the cross-sections will remain planar and not warp.

We have presented in this chapter some fundamental cases regarding three-dimensional elasticity of sandwich structures. Elasticity solutions for other cases, e.g. a hollow orthotropic sandwich cylinder loaded by bending moments applied at the ends, or an orthotropic sandwich curved bar, loaded by couples or terminal forces, can be found by extending these solutions.

Chapter 6
High-Order Sandwich Panel Theories

The effects of transverse shear and core compressibility are of high importance in sandwich structures, having an influence on the entire structural behavior including bending, buckling and vibrations. The unusually large transverse shear effects arise due to the very low shear modulus of the core in relation to the extensional modulus of the face sheets. The compressibility effects arise due to the soft nature of the core. This chapter presents two one-dimensional high-order core shear theories for sandwich beams or wide plates, namely the "High-Order Sandwich Panel Theory" (HSAPT), and the more recent "Extended High-Order Sandwich Panel Theory" (EHSAPT). It should be noted that the basic assumptions regarding the face sheets kinematics and face sheet constitutives are the same in all theories and the differences are in dealing with the core kinematics and constitutive relations. In addition, although these theories are presented for the simpler one-dimensional beam configuration, they can be easily extended to the two-dimensional plate or shell geometries. Other high-order theories for sandwich structures available in the literature are briefly outlined.

6.1 Basic Assumptions

In the following, we consider a sandwich beam of length a with a core of thickness $2c$ and top and bottom face sheets of thicknesses f_t and f_b, respectively. as shown in Figure 6.1. A Cartesian coordinate system (x, y, z) is defined at one end of the beam and its origin is placed at the middle of the core. Only loading in the x–z plane is considered to act on the beam which solely causes displacements in the x and z directions designated by u and w, respectively.

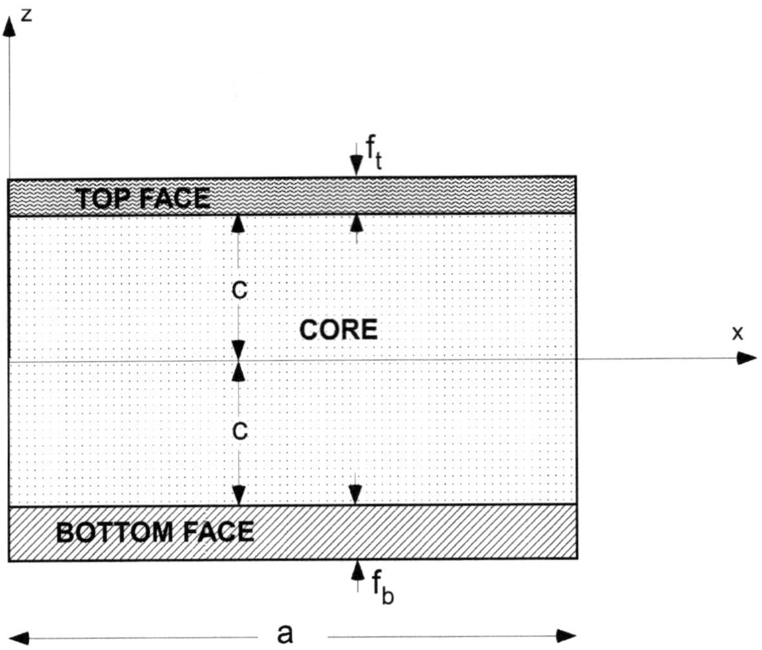

Figure 6.1 Definition of the sandwich configuration.

The superscripts t, b, and c refer to the top face sheet, bottom face sheet, and core, respectively. The subscript 0 refers to the middle surface of the corresponding layer. For example, u_0^t denotes the mid-surface displacement in the x direction of the top face sheet. We should also note that the rigidities and all applied loadings are per unit width.

The displacements of the top and bottom face sheets are assumed to satisfy the Euler–Bernoulli assumptions: Therefore, the displacement field for the top face sheet, $c \leq z \leq c + f_t$, is

$$w^t(x,z) = w_0^t(x); \quad u^t(x,z) = u_0^t(x) - \left(z - c - \frac{f_t}{2}\right) w_{0,x}^t(x), \quad (6.1a)$$

and for the bottom face sheet, $-(c + f_b) \leq z \leq -c$:

$$w^b(x,z) = w_0^b(x); \quad u^b(x,z) = u_0^b(x) - \left(z + c + \frac{f_b}{2}\right) w_{0,x}^b(x). \quad (6.1b)$$

The only non-zero strain in the faces is the axial strain which, in the general nonlinear case (necessary, for example, for buckling), is

$$\varepsilon_{xx}^{t,b}(x,z) = u_{,x}^{t,b}(x,z) + \frac{1}{2}\left[w_{0,x}^{t,b}(x)\right]^2. \quad (6.1c)$$

If a linear analysis is pursued, the second (squared) term in (6.1c) is neglected.

We assume plane stress in the orthotropic face sheets, thus the non-zero stresses for the faces are

$$\sigma_{xx}^{t,b} = C_{11}^{t,b}\varepsilon_{xx}^{t,b}; \quad \sigma_{zz}^{t,b} = C_{13}^{t,b}\varepsilon_{xx}^{t,b}. \tag{6.2}$$

6.2 The "High-Order Sandwich Panel Theory" (HSAPT)

This theory, proposed by Frostig et al. (1992), was developed for sandwich panels, i.e. beams and wide beams in cylindrical bending, and plates and shells. It has been applied successfully for linear and nonlinear cases. In this theory, the displacement functions are a result of the solution of the governing equations of the theory rather than pre-assumed patterns. The theory yields a transverse displacement in the core that is quadratic in z, and an in-plane displacement that is cubic. The resulting shear stress in the core τ^c does not depend on z, i.e. it is only a function of x.

This theory is expressed in terms of five generalized coordinates, the $u_0^t(x)$, $w_0^t(x)$, $u_0^b(x)$, $w_0^b(x)$ and $\tau^c(x)$.

In the following, the equations of the theory have been adapted from Frostig et al. (1992) for the present sandwich geometric and coordinate configuration.

In this theory, the displacement field of the core depends on the generalized coordinates in the following way:

$$w^c(x,z) = w_0^b(x) + \frac{z+c}{2c}\left[w_0^t(x) - w_0^b(x)\right] - \frac{z^2-c^2}{2C_{33}^c}\tau_{,x}^c(x), \tag{6.3a}$$

and

$$u^c(x,z) = u_0^b(x) + \frac{z+c}{C_{55}^c}\tau^c(x) - \frac{1}{2C_{33}^c}\left[(z+c)^2 c - \frac{(z+c)^3}{3}\right]\tau_{,xx}^c(x)$$
$$- \frac{(z+c)^2}{4c}w_{0,x}^t(x) - \left[\frac{f_t}{2} - \frac{(z+c)^2}{4c} + (z+c)\right]w_{0,x}^b(x), \tag{6.3b}$$

where $C_{33}^c = E_c$ and $C_{55}^c = G_c$. Notice that in order to define the displacement patterns explicitly in the core, three compatibility conditions are required: two at the upper face-core interface and a single one in the vertical direction at the lower interface.

The stresses in the core are

$$\sigma_{xx}^c \simeq 0; \quad \sigma_{zz}^c = E_c \varepsilon_{zz}^c; \quad \tau_{xz}^c = \tau^c = G_c \gamma_{xz}^c. \tag{6.3c}$$

The governing equations are formulated in terms of five unknowns: the two displacements in each face, w_0^t, u_0^t and w_0^b, u_0^b and the shear stress in the core, τ^c. Referring to the configuration in Figure 6.1, the governing equations in the case of a sandwich beam with isotropic face sheets read for the linear case (i.e. neglecting the second term in the strains (6.1c)):

$$\widetilde{(EA)}_b \frac{d^2 u_0^b(x)}{dx^2} + \tau^c(x) + \widetilde{p}_b = 0, \tag{6.4a}$$

$$\begin{aligned} -\left(c + \frac{f_b}{2}\right) \frac{d\tau^c(x)}{dx} + \frac{E_c}{2c}\left[w_0^b(x) - w_0^t(x)\right] \\ + \widetilde{(EI)}_b \frac{d^4 w_0^b(x)}{dx^4} + \frac{d\widetilde{m}_b(x)}{dx} - \widetilde{q}_b = 0, \end{aligned} \tag{6.4b}$$

$$\widetilde{(EA)}_t \frac{d^2 u_0^t(x)}{dx^2} - \tau^c(x) + \widetilde{p}_t = 0, \tag{6.4c}$$

$$\begin{aligned} -\left(c + \frac{f_t}{2}\right) \frac{d\tau^c(x)}{dx} + \frac{E_c}{2c}\left[w_0^t(x) - w_0^b(x)\right] \\ + \widetilde{(EI)}_t \frac{d^4 w_0^t(x)}{dx^4} + \frac{d\widetilde{m}_t(x)}{dx} - \widetilde{q}_t = 0, \end{aligned} \tag{6.4d}$$

and

$$\begin{aligned} -\frac{2c^3}{3E_c} \frac{d^2\tau^c(x)}{dx^2} + 2c\frac{\tau^c(x)}{G_c} - \left(c + \frac{f_t}{2}\right) \frac{dw_0^t(x)}{dx} \\ - \left(c + \frac{f_b}{2}\right) \frac{dw_0^b(x)}{dx} + u_0^b(x) - u_0^t(x) = 0, \end{aligned} \tag{6.4e}$$

where $\widetilde{q}_i$, $\widetilde{p}_i$ $(i = t, b)$ are the distributed transverse (vertical), and in-plane (horizontal) loads per unit width, and $\widetilde{m}_i$ the distributed bending moments per unit width at the upper and lower face sheets. Moreover, $\widetilde{(EA)}_{t,b}$ and $\widetilde{(EI)}_{t,b}$ are the axial and flexural rigidities of the top and bottom face sheets per unit width, i.e. $\widetilde{(EA)}_{t,b} = (EA)_{t,b}/b$ and $\widetilde{(EI)}_{t,b} = (EI)_{t,b}/b$, where b is the width.

The first four equations (6.4a–d) are due to equilibrium in the longitudinal and transverse directions of the two face sheets and the last one (6.4e) is a result of the compatibility condition at the lower face-core interface in the longitudinal direction.

Although the equations are coupled, the linearity of the equations allow the use of well-known solution procedures for ordinary differential equations.

6.3 The "Extended High-Order Sandwich Panel Theory" (EHSAPT)

The HSAPT sandwich theory is very effective in dealing with sandwich configurations where the assumption of negligible in-plane rigidity of the core is justified, such as a weak core. However, as it will be shown in Section 6.4, for applications involving stiffer cores the HSAPT cannot accurately predict the shear and axial stress distributions in the core. Thus, in this section we present the "Extended High-Order Sandwich Panel Theory" (EHSAPT) panel theory which, as will be shown in the following, is applicable to all cores, either weak or strong.

In this theory, recently formulated by Frostig (2010) and Phan et al. (2010), the in-plane rigidity of the core is considered in addition to the compressibility of the core in the transverse direction. The displacement field of the core is as in the HSAPT, i.e. the transverse displacement in the core is of second order in the transverse coordinate and the in-plane displacements are of third order in the transverse coordinate. The novelty of this theory is that it allows for three generalized coordinates in the core (the axial and transverse displacements at the centroid of the core, and the rotation at the centroid of the core) instead of just one (mid-point transverse displacement) commonly adopted in other available theories. It will proven in Section 6.4, by comparison to the elasticity solution, that this approach results in superior accuracy, especially for the cases of stiffer cores, for which cases the other available sandwich computational models cannot correctly predict the stress fields. Thus, this theory can be used with any combinations of core and face sheets and not only the very "soft" cores that the HSAPT demands. The theory is derived so that all core/facesheet displacement continuity conditions are fulfilled. The full nonlinear version of the theory is outlined.

While the face sheets can change their length only, the core can change both its thickness and length. The displacement fields considered for the core follow the resulting fields of the HSAPT model (see Frostig et al., 1992) and they read

$$w^c(x,z) = w_0^c(x) + w_1^c(x)z + w_2^c(x)z^2, \tag{6.5a}$$

$$u^c(x,z) = u_0^c(x) + \phi_0^c(x)z + u_2^c(x)z^2 + u_3^c(x)z^3, \tag{6.5b}$$

where w_0^c and u_0^c are the transverse and in-plane displacements, respectively, ϕ_0^c is the slope at the centroid of the core, while w_1^c, w_2^c and u_2^c, u_3^c are the transverse the in-plane unknown functions to be determined by the transverse and the in-plane compatibility conditions applied at the upper ($z = c$) and lower ($z = -c$) face-core interfaces,

Hence, using the compatibility condition in the transverse direction (z-axis) at the upper and the lower face core interfaces

$$w^c(x,c) = w_0^t(x); \quad w^c(x,-c) = w_0^b(x), \tag{6.5c}$$

yields the following distribution of the transverse displacement:

$$w^c(x,z) = \left(-\frac{z}{2c} + \frac{z^2}{2c^2}\right) w_0^b(x) + \left(1 - \frac{z^2}{c^2}\right) w_0^c(x) + \left(\frac{z}{2c} + \frac{z^2}{2c^2}\right) w_0^t(x). \tag{6.6}$$

The axial displacement of the core, $u^c(x,z)$, is determined through the fulfillment of the compatibility conditions in the in-plane direction at $z = c$ and $-c$, i.e.,

$$u^c(x,c) = u_0^t(x) + \frac{f_t}{2} w_{0,x}^t(x), \tag{6.7a}$$

and

$$u^c(x,-c) = u_0^b(x) - \frac{f_b}{2} w_{0,x}^b(x). \tag{6.7b}$$

Hence, after some algebraic manipulation:

$$\begin{aligned} u^c(x,z) = {} & z\left(1 - \frac{z^2}{c^2}\right)\phi_0^c(x) + \frac{z^2}{2c^2}\left(1 - \frac{z}{c}\right)u_0^b + \left(1 - \frac{z^2}{c^2}\right)u_0^c \\ & + \frac{z^2}{2c^2}\left(1 + \frac{z}{c}\right)u_0^t + \frac{f_b z^2}{4c^2}\left(-1 + \frac{z}{c}\right)w_{0,x}^b + \frac{f_t z^2}{4c^2}\left(1 + \frac{z}{c}\right)w_{0,x}^t. \end{aligned} \tag{6.8}$$

Therefore, this theory is expressed in terms of seven generalized coordinates (unknown functions of x): two for the top face sheet, w_0^t, u_0^t, two for the bottom face sheet, w_0^b, u_0^b, and three for the core, w_0^c, u_0^c and ϕ_0^c.

The strains can be obtained from the displacements using the linear strain-displacement relations. For the core, the transverse normal strain is

$$\varepsilon_{zz}^c = \frac{\partial w^c}{\partial z} = \left(\frac{z}{c^2} - \frac{1}{2c}\right) w_0^b - \frac{2z}{c^2} w_0^c + \left(\frac{z}{c^2} + \frac{1}{2c}\right) w_0^t, \tag{6.9}$$

and the shear strain

$$\begin{aligned}\gamma_{zx}^c &= \frac{\partial u^c}{\partial z} + \frac{\partial w^c}{\partial x} = \left(1 - \frac{3z^2}{c^2}\right) \phi_0^c + \left(\frac{z}{c^2} - \frac{3z^2}{2c^3}\right) u_0^b \\ &- \left(\frac{2z}{c^2}\right) u_0^c + \left(\frac{z}{c^2} + \frac{3z^2}{2c^3}\right) u_0^t \\ &+ \left[-\left(\frac{c+f_b}{2c^2}\right) z + \left(\frac{2c+3f_b}{4c^3}\right) z^2\right] w_{0,x}^b + \left(1 - \frac{z^2}{c^2}\right) w_{0,x}^c \\ &+ \left[\left(\frac{c+f_t}{2c^2}\right) z + \left(\frac{2c+3f_t}{4c^3}\right) z^2\right] w_{0,x}^t.\end{aligned} \tag{6.10}$$

There is also a non-zero linear axial strain in the core $\varepsilon_{xx}^c = du^c/dx$, which has the same structure as Equation (6.8), but with the generalized function coordinates replaced by one-order higher derivative with respect to x.

We assume an orthotropic core with stress-strain relations

$$\begin{bmatrix} \sigma_{xx}^c \\ \sigma_{zz}^c \\ \tau_{xz}^c \end{bmatrix} = \begin{bmatrix} C_{11}^c & C_{13}^c & 0 \\ C_{13}^c & C_{33}^c & 0 \\ 0 & 0 & C_{55}^c \end{bmatrix} \begin{bmatrix} \varepsilon_{xx}^c \\ \varepsilon_{zz}^c \\ \gamma_{xz}^c \end{bmatrix}, \tag{6.11}$$

where $C_{ij}^{t,b,c}$ are the corresponding stiffness constants and we have used the notation $1 \equiv x$, $3 \equiv z$, and $55 \equiv xz$.

The governing equations and boundary conditions are derived from the variational principle

$$\delta(U + V) = 0, \tag{6.12}$$

where U is the strain energy of the sandwich beam, and V is the potential due to the applied loading. The first variation of the strain energy per unit width of the sandwich beam is

$$\begin{aligned}\delta U = \int_0^a \Bigg[&\int_{-c+f_b}^{-c} \sigma_{xx}^b \delta\varepsilon_{xx}^b dz + \int_{-c}^{c} \left(\sigma_{xx}^c \delta\varepsilon_{xx}^c + \sigma_{zz}^c \delta\varepsilon_{zz}^c + \tau_{zx}^c \delta\gamma_{zx}^c\right) dz \\ &+ \int_c^{c+f_t} \sigma_{xx}^t \delta\varepsilon_{xx}^t dz \Bigg] dx,\end{aligned} \tag{6.13a}$$

and the first variation of the external potential per unit width due to several general loading conditions is

$$\delta V = -\int_0^a \left(\widetilde{p}^t \delta u_0^t + \widetilde{p}^b \delta u_0^b + \widetilde{q}^t \delta w_0^t + \widetilde{q}^b \delta w_0^b + \widetilde{m}^t \delta w_{0,x}^t + \widetilde{m}^b \delta w_{0,x}^b\right) dx$$
$$- \widetilde{N}^t \delta u_0^t \Big|_{x=0}^a - \widetilde{N}^b \delta u_0^b \Big|_{x=0}^a - \widetilde{V}^t \delta w_0^t \Big|_{x=0}^a - \widetilde{V}^b \delta w_0^b \Big|_{x=0}^a$$
$$- \widetilde{M}^t \delta w_{0,x}^t \Big|_{x=0}^a - \widetilde{M}^b \delta w_{0,x}^b \Big|_{x=0}^a - \left(\int_{-c}^{c} \widetilde{n}^c \delta u^c + \widetilde{v}^c \delta w^c\right) dz \Big|_{x=0}^a, \tag{6.13b}$$

where $\widetilde{p}^{t,b}$ is the distributed in-plane force (along x) per unit width, $\widetilde{q}^{t,b}$ is the distributed transverse (along z) force per unit width and $\widetilde{m}^{t,b}$ is the distributed moment per unit width on the top and bottom face sheets. Moreover, $\widetilde{N}^{t,b}$ is the end axial force per unit width, $\widetilde{V}^{t,b}$ is the end shear force per unit width and $\widetilde{M}^{t,b}$ is the end moment per unit width at the top and bottom face sheets, at the ends $x = 0, a$. In addition, $\widetilde{n}^c$ is the end axial force per unit width and $\widetilde{v}^c$ is the end shear force per unit width in the core at the ends $x = 0, a$.

In the following we assume that $\widetilde{n}^c$ and $\widetilde{v}^c$ are constants. In this case,

$$\int_{-c}^{c} \widetilde{n}^c \delta u^c dz = \widetilde{n}^c c \left[\frac{1}{3}\left(\delta u_0^b + \delta u_0^t\right) + \frac{4}{3}\left(\delta u_0^c\right) - \frac{f_b}{6}\delta w_{0,x}^b + \frac{f_t}{6}\delta w_{0,x}^t\right], \tag{6.13c}$$

$$\int_{-c}^{c} \widetilde{v}^c \delta w^c dz = \widetilde{v}^c c \left(\frac{1}{3}\delta w_0^b + \frac{4}{3}\delta w_0^c + \frac{1}{3}\delta w_0^t\right). \tag{6.13d}$$

Of course, the theory can admit any variation of $\widetilde{n}^c$ and $\widetilde{v}^c$ along z; for example, a bending moment on the core would correspond to a linear variation of $\widetilde{n}^c$ with respect to z. However, for most practical purposes, loads are applied to the face sheets and not to the core.

For sandwich panels made from orthotropic materials, we can substitute the stresses in terms of the strains from the constitutive relations, and then the strains in terms of the displacements and the displacement profiles, and finally apply the variational principle. Thus, we can write a set of nonlinear governing differential equations (DEs) in terms of the seven unknown generalized coordinates as follows:

Top face sheet DEs (two nonlinear):

$$\delta u_0^t : \ -\left(\frac{4}{5}C_{55}^c + \frac{2c^2}{35}C_{11}^c\frac{\partial^2}{\partial x^2}\right)\phi_0^c - \left(\frac{7}{30c}C_{55}^c + \frac{c}{35}C_{11}^c\frac{\partial^2}{\partial x^2}\right)u_0^b$$
$$-\left(\frac{4}{3c}C_{55}^c + \frac{2c}{15}C_{11}^c\frac{\partial^2}{\partial x^2}\right)u_0^c + \left(\frac{47}{30c}C_{55}^c - \alpha_1^t\frac{\partial^2}{\partial x^2}\right)u_0^t$$

$$-\left(\alpha_2^b\frac{\partial}{\partial x}-\frac{cf_b}{70}C_{11}^c\frac{\partial^3}{\partial x^3}\right)w_0^b+\left(\beta_1\frac{\partial}{\partial x}\right)w_0^c$$

$$+\left(\alpha_3^t\frac{\partial}{\partial x}-\frac{3cf_t}{35}C_{11}^c\frac{\partial^3}{\partial x^3}\right)w_0^t=\widetilde{p}^t+F_u^t, \tag{6.14a}$$

where F_u^t is the nonlinear term

$$F_u^t=C_{11}^t f_t w_{0,x}^t w_{0,xx}^t, \tag{6.14b}$$

and $\widetilde{p}^t$ is the distributed in-plane force (along x) per unit width at the top face; and

$$\delta w_0^t\colon \left(\alpha_4^t\frac{\partial}{\partial x}+\frac{c^2 f_t}{35}C_{11}^c\frac{\partial^3}{\partial x^3}\right)\phi_0^c+\left(\alpha_5^t\frac{\partial}{\partial x}+\frac{cf_t}{70}C_{11}^c\frac{\partial^3}{\partial x^3}\right)u_0^b$$

$$+\left(\alpha_6^t\frac{\partial}{\partial x}+\frac{cf_t}{15}C_{11}^c\frac{\partial^3}{\partial x^3}\right)u_0^c+\left(-\alpha_3^t\frac{\partial}{\partial x}+\frac{3cf_t}{35}C_{11}^c\frac{\partial^3}{\partial x^3}\right)u_0^t$$

$$+\left(\frac{1}{6c}C_{33}^c+\beta_2\frac{\partial^2}{\partial x^2}-\frac{cf_bf_t}{140}C_{11}^c\frac{\partial^4}{\partial x^4}\right)w_0^b+\left(-\frac{4}{3c}C_{33}^c+\alpha_7^t\frac{\partial^2}{\partial x^2}\right)w_0^c$$

$$+\left(\frac{7}{6c}C_{33}^c+\alpha_8^t\frac{\partial^2}{\partial x^2}+\alpha_9^t\frac{\partial^4}{\partial x^4}\right)w_0^t=\widetilde{q}^t-\widetilde{m}_{,x}^t+F_w^t. \tag{6.15a}$$

where F_w^t is the nonlinear term

$$F_w^t=C_{11}^t f_t\left[w_{0,x}^t u_{0,xx}^t+u_{0,x}^t w_{0,xx}^t+\frac{3}{2}(w_{0,x}^t)^2 w_{0,xx}^t\right], \tag{6.15b}$$

and $\widetilde{q}^t$ is the distributed transverse (along z) force per unit width, and $\widetilde{m}^t$ is the distributed moment per unit width on the top face sheet.

Core DEs (three linear):

$$\delta u_0^c\colon -\left(\frac{4}{3c}C_{55}^c+\frac{2c}{15}C_{11}^c\frac{\partial^2}{\partial x^2}\right)u_0^b+\left(\frac{8}{3c}C_{55}^c-\frac{16c}{15}C_{11}^c\frac{\partial^2}{\partial x^2}\right)u_0^c$$

$$-\left(\frac{4}{3c}C_{55}^c+\frac{2c}{15}C_{11}^c\frac{\partial^2}{\partial x^2}\right)u_0^t+\left(\alpha_6^b\frac{\partial}{\partial x}+\frac{cf_b}{15}C_{11}^c\frac{\partial^3}{\partial x^3}\right)w_0^b$$

$$-\left(\alpha_6^t\frac{\partial}{\partial x}+\frac{cf_t}{15}C_{11}^c\frac{\partial^3}{\partial x^3}\right)w_0^t=0, \tag{6.16a}$$

$$\delta\phi_0^c: \left(\frac{8c}{5}C_{55}^c - \frac{16c^3}{105}C_{11}^c\frac{\partial^2}{\partial x^2}\right)\phi_0^c + \left(\frac{4}{5}C_{55}^c + \frac{2c^2}{35}C_{11}^c\frac{\partial^2}{\partial x^2}\right)u_0^b$$
$$- \left(\frac{4}{5}C_{55}^c + \frac{2c^2}{35}C_{11}^c\frac{\partial^2}{\partial x^2}\right)u_0^t - \left(\alpha_4^b\frac{\partial}{\partial x} + \frac{c^2 f_b}{35}C_{11}^c\frac{\partial^3}{\partial x^3}\right)w_0^b$$
$$+ \left(\beta_3\frac{\partial}{\partial x}\right)w_0^c - \left(\alpha_4^t\frac{\partial}{\partial x} + \frac{c^2 f_t}{35}C_{11}^c\frac{\partial^3}{\partial x^3}\right)w_0^t = 0, \tag{6.16b}$$

and

$$\delta w_0^c: -\left(\beta_3\frac{\partial}{\partial x}\right)\phi_0^c + \left(\beta_1\frac{\partial}{\partial x}\right)u_0^b - \left(\beta_1\frac{\partial}{\partial x}\right)u_0^t$$
$$+ \left(-\frac{4}{3c}C_{33}^c + \alpha_7^b\frac{\partial^2}{\partial x^2}\right)w_0^b + \left(\frac{8}{3c}C_{33}^c - \frac{16c}{15}C_{55}^c\frac{\partial^2}{\partial x^2}\right)w_0^c$$
$$+ \left(-\frac{4}{3c}C_{33}^c + \alpha_7^t\frac{\partial^2}{\partial x^2}\right)w_0^t = 0. \tag{6.16c}$$

Bottom face sheet DEs (two nonlinear):

$$\delta u_0^b: \left(\frac{4}{5}C_{55}^c + \frac{2c^2}{35}C_{11}^c\frac{\partial^2}{\partial x^2}\right)\phi_0^c + \left(\frac{47}{30c}C_{55}^c - \alpha_1^b\frac{\partial^2}{\partial x^2}\right)u_0^b$$
$$- \left(\frac{4}{3c}C_{55}^c + \frac{2c}{15}C_{11}^c\frac{\partial^2}{\partial x^2}\right)u_0^c - \left(\frac{7}{30c}C_{55}^c + \frac{c}{35}C_{11}^c\frac{\partial^2}{\partial x^2}\right)u_0^t$$
$$+ \left(-\alpha_3^b\frac{\partial}{\partial x} + \frac{3cf_b}{35}C_{11}^c\frac{\partial^3}{\partial x^3}\right)w_0^b - \left(\beta_1\frac{\partial}{\partial x}\right)w_0^c$$
$$+ \left(\alpha_2^t\frac{\partial}{\partial x} - \frac{cf_t}{70}C_{11}^c\frac{\partial^3}{\partial x^3}\right)w_0^t = \widetilde{p}^b + \hat{F}_u^b, \tag{6.17a}$$

where F_u^b is the nonlinear term

$$F_u^b = C_{11}^b f_b w_{0,x}^b w_{0,xx}^b, \tag{6.17b}$$

and $\widetilde{p}^b$ is the distributed in-plane force (along x) per unit width at the bottom face; and

$$\delta w_0^b: \left(\alpha_4^b\frac{\partial}{\partial x} + \frac{c^2 f_b}{35}C_{11}^c\frac{\partial^3}{\partial x^3}\right)\phi_0^c + \left(\alpha_3^b\frac{\partial}{\partial x} - \frac{3cf_b}{35}C_{11}^c\frac{\partial^3}{\partial x^3}\right)u_0^b$$
$$- \left(\alpha_6^b\frac{\partial}{\partial x} + \frac{cf_b}{15}C_{11}^c\frac{\partial^3}{\partial x^3}\right)u_0^c - \left(\alpha_5^b\frac{\partial}{\partial x} + \frac{cf_b}{70}C_{11}^c\frac{\partial^3}{\partial x^3}\right)u_0^t$$

$$+\left(\frac{7}{6c}C_{33}^c+\alpha_8^b\frac{\partial^2}{\partial x^2}+\alpha_9^b\frac{\partial^4}{\partial x^4}\right)w_0^b+\left(-\frac{4}{3c}C_{33}^c+\alpha_7^b\frac{\partial^2}{\partial x^2}\right)w_0^c$$

$$+\left(\frac{1}{6c}C_{33}^c+\beta_2\frac{\partial^2}{\partial x^2}-\frac{cf_bf_t}{140}\frac{\partial^4}{\partial x^4}\right)w_0^t=\widetilde{q}^b-\widetilde{m}^b_{,x}+F_w^b, \quad (6.18a)$$

where F_w^b is the nonlinear term:

$$F_w^b=C_{11}^bf_b\left[w_{0,x}^bu_{0,xx}^b+u_{0,x}^bw_{0,xx}^b+\frac{3}{2}(w_{0,x}^b)^2w_{0,xx}^b\right]. \quad (6.18b)$$

and $\widetilde{q}^b$ is the distributed transverse (along z) force per unit width and $\widetilde{m}^b$ is the distributed moment per unit width on the bottom face sheet.

In the above expressions, the following constants are defined for $i=t,b$:

$$\alpha_1^i=\frac{6c}{35}C_{11}^c+f_iC_{11}^i;\quad \alpha_2^i=\frac{1}{30}C_{13}^c+\left(\frac{1}{30}-\frac{7f_i}{60c}\right)C_{55}^c, \quad (6.19a)$$

$$\alpha_3^i=-\frac{11}{30}C_{13}^c+\left(\frac{19}{30}+\frac{47f_i}{60c}\right)C_{55}^c;\quad \alpha_4^i=\frac{4c}{15}C_{13}^c+\left(\frac{4c}{15}+\frac{2f_i}{5}\right)C_{55}^c, \quad (6.19b)$$

$$\alpha_5^i=-\frac{1}{30}C_{13}^c+\left(-\frac{1}{30}+\frac{7f_i}{60c}\right)C_{55}^c;\quad \alpha_6^i=\frac{2}{3}C_{13}^c+\left(\frac{2}{3}+\frac{2f_i}{3c}\right)C_{55}^c, \quad (6.19c)$$

$$\alpha_7^i=-\frac{f_i}{5}C_{13}^c-\left(\frac{2c}{15}+\frac{f_i}{5}\right)C_{55}^c, \quad (6.19d)$$

$$\alpha_8^i=\frac{11f_i}{30}C_{13}^c-\left(\frac{4c}{15}+\frac{19f_i}{30}+\frac{47f_i^2}{120c}\right)C_{55}^c, \quad (6.19e)$$

$$\alpha_9^i=\frac{f_i^3}{12}C_{11}^i+\frac{3cf_i^2}{70}C_{11}^c, \quad (6.19f)$$

and

$$\beta_1=\frac{2}{5}(C_{13}^c+C_{55}^c), \quad (6.19g)$$

$$\beta_2=\frac{f_b+f_t}{60}C_{13}^c+\left(\frac{c}{15}+\frac{f_b+f_t}{60}-\frac{7f_bf_t}{120c}\right)C_{55}^c, \quad (6.19h)$$

$$\beta_3=\frac{8c}{15}(C_{13}^c+C_{55}^c). \quad (6.19i)$$

Boundary conditions (BCs) at $x=0,a$ are either "natural" or "kinematic", referred to as "either" and "or" in the remainder of this section. At each end there are nine boundary conditions, three for each face sheet and three for the core:

Top face sheet BCs (three):

(i) *Either* $\delta u_0^t = 0$ *or*

$$\left(\frac{2c^2}{35}C_{11}^c\frac{\partial}{\partial x}\right)\phi_0^c + \left(\frac{c}{35}C_{11}^c\frac{\partial}{\partial x}\right)u_0^b + \left(\frac{2c}{15}C_{11}^c\frac{\partial}{\partial x}\right)u_0^c$$
$$+\left[\left(\frac{6c}{35}C_{11}^c + f_t C_{11}^t\right)\frac{\partial}{\partial x}\right]u_0^t + \left(\frac{1}{30}C_{13}^c - \frac{cf_b}{70}C_{11}^c\frac{\partial^2}{\partial x^2}\right)w_0^b$$
$$-\left(\frac{2}{5}C_{13}^c\right)w_0^c + \left(\frac{11}{30}C_{13}^c + \frac{3cf_t}{35}C_{11}^c\frac{\partial^2}{\partial x^2}\right)w_0^t$$
$$= \widetilde{N}^t + \frac{\widetilde{n}^c c}{3} + B_u^t, \tag{6.20a}$$

where $\widetilde{N}^t$ is the end axial force per unit width at the top face and $\widetilde{n}^c$ is the end axial force per unit width at the the core (at the end $x = 0$ or $x = a$) and the nonlinear term

$$B_u^t = -\frac{f_t}{2}C_{11}^t(w_{0,x}^t)^2. \tag{6.20b}$$

(ii) *Either* $\delta w_0^t = 0$ *or*

$$-\left[\frac{2(2c+3f_t)}{15}C_{55}^c + \frac{c^2 f_t}{35}C_{11}^c\frac{\partial^2}{\partial x^2}\right]\phi_0^c$$
$$+\left[\frac{(2c-7f_t)}{60c}C_{55}^c - \frac{cf_t}{70}C_{11}^c\frac{\partial^2}{\partial x^2}\right]u_0^b$$
$$-\left[\frac{2(c+f_t)}{3c}C_{55}^c + \frac{cf_t}{15}C_{11}^c\frac{\partial^2}{\partial x^2}\right]u_0^c$$
$$+\left[\frac{(38c+47f_t)}{60c}C_{55}^c - \frac{3cf_t}{35}C_{11}^c\frac{\partial^2}{\partial x^2}\right]u_0^t$$
$$+\left[\left(\frac{f_b}{60}C_{13}^c - \beta_2\right)\frac{\partial}{\partial x} + \frac{cf_b f_t}{140}C_{11}^c\frac{\partial^3}{\partial x^3}\right]w_0^b$$
$$-\left(\alpha_7^t\frac{\partial}{\partial x}\right)w_0^c + \left[\left(\frac{11f_t}{60}C_{13}^c - \alpha_8^t\right)\frac{\partial}{\partial x} - \alpha_9^t\frac{\partial^3}{\partial x^3}\right]w_0^t$$
$$= \widetilde{V}^t + \widetilde{m}^t + \frac{\widetilde{v}^c c}{3} + B_w^t, \tag{6.21a}$$

where B_w^t is the nonlinear term

$$B_w^t = -\frac{f_t}{2} C_{11}^t w_{0,x}^t \left[2u_{0,x}^t + (w_{0,x}^t)^2\right], \tag{6.21b}$$

and $\widetilde{V}^t$ is the end shear force per unit width in the top face and $\widetilde{v}^c$ is the end shear force per unit width in the core (at the end $x = 0$ or $x = a$).

(iii) Either $\delta w_{0,x}^t = 0$ or

$$\left(\frac{c^2 f_t}{35} C_{11}^c \frac{\partial}{\partial x}\right) \phi_0^c + \left(\frac{c f_t}{70} C_{11}^c \frac{\partial}{\partial x}\right) u_0^b + \left(\frac{c f_t}{15} C_{11}^c \frac{\partial}{\partial x}\right) u_0^c$$
$$+ \left(\frac{3 c f_t}{35} C_{11}^c \frac{\partial}{\partial x}\right) u_0^t + \left(\frac{f_t}{60} C_{13}^c - \frac{c f_b f_t}{140} C_{11}^c \frac{\partial^2}{\partial x^2}\right) w_0^b$$
$$- \left(\frac{f_t}{5} C_{13}^c\right) w_0^c + \left(\frac{11 f_t}{60} C_{13}^c + \alpha_9^t \frac{\partial^2}{\partial x^2}\right) w_0^t = \widetilde{M}^t + \frac{\widetilde{n}^c c f_t}{6}, \tag{6.22}$$

where $\widetilde{M}^t$ is the end moment per unit width at the top face (at the end $x = 0$ or $x = a$).

Core BCs (three):

(i) *Either* $\delta u_0^c = 0$ *or*,

$$\left(\frac{2c}{15} C_{11}^c \frac{\partial}{\partial x}\right) u_0^b + \left(\frac{16c}{15} C_{11}^c \frac{\partial}{\partial x}\right) u_0^c + \left(\frac{2c}{15} C_{11}^c \frac{\partial}{\partial x}\right) u_0^t$$
$$- \left(\frac{2}{3} C_{13}^c + \frac{c f_b}{15} C_{11}^c \frac{\partial^2}{\partial x^2}\right) w_0^b + \left(\frac{2}{3} C_{13}^c + \frac{c f_t}{15} C_{11}^c \frac{\partial^2}{\partial x^2}\right) w_0^t = \frac{4 \widetilde{n}^c c}{3}. \tag{6.23a}$$

(ii) *Either* $\delta \phi_0^c = 0$ *or*

$$\left(\frac{16c^3}{105} C_{11}^c \frac{\partial}{\partial x}\right) \phi_0^c - \left(\frac{2c^2}{35} C_{11}^c \frac{\partial}{\partial x}\right) u_0^b + \left(\frac{2c^2}{35} C_{11}^c \frac{\partial}{\partial x}\right) u_0^t$$
$$+ \left(\frac{4c}{15} C_{13}^c + \frac{c^2 f_b}{35} C_{11}^c \frac{\partial^2}{\partial x^2}\right) w_0^b - \left(\frac{8c}{15} C_{13}^c\right) w_0^c$$
$$+ \left(\frac{4c}{15} C_{13}^c + \frac{c^2 f_t}{35} C_{11}^c \frac{\partial^2}{\partial x^2}\right) w_0^t = 0. \tag{6.23b}$$

(iii) *Either* $\delta w_0^c = 0$ *or*

$$C_{55}^c \left[\frac{8c}{15} \phi_0^c - \frac{2}{5} u_0^b + \frac{2}{5} u_0^t + \frac{(2c + 3 f_b)}{15} w_{0,x}^b\right.$$
$$\left. + \frac{16c}{15} w_{0,x}^c + \frac{(2c + 3 f_t)}{15} w_{0,x}^t\right] = \frac{4}{3} \widetilde{v}^c c. \tag{6.23c}$$

Bottom face sheet BCs (three):

(i) *Either* $\delta u_0^b = 0$ *or*

$$-\left(\frac{2c^2}{35}C_{11}^c\frac{\partial}{\partial x}\right)\phi_0^c+\left[\left(\frac{6c}{35}C_{11}^c+f_bC_{11}^b\right)\frac{\partial}{\partial x}\right]u_0^b$$
$$+\left(\frac{2c}{15}C_{11}^c\frac{\partial}{\partial x}\right)u_0^c+\left(\frac{c}{35}C_{11}^c\frac{\partial}{\partial x}\right)u_0^t$$
$$-\left(\frac{11}{30}C_{13}^c+\frac{3cf_b}{35}C_{11}^c\frac{\partial^2}{\partial x^2}\right)w_0^b+\left(\frac{2}{5}C_{13}^c\right)w_0^c$$
$$+\left(-\frac{1}{30}C_{13}^c+\frac{cf_t}{70}C_{11}^c\frac{\partial^2}{\partial x^2}\right)w_0^t=\widetilde{N}^b+\frac{\widetilde{n}^c c}{3}+B_u^b, \quad (6.24a)$$

where $\widetilde{N}^b$ is the end axial force per unit width at the bottom face and the nonlinear term,

$$B_u^b=-\frac{f_b}{2}C_{11}^b(w_{0,x}^b)^2. \quad (6.24b)$$

(ii) *Either* $\delta w_0^b = 0$ *or*

$$-\left[\frac{2(2c+3f_b)}{15}C_{55}^c+\frac{c^2f_b}{35}C_{11}^c\frac{\partial^2}{\partial x^2}\right]\phi_0^c$$
$$+\left[-\frac{(38c-47f_b)}{60c}C_{55}^c+\frac{3cf_b}{35}C_{11}^c\frac{\partial^2}{\partial x^2}\right]u_0^b$$
$$+\left[\frac{2(c+f_b)}{3c}C_{55}^c+\frac{cf_b}{15}C_{11}^c\frac{\partial^2}{\partial x^2}\right]u_0^c$$
$$+\left[\frac{(-2c+7f_b)}{60c}C_{55}^c+\frac{cf_b}{70}C_{11}^c\frac{\partial^2}{\partial x^2}\right]u_0^t$$
$$+\left[\left(\frac{11f_b}{60}C_{13}^c-\alpha_8^b\right)\frac{\partial}{\partial x}-\alpha_9^b\frac{\partial^3}{\partial x^3}\right]w_0^b-\left(\alpha_7^b\frac{\partial}{\partial x}\right)w_0^c$$
$$+\left[\left(\frac{f_t}{60}C_{13}^c-\beta_2\right)\frac{\partial}{\partial x}+\frac{cf_bf_t}{140}C_{11}^c\frac{\partial^3}{\partial x^3}\right]w_0^t$$
$$=\widetilde{V}^b+\widetilde{m}^b+\frac{\widetilde{v}^c c}{3}+B_w^b, \quad (6.25a)$$

where B_w^b is the nonlinear term

$$B_w^b=-\frac{f_b}{2}C_{11}^bw_{0,x}^b\left[2u_{0,x}^b+(w_{0,x}^b)^2\right]. \quad (6.25b)$$

and $\widetilde{V}^b$ is the end shear force per unit width at bottom face.

(iii) *Either* $\delta w^b_{0,x} = 0$ *or*

$$\left(\frac{c^2 f_b}{35} C^c_{11} \frac{\partial}{\partial x}\right) \phi^c_0 - \left(\frac{3cf_b}{35} C^c_{11} \frac{\partial}{\partial x}\right) u^b_0 - \left(\frac{cf_b}{15} C^c_{11} \frac{\partial}{\partial x}\right) u^c_0$$
$$- \left(\frac{cf_b}{70} C^c_{11} \frac{\partial}{\partial x}\right) u^t_0 + \left(\frac{11 f_b}{60} C^c_{13} + \alpha^b_9 \frac{\partial^2}{\partial x^2}\right) w^b_0$$
$$- \left(\frac{f_b}{5} C^c_{13}\right) w^c_0 + \left(\frac{f_b}{60} C^c_{13} - \frac{cf_b f_t}{140} C^c_{11} \frac{\partial^2}{\partial x^2}\right) w^t_0$$
$$= \widetilde{M}^b - \frac{\widetilde{n}^c c f_b}{6}. \tag{6.26}$$

where $\widetilde{M}^b$ is the end moment per unit width at the bottom face.

In the above equations the tilde denotes the known external boundary values.

6.4 Application to a Simply Supported Sandwich Beam

In this section we shall study the response of a simply supported sandwich beam under a transversely applied load of the form:

$$\widetilde{q}^t(x) = q_0 \sin \frac{\pi x}{a}. \tag{6.27}$$

The numerical results for several typical sandwich plate configurations with orthotropic layers will be compared with the results using the elasticity solution. The elasticity solution for sandwich plates is outlined in Chapter 5. For beams, a similar solution can be derived. Indeed, Pagano (1969) presented a three-dimensional elasticity solution for a laminated or sandwich beam for the case of a positive discriminant of the quadratic characteristic equation, which is formed from the orthotropic material constants, and only when these two real roots are positive. The isotropic case, in which there are two equal real roots, was also outlined. Recently, Kardomateas and Phan (2010) extended the Pagano (1969) solution to the case of (i) negative discriminant, which results in two complex conjugate roots of the quadratic equation, and (ii) positive discriminant but with real negative roots. The case of a negative discriminant is actually frequently encountered in sandwich construction where the orthotropic core is stiffer in the transverse direction than in-plane. Notice that we here refer to the elasticity solution

Table 6.1 Material properties. Moduli data are in GPa

	Carbon/ Epoxy Face	E-Glass/ Polyester Face	Balsa Wood Core	Glass-Phenolic Honeycomb Core
E_1	181	40.0	0.671	0.032
E_2	10.3	10.0	0.158	0.032
E_3	10.3	10.0	7.72	0.300
G_{23}	5.96	3.5	0.312	0.048
G_{13}	7.17	4.5	0.312	0.048
G_{12}	7.17	4.5	0.200	0.013
ν_{32}	0.40	0.40	0.49	0.25
ν_{31}	0.016	0.260	0.23	0.25
ν_{12}	0.277	0.065	0.66	0.25

for a sandwich beam (1D) which is different from the one for a sandwich plate (2D) as outlined in Chapter 5.

Thus, the results from the HSAPT and the EHSAPT theories will be compared with the elasticity results from Kardomateas and Phan (2010). In addition, the classical model and the first-order shear deformation theory (FOSD) will be examined.

For the EHSAPT theory, boundary conditions for $x = 0, a$ (Figure 6.1) utilize the seven kinematic conditions

$$u_0^t = u_0^b = u_0^c = \phi_0^c = 0; \quad w_0^t = w_0^b = w_0^c = 0, \tag{6.28}$$

and the two natural boundary conditions (6.22) and (6.26).

All these are satisfied by displacements in the form

$$u_0^t = U_0^t \cos\frac{\pi x}{a}; \quad u_0^c = U_0^c \cos\frac{\pi x}{a};$$

$$\phi_0^c = \Phi_0^c \cos\frac{\pi x}{a}; \quad u_0^b = U_0^b \cos\frac{\pi x}{a}, \tag{6.29a}$$

$$w_0^t = W_0^t \sin\frac{\pi x}{a}; \quad w_0^c = W_0^c \sin\frac{\pi x}{a}; \quad w_0^b = W_0^b \sin\frac{\pi x}{a}. \tag{6.29b}$$

We consider the linear problem, which means that the nonlinear terms $F_{u,w}^{t,b}$ in the governing differential equations and the nonlinear terms $B_{u,w}^{t,b}$ in the boundary conditions are neglected.

Substituting Equation (6.29) into the governing equations (6.12–6.19) results in a system of seven linear equations for the seven unknown constants U_0^t, U_0^c, Φ_0^c, U_0^b, W_0^t, W_0^c, W_0^b.

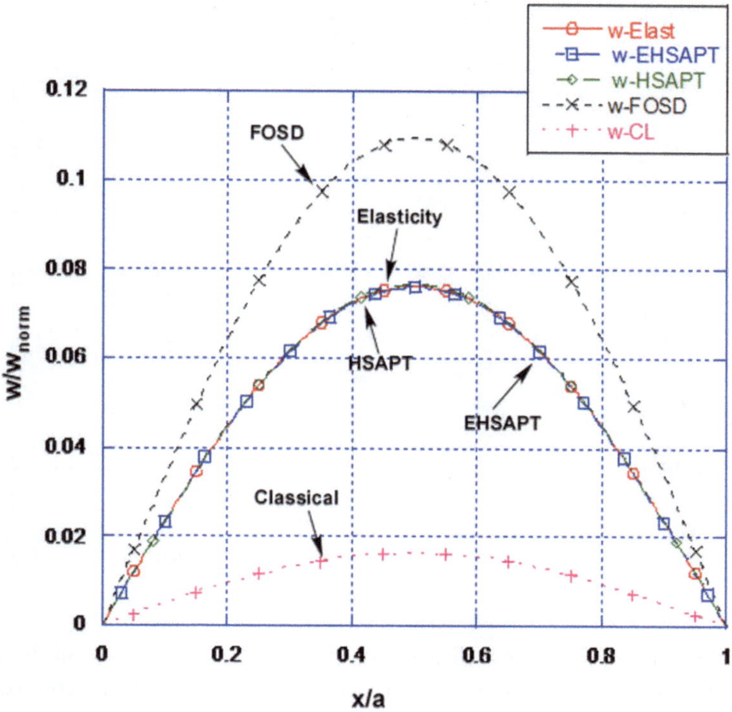

Figure 6.2 Transverse displacement, w, at the top, $z = c + f$, for a sandwich beam with carbon/epoxy faces and glass phenolic honeycomb core.

We will consider sandwich configurations with faces made from either carbon/epoxy or E-glass/polyester unidirectional composite and core made from either hexagonal glass/phenolic honeycomb or balsa wood. The properties for these materials are listed in Table 6.1.

The two face sheets are assumed identical with thickness $f_t = f_b = f = 2$ mm. The core thickness is $2c = 16$ mm. The total thickness of the beam is $h = 2f + 2c = 20$ mm and the span length of the beam is $a = 20h = 400$ mm.

In the following, the displacements are normalized with

$$w_{\text{norm}} = \frac{3q_0 a^4}{2\pi^4 E_1^f f^3}. \tag{6.30}$$

and the stresses with q_0.

The normalized displacement of the top face sheet is plotted as a function of x, for a sandwich beam with carbon/epoxy faces and glass phenolic honeycomb core in Figure 6.2. In this figure, we also show the predictions from

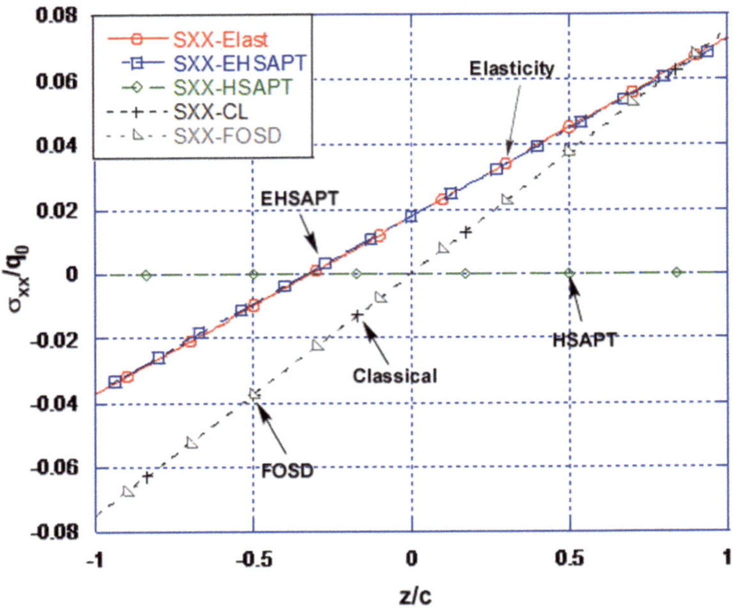

Figure 6.3 Through-thickness distribution of the axial stress, σ_{xx}, in the core at $x = a/2$ for the case of carbon/epoxy faces and glass phenolic honeycomb core.

classical beam theory, which does not include transverse shear, as well as the First-Order Shear Deformation (FOSD) theory. From Figure 6.2, we can see that both the classical and first-order shear approaches appear inadequate. The classical theory is too unconservative and the first-order shear theory is too conservative; this clearly demonstrates the need for higher-order theories in dealing with sandwich structures. In this regard, both the HSAPT and the EHSAPT give a displacement profile which is essentially identical to the elasticity solution. In Figure 6.2 we can also readily observe the large effect of transverse shear, which is an important feature of sandwich structures.

The distribution of the axial stress in the core, σ_{xx}, as a function of z at the midspan location, $x = a/2$ (where the bending moment is maximum), is plotted in Figure 6.3, again for the case of carbon/epoxy faces and glass phenolic honeycomb core. The EHSAPT predicts a stress profile practically identical to the Elasticity. Note that the HSAPT neglects the in-plane rigidity of the core, and this yields a zero axial stress. The classical and first-order shear theories give practically identical predictions but they are in appreciable error by comparison to the elasticity, the error increasing towards the lower end of the core ($z = -c$). All curves are linear.

Notice also that for the elasticity and the extended high-order theory there is no symmetry with regard to the mid-plane ($z = 0$) unlike in the classical and first-order shear theories.

The through-thickness distribution of the transverse normal stress in the core, σ_{zz}, at the midspan location, $x = a/2$, is shown in Figure 6.4a for the case of E-glass/polyester faces and balsa wood core.

Only the profiles using elasticity, the HSAPT and the EHSAPT are presented, since the FOSD theory and the classical theories consider the core to be incompressible. Both high-order theories are practically coinciding with the elasticity curve and all are nearly linear. However, the theories differ when the transverse normal strain is examined in Figure 6.4b with the present extended high-order theory being very close to the elasticity theory.

Figures 6.5 and 6.6 show the through-thickness distribution of the transverse shear stress in the core, τ_{xz}, at $x = a/10$, i.e. near the ends where shear is expected to be significant, for the cases of carbon/epoxy faces and glass phenolic honeycomb core (Figure 6.5) and E-glass/polyester faces and balsa wood core (Figure 6.6).

For the very soft core case of Figure 6.5, the shear stress is nearly constant and all theories practically coincide with the elasticity. The elasticity data show that the range of the shear stress variation is about 0.05% of the maximum value, i.e. the shear stress is practically constant. This case of a very soft core would justify the neglect of the in-plane rigidity of the core made in the HSAPT.

For the case of the E-glass/polyester faces and balsa wood core, however, the shear stress shows a noticeable distribution (about 5%) through the thickness, which is very nicely captured by the EHSAPT, which is very close to the elasticity (Figure 6.6). For this sandwich configuration, it is obvious that a theory based on a constant shear stress assumption would not capture this distribution.

This issue is further explored by considering a sandwich construction in which both the face sheets and the core are isotropic. By varying the face/core modulus ratio, we can change the shear stress in the core. Thus, we assume that the face sheets are made from an isotropic alloy with $E_f = 100$ GPa and the core is made from isotropic material having a modulus E_c such that the ratio E_f/E_c assumes the values of 50, 5 and 2. The Poisson ratios are assumed to be $\nu_f = \nu_c = 0.30$. Figure 6.7 shows the shear stress distribution through the thickness. For a modulus ratio of 2, the maximum over minimum shear stress ratio is about 2. On the contrary, for a modulus ratio of 50, the shear stress range is very small, with the corresponding maximum over minimum shear stress ratio being only about 1.04. The

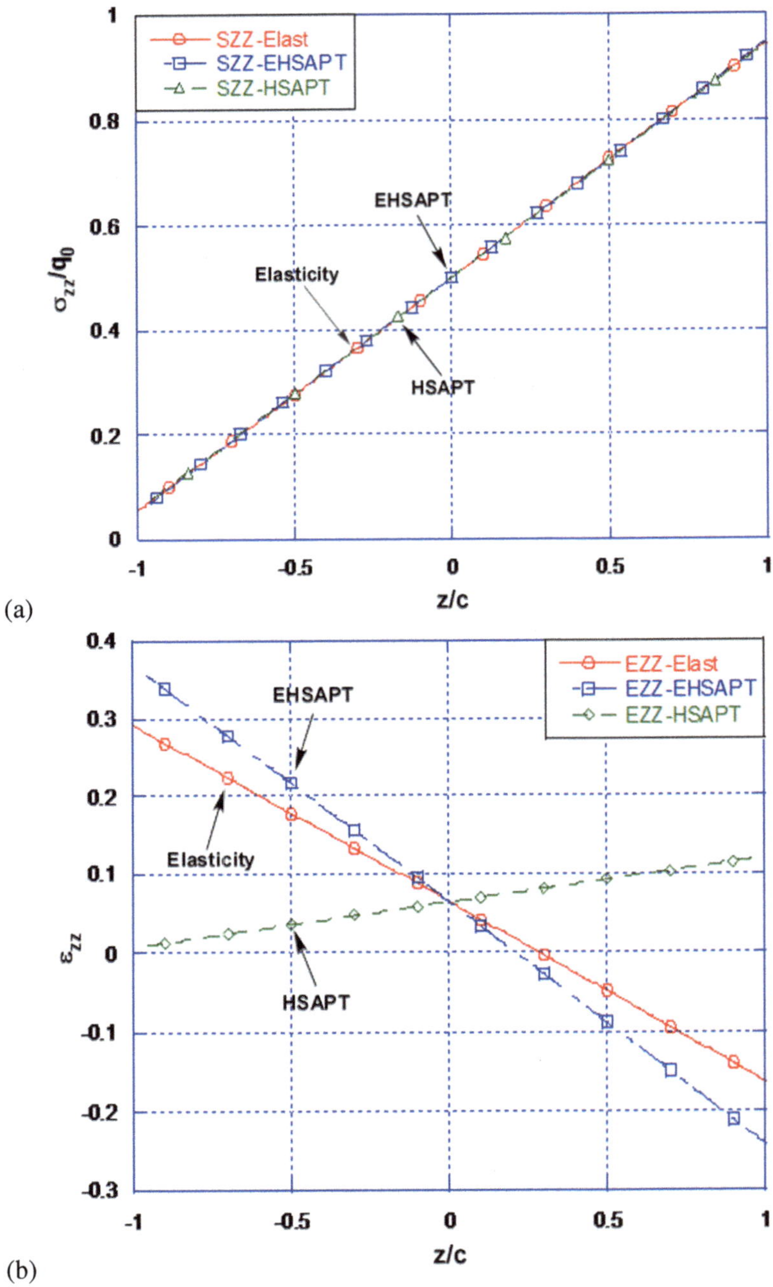

Figure 6.4 Through-thickness distribution of normal stress, σ_{zz} (a) and strain, ε_{zz} (b) in the core at $x = a/2$ for the case of E-glass/polyester faces and balsa wood core.

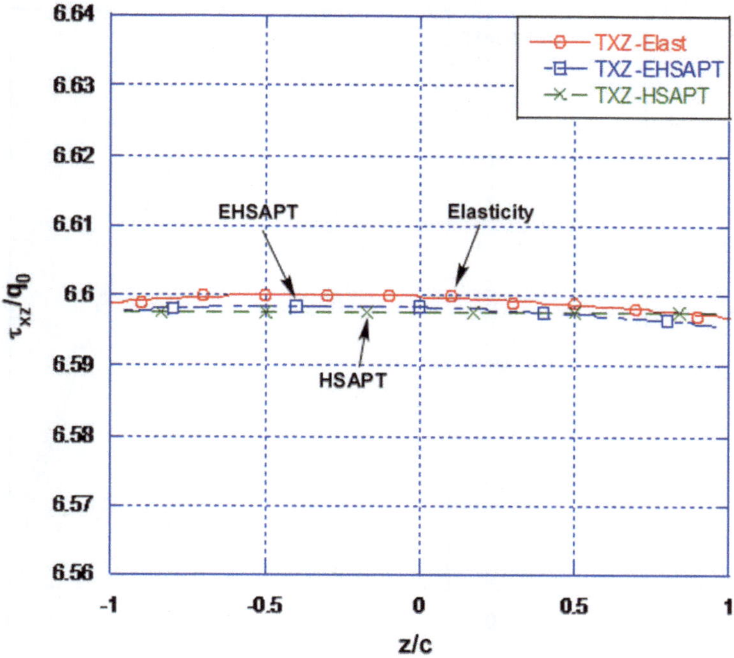

Figure 6.5 Through-thickness distribution of transverse shear stress, τ_{xz}, in the core at $x = a/10$ for the case of carbon/epoxy faces and glass phenolic honeycomb core.

EHSAPT is capable of capturing the shear stress profile in all cases, even in the most demanding case of $E_f/E_c = 2$, and in all cases it is very close to the elasticity solution. On the contrary, a constant shear stress assumption, as in the HSAPT, would be applicable only for sandwich beams with large ratios of E_f/E_c.

6.5 Other High-Order Sandwich Panel Theories

There are a few two-dimensional high-order theories, formulated for sandwich plates and shells, which will be briefly outlined. In the Hohe et al. (2006) plate theory, the transverse displacement in the core, w^c, is assumed to be a linear function of the thickness coordinate, z, and the in-plane displacements, u^c and v^c are quadratic in z. Accordingly, one can easily see that for this theory the strain in the transverse direction is constant, i.e., $\varepsilon^c_{zz} = (w^b - w^t)/(2c)$, and that the transverse displacement of the middle of the core is equal to the average of the transverse displacements of

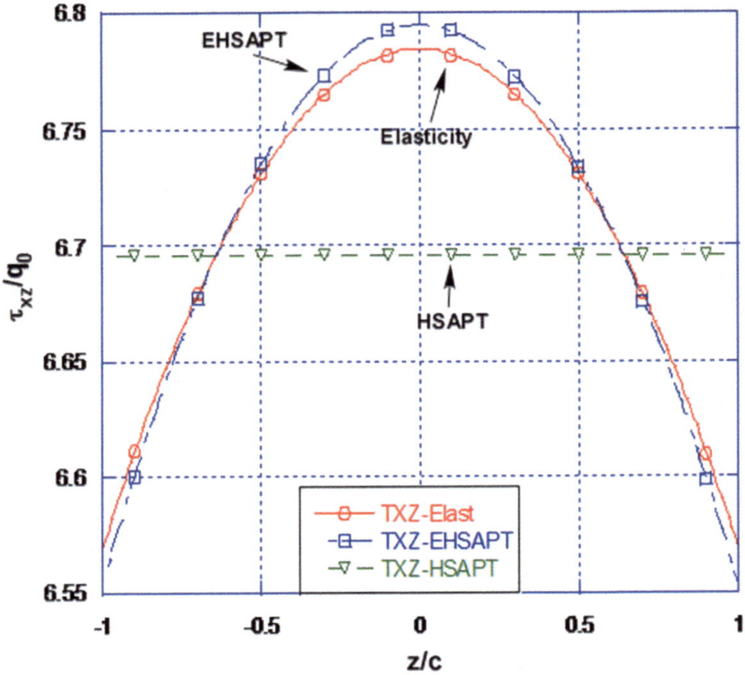

Figure 6.6 Through-thickness distribution of the transverse shear stress, τ_{xz}, in the core at $x = a/10$ for the case of E-glass/polyester faces and balsa wood core.

the top and bottom face sheets $w^c(x, y, 0) = (w^t + w^b)/2$. Therefore, this theory is in terms of six generalized coordinates: u_0^t, v_0^t, w^t; u_0^b, v_0^b and w^b. Thus, this theory cannot capture the nearly linear distribution of transverse normal stress and strain through the thickness, which is predicted from elasticity; instead this theory would result in a nearly constant transverse normal stress.

In the Li and Kardomateas (2008) high-order plate/shell theory, the transverse displacement in the core w^c is of fourth order in the transverse coordinate z, and the in-plane displacements u^c and v^c are of fifth order. In this theory there are seven generalized coordinates: u_0^t, v_0^t, w^t, w_0^c, u_0^b, v_0^b, w^b, i.e. one more than in the Hohe et al. (2006) theory, namely, the mid-plane transverse displacement in the core w_0^c. This theory produces distributions of tranverse normal stress and strain which are close to the elasticity results, but the profile is nonlinear unlike the nearly linear profile predicted by elasticity.

It should be noted that when it comes to the transverse displacement profiles, both the Hohe et al. (2006) and the Li and Kardomateas (2008) theories show an excellent agreement and for all practical purposes identical to the

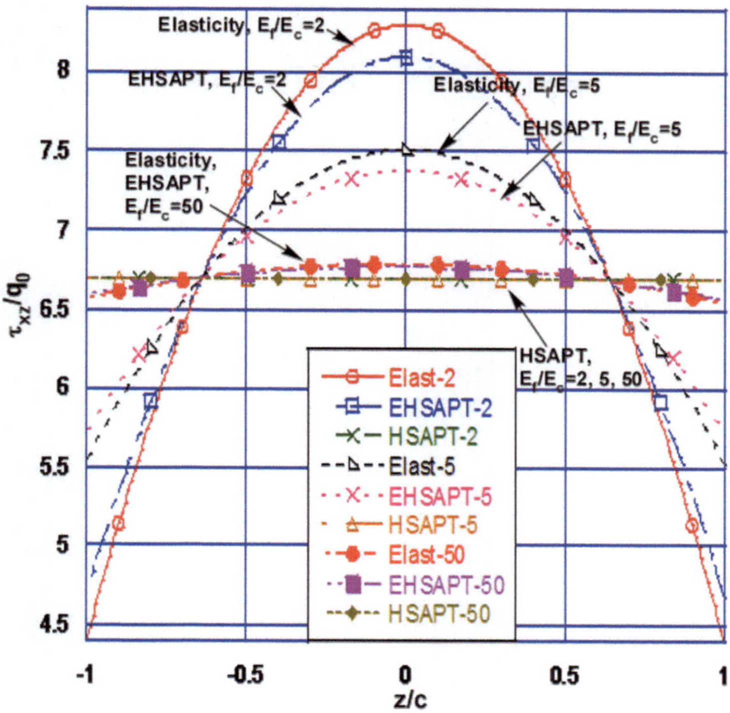

Figure 6.7 Through-thickness shear-stress distribution in the core at $z = a/10$ for the case of E-glass/polyester faces and balsa wood core.

exact elasticity solution. However, as has been demonstrated in this chapter, the theories may differ in the prediction of stresses and may be be significantly off the corresponding elasticity solution.

Chapter 7
Global Buckling of Sandwich Columns and Wide Panels

The most important issue regarding buckling of sandwich structures is the effect of transverse shear which can significantly reduce the Euler critical load. Simply put, the effect of transverse shear absolutely cannot be neglected. Therefore, all formulas for sandwich buckling are essentially ways to include this effect into the Euler formulas. Two basic ways for including transverse shear in column buckling are the Engesser (1891) and the Haringx (1948, 1949) approaches. Both of these approaches are also outlined by Timoshenko (1936).

In this chapter, some of the most widely used column buckling approaches will be presented and compared, followed by the application of first-order shear analysis to the buckling of wide panels and simply supported rectangular panels. Panel compression test methods and data reduction analysis for the evaluation of the critical load will be presented along with examples and collapse strength estimates.

7.1 The Engesser Approach

The Engesser approach is based on considering the additional slope and hence the additional curvature produced by the shear force. Consider the fixed-free sandwich column shown in Figure 7.1a, which is loaded axially by a compressive load P . When buckling occurs, there will be shear forces acting on the cross-sections of the column (Figure 7.1b). The magnitude of the shear force Q acting at a cross-section *mm* can be found from Figure 7.1c:

$$Q \simeq P\frac{dw}{dx}. \tag{7.1}$$

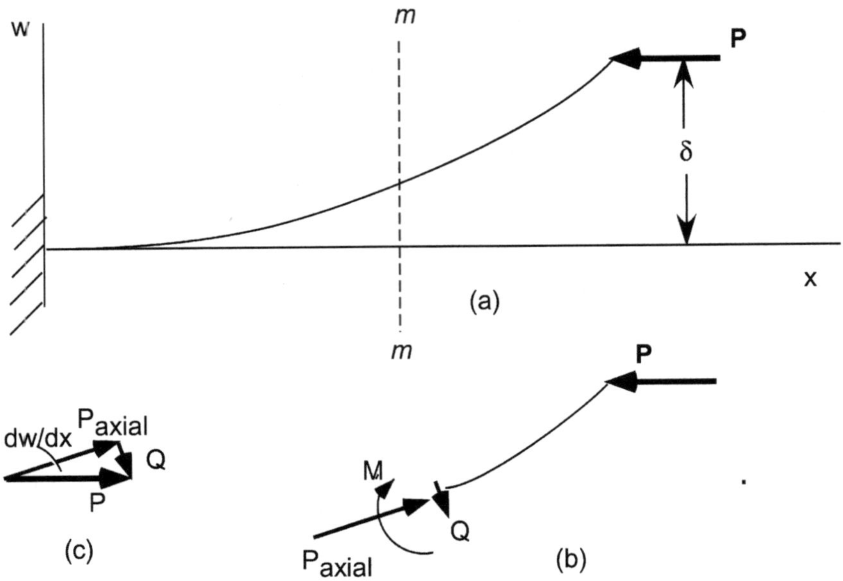

Figure 7.1 Forces and moments acting on a buckled sandwich column.

Note that at the section mm there will also be an axial force, $P_{\text{axial}} \simeq P$.

The change in slope of the deflection curve produced by the shear force is

$$\frac{\beta Q}{AG_{\text{eq}}}, \tag{7.2}$$

where A is the total cross-sectional area of the column, G_{eq} is the "equivalent" or "effective" modulus in shear, and β is a correction factor depending on the shape of the cross-section, which accounts for the fact that shear is not distributed uniformly throughout the section. If the section is rectangular and the column homogeneous isotropic, then $G_{\text{eq}} = G =$ shear modulus of the homogeneous material and $\beta = 1.2$. We shall discuss both G_{eq} and β for sandwich construction later in this section.

The rate of change of slope produced by the shear force Q represents the additional curvature due to shear and, from Equation (7.1), this is equal to

$$\frac{\beta}{AG_{\text{eq}}}\frac{dQ}{dx} = \frac{\beta P}{AG_{\text{eq}}}\frac{d^2w}{dx^2}.$$

The total curvature of the deflection curve is now obtained by adding the curvature produced by the shear force to the curvature produced by the bending moment. Thus, for the column of Figure 7.1a, the differential equation for the deflection curve becomes:

$$\frac{d^2w}{dx^2} = \frac{M}{(EI)_{\text{eq}}} + \frac{\beta}{AG_{\text{eq}}}\frac{dQ}{dx}, \tag{7.3a}$$

or

$$\frac{d^2w}{dx^2} = \frac{P(\delta - w)}{(EI)_{\text{eq}}} + \frac{\beta P}{AG_{\text{eq}}}\frac{d^2w}{dx^2}, \tag{7.3b}$$

where $(EI)_{\text{eq}}$ is the "equivalent" or "effective" bending rigidity of the sandwich cross-section (see Chapter 4 and discussion later).

Equation (7.3b) can be written as

$$\frac{d^2w}{dx^2} = \frac{P}{(EI)_{\text{eq}}[1 - \beta P/(AG_{\text{eq}})]}(\delta - w) . \tag{7.4}$$

If we set

$$k^2 = \frac{P}{(EI)_{\text{eq}}[1 - \beta P/(AG_{\text{eq}})]}, \tag{7.5a}$$

we can write (7.4) in the form

$$\frac{d^2w}{dx^2} + k^2 w = k^2 \delta. \tag{7.5b}$$

The general solution of this equation is

$$w = A_1 \cos kx + A_2 \sin kx + \delta , \tag{7.5c}$$

in which A_1 and A_2 are constants of integration. These constants are determined from the fixed end conditions:

$$w = \frac{dw}{dx} = 0 \quad \text{at } x = 0. \tag{7.5d}$$

These two conditions are fulfilled if

$$A_1 = -\delta, \quad A_2 = 0 \tag{7.5e}$$

and then

$$w = \delta(1 - \cos kx). \tag{7.5f}$$

The condition at the free end of the column requires that

$$w = \delta \quad \text{at } x = L, \tag{7.5g}$$

which is satisfied if

$$\delta \cos kL = 0. \tag{7.6}$$

For a non-zero δ, the smallest value of kl which satisfies Equation (7.6) is $kl = \pi/2$, which when combined with (7.5) gives

$$\frac{P}{(EI)_{\text{eq}}[1 - \beta P/(AG_{\text{eq}})]} = \frac{\pi^2}{4L^2}. \tag{7.7}$$

Solving for P gives the critical load:

$$P_{\text{cr}} = \frac{P_E}{1 + \beta P_E/(AG_{\text{eq}})}, \tag{7.8}$$

where $P_E = \pi^2 (EI)_{\text{eq}}/(4L^2)$ represents the Euler critical load for this case.

7.2 The Haringx Approach

In this approach, due to the shear strain, γ, there is an additional slope measured from the normal to the section to the tangent to the axis of the deflected column. This additional slope is added to the slope, θ, due to the bending moment, measured from the x axis to the normal to the cross-section. Thus the slope of the deflected curve is by use of Equation (7.2):

$$\frac{dw}{dx} = \theta + \gamma = \theta + \frac{\beta Q}{AG_{\text{eq}}}. \tag{7.9}$$

The axial force P has a component in the direction normal to the section equal to $P\cos\theta \simeq P$ and a component

$$Q = P\sin\theta \simeq P\theta. \tag{7.10}$$

Substituting in Equation (7.9), the slope becomes

$$\frac{dw}{dx} = \theta + \frac{\beta P\theta}{AG_{\text{eq}}} = \theta\left(1 + \frac{\beta P}{AG_{\text{eq}}}\right). \tag{7.11}$$

Observing that

$$\frac{d\theta}{dx} = \frac{M}{(EI)_{\text{eq}}} = \frac{P(\delta - w)}{(EI)_{\text{eq}}},$$

we obtain from Equation (7.11) the following expression for the curvature:

$$\frac{d^2w}{dx^2} = \frac{P(\delta - w)}{(EI)_{\text{eq}}}\left(1 + \frac{\beta P}{AG_{\text{eq}}}\right). \tag{7.12}$$

The difference between Equation (7.12) and the previous equation (7.4) is due to the fact that in the derivation of Equation (7.4) the shear force is calculated from the total slope dw/dx of the deflection curve (see Equation (7.1)), whereas in the derivation of Equation (7.12), only the angle of rotation of the cross-section is used (see Equation (7.10)).

Solving the differential equation (7.12) in the same manner as before we find that the critical load is

$$P_{\text{cr}} = \frac{\sqrt{1 + 4\beta P_E/(AG_{\text{eq}})} - 1}{2\beta/(AG_{\text{eq}})} . \tag{7.13}$$

Now one important note regarding Haringx's formula: In a recent paper, Bazant and Beghini (2006) showed that the Engesser and Haringx-type theories are equivalent (i.e., one to follow from the other) provided that a proper transformation of the shear modulus of the core, G_c, is made. However, this transfomation implies that G_c of the soft core is a function of the axial stress in the stiff face sheets. This paradox was clarified by showing that the energetic variational analysis merely requires that the shear stiffness of the cross-section, characterized by G_c of the core, to be a function of the axial force in the face sheets. In other words, if the Haringx-type theory was used with a constant shear modulus, results as in Equation (7.13) would be obtained. However, if the shear modulus is updated as a function of the axial load, then the results are expected to agree with Engesser's formula.

Equations (7.8) and (7.13), the first (Engesser's) are the most widely used. The Haringx formula is expected to have accuracy issues if a constant shear modulus is used (this will be confirmed in Section 7.6). In fact, the global buckling formulas for sandwich columns in the literature provide ways of defining the G_{eq} and $(EI)_{\text{eq}}$ for use in the Engesser's formula (7.8). In the following, we outline the Allen (1969), the Bazant and Cedolin (1991), and the Huang and Kardomateas (2002) approaches for defining these quantities.

7.3 Allen's Formulas

7.3.1 *Thin Faces*

With regard to the cross-section in Figure 7.2, Allen's formula for thin faces assumes that the equivalent bending rigidity is due to the face sheets only and the face sheets are considered as two areas fb, where b is the width of

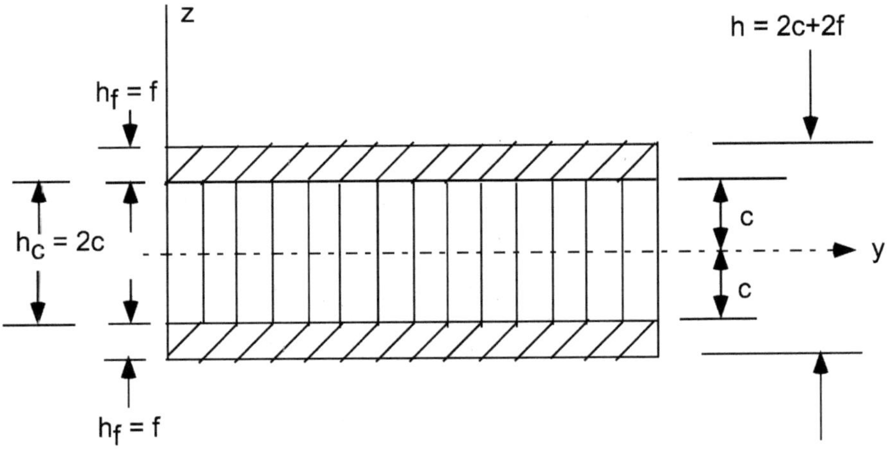

Figure 7.2 Sandwich cross-section.

the beam, located at a distance $c + f/2$ from the mid-axis (the neutral axis of the section), i.e.

$$(EI)_{\text{eq}} = 2E_f bf \left(c + \frac{f}{2}\right)^2. \tag{7.14a}$$

The equivalent shear modulus of the section is just the shear modulus of the core, and the shear correction factor in Equation (7.8) becomes

$$\frac{\beta}{AG_{\text{eq}}} = \frac{1}{b(2c+f)G_c}. \tag{7.14b}$$

Then, Equation (7.8) for a simply-supported column becomes

$$P_{\text{cr}} = \frac{P_E}{1 + P_E/(AG_c)}; \quad P_E = \frac{\pi^2}{L^2} 2E_f bf \left(c + \frac{f}{2}\right)^2. \tag{7.14c}$$

An alternative form of the above equation is

$$\frac{1}{P_{\text{cr}}} = \frac{1}{P_E} + \frac{1}{AG_c}, \quad A = b(2c+f). \tag{7.15}$$

This formula shows that when the sandwich construction involves a core of very low shear modulus, the critical load would be dominated by the second term of Equation (7.15), i.e. by the core, and it would approach the value AG_c. On the other hand, if the core shear modulus is very high, the critical load would be dominated by the first term of Equation (7.15), and it would approach the Euler load P_E.

7.3.2 Thick Faces

When the faces are thick, see Section 1.1, the bending rigidity of the faces about their own separate centroidal axes cannot be neglected. Therefore, the equivalent rigidity is now

$$(EI)_{\text{eq}} = E_f I = E_f(I_1 + I_f), \quad I_1 = 2bf\left(c + \frac{f}{2}\right)^2, \quad I_f = \frac{bf^3}{6}. \tag{7.16}$$

Note that again the bending rigidity of the core is neglected.

The Allen approach consists of considering that, at the buckled state, there occur two superimposed displacements, w_1 (the ordinary bending displacement) and w_2, an additional displacement associated with the shear deformation of the core.

The interaction between the bending stiffness of the faces and the shear stiffness of the core can be seen most easily if we first consider a sandwich with a core which is rigid in shear ($G_c = \infty$). A deflection w_1 occurs in accordance with ordinary bending theory. This deflection is associated with a bending moment M_1 and a shear force Q_1, the latter being

$$-Q_1 = (EI)_{\text{eq}}\frac{d^3w_1}{dx^3} = E_f I_1\frac{d^3w_1}{dx^3} + E_f I_f\frac{d^3w_1}{dx^3}. \tag{7.17}$$

The first term on the right-hand side of Equation (7.17) represents the shear force carried by the beam as a whole, supposing the faces to undergo only uniform extension or contraction without bending locally. In this state the shear stress τ is uniform across the thickness of the core and diminishes linearly to zero across the thickness of each face. The first term may therefore be replaced by $-b(2c + f)\tau$ where τ is the shear stress in the core:

$$-Q_1 = -b(2c + f)\tau + E_f I_f\frac{d^3w_1}{dx^3}. \tag{7.18}$$

As a result of the shear stress τ, the core undergoes a shear strain $\gamma_c = \tau/G_c$ which corresponds to an additional beam deflection w_2.

The shear deformation is illustrated in Figure 7.3, which shows a simply supported beam under three point bending. The points a, b, ... lie on the mid-lines of the faces and do not move horizontally (as in the ordinary bending w_1, case) but instead are displaced just vertically by w_2. The faces and the longitudinal center-line of the beam tilt, and the relationship between this additional slope of the beam dw_2/dx and the core shear strain can be obtained from Figure 7.4:

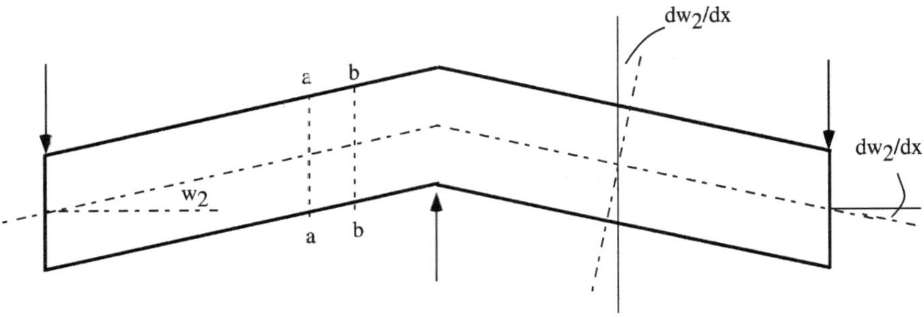

Figure 7.3 Additional beam deflection, w_2, due to transverse shear.

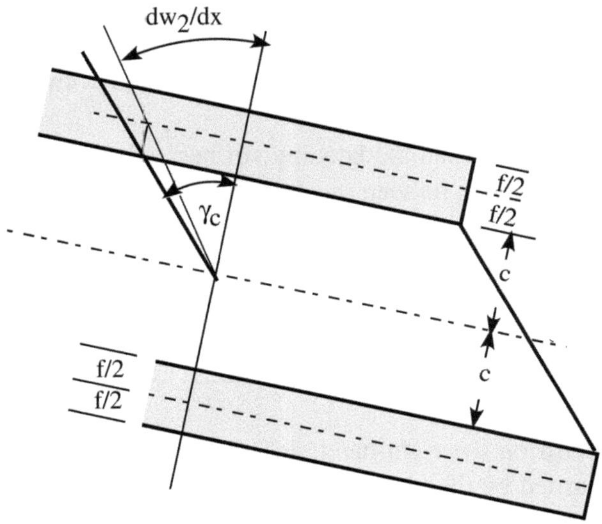

Figure 7.4 Schematic of the relationship between additional slope of the beam due to transverse shear and core shear strain.

$$c\gamma_c = \left(c + \frac{f}{2}\right)\frac{dw_2}{dx}, \tag{7.19a}$$

therefore

$$\tau = \left(1 + \frac{f}{2c}\right)G_c\frac{dw_2}{dx}. \tag{7.19b}$$

Substitution in Equation (7.18) yields

$$-Q_1 = -AG_c\frac{dw_2}{dx} + E_f I_f \frac{d^3 w_1}{dx^3}, \quad \text{where} \quad A = \frac{b(2c+f)^2}{2c}. \tag{7.20}$$

Rearranging the above equation and substituting $Q_1 = -(EI)_{\mathrm{eq}} d^3 w_1/dx^3$ gives

$$\frac{dw_2}{dx} = -\frac{(EI)_{\mathrm{eq}}}{AG_c}\left(1 - \frac{I_f}{I}\right)\frac{d^3w_1}{dx^3} = \frac{Q_1}{AG_c}\left(1 - \frac{I_f}{I}\right). \tag{7.21}$$

The additional transverse deflection w_2 corresponds to an additional shear force Q_2, since the faces must share this extra deflection and, in order to do so, they must be subjected to an additional bending moment and shear force, hence the total shear force is

$$Q = Q_1 + Q_2 = Q_1 - E_f I_f \frac{d^3 w_2}{dx^3}.$$

Substitution of d^3w_2/dx^3 from (7.21) provides a differential equation for Q_1:

$$\frac{d^2Q_1}{dx^2} - \lambda^2 Q_1 = -\lambda^2 Q, \tag{7.22a}$$

where

$$\lambda^2 = \frac{AG_c}{E_f I_f (1 - I_f/I)}. \tag{7.22b}$$

Now the total shear force is (Equation (7.1))

$$Q = P\frac{dw}{dx} = P\left(\frac{dw_1}{dx} + \frac{dw_2}{dx}\right).$$

Using (7.21) for the slope dw_2/dx in terms of Q_1, we obtain the following differential equation for Q_1:

$$\frac{d^2Q_1}{dx^2} - \left(\lambda^2 - \frac{P}{E_f I_f}\right) Q_1 = -\lambda^2 P \frac{dw_1}{dx}. \tag{7.23}$$

Substituting $Q_1 = -(EI)_{\mathrm{eq}} d^3w_1/dx^3$ gives a differential equation for w_1:

$$\frac{d^5w_1}{dx^5} - \left(\lambda^2 - \frac{P}{E_f I_f}\right)\frac{d^3w_1}{dx^3} = -\frac{\lambda^2 P}{(EI)_{\mathrm{eq}}}\frac{dw_1}{dx}. \tag{7.24}$$

Consider the case of simply-supported ends; then the boundary conditions are $w_1 = d^2w_1/dx^2 = 0$ at $x = 0, L$. These conditions are fulfilled by a sinusoidal displacement:

$$w_1 = a_1 \sin\frac{\pi x}{L}.$$

Substitution into (7.24) gives the following:

$$\left[\frac{\pi^4}{L^4} + \left(\lambda^2 - \frac{P}{E_f I_f}\right)\frac{\pi^2}{L^2} - \frac{\lambda^2 P}{E_f I}\right] a_1 = 0,$$

from which we obtain the critical load:

$$P_{\text{cr}} = \frac{\frac{\pi^4}{L^4} + \frac{\lambda^2 \pi^2}{L^2}}{\frac{\pi^2}{L^2 E_f I_f} + \frac{\lambda^2}{E_f I}}. \tag{7.25}$$

This can be expressed in a more general form by using the following definitions:

$$P_E = \frac{\pi^2 E_f I}{L^2}, \quad P_{Ef} = \frac{\pi^2 E_f I_f}{L^2}, \quad P_c = AG_c = \frac{b(2c+f)^2}{2c} G_c, \tag{7.26}$$

where P_E is the Euler load of the entire sandwich column, P_{Ef} is the Euler load of the two faces when they buckle as independent struts, and P_c may be described as the shear buckling load, which is essentially numerically equal to the shear stiffness AG_c. In terms of these quantities, the critical load from Equation (7.25) can be expressed in the general form

$$P_{\text{cr}} = P_E \left[\frac{1 + \frac{P_{Ef}}{P_c} - \frac{P_{Ef}^2}{P_c P_E}}{1 + \frac{P_E}{P_c} - \frac{P_{Ef}}{P_c}} \right]. \tag{7.27}$$

When the faces are very thin, $P_{Ef} \to 0$ and Equation (7.27) coincides with the thin face formula (7.15).

One additional note: Allen's thick face formula, (7.27), turns out to be the prediction from the high-order sandwich panel theory (HSAPT) discussed in Section 6.2 in the limit when the core modulus approaches infinity ($E_c \to \infty$) (Frostig, 2010, personal communication).

7.4 Bazant and Cedolin's Formula

In this formula, the shear correction is defined as follows. Keeping the same notation, the rotation dw_1/dx of the cross-section is defined by the longitudinal displacements of the face centroids, which differs slightly from the rotation of the core cross-section (Figure 7.4). Denoting by γ_c the shear strain in the core and by γ the average shear strain (which is dw_2/dx in the previous section), we can see from Figure 7.4:

$$\left(c + \frac{f}{2} \right) \gamma = c\gamma_c. \tag{7.28a}$$

This is the same as the relation (7.19a) derived in the previous section. Therefore,

$$\gamma_c = \frac{Q}{G_c(2cb)} = \left(1 + \frac{f}{2c}\right)\gamma. \tag{7.28b}$$

Solving for γ, which is the difference between the slope dw/dx of the deflected beam axis and the rotation dw_1/dx of the cross-section, gives

$$\frac{dw}{dx} - \frac{dw_1}{dx} = \gamma = \frac{Q}{G_c A_1}, \quad A_1 = (2c + f)b. \tag{7.29}$$

The axial strain in the face sheet at the mid-face location is $-(c + f/2)d^2w_1/dx^2$ (from bending theory). Therefore the resultant axial forces at the faces (compressive at the upper face and tensile at the lower) are

$$P_f = E_f(fb)\left(c + \frac{f}{2}\right)\frac{d^2w_1}{dx^2}. \tag{7.30a}$$

The bending moment can now be written as

$$M = P_f(2c + f) + 2M_f \quad \text{where} \quad M_f = E_f\frac{bf^3}{12}\frac{d^2w}{dx^2}. \tag{7.30b}$$

Using (7.30a) gives

$$M = E_f I_1 \frac{d^2w_1}{dx^2} + E_f I_f \frac{d^2w}{dx^2} \quad \text{where} \quad I_1 = bf\frac{(2c+f)^2}{2}; \quad I_f = \frac{bf^3}{6}. \tag{7.31}$$

Now differentiating Equation (7.29), expressing from this d^2w_1/dx^2 and substituting it into Equation (7.31), gives

$$\frac{d^2w}{dx^2} = \frac{M}{(EI)_{\text{eq}}} + \frac{1}{G_c A_1(1 + I_f/I_1)}\frac{dQ}{dx}, \tag{7.32a}$$

where

$$(EI)_{\text{eq}} = E_f(I_1 + I_f), \tag{7.32b}$$

i.e., the equivalent bending rigidity is again due to the face sheets only but the bending rigidity of the faces about their own separate centroidal axes is included.

This equation essentially means that the total curvature d^2w/dx^2 is the sum of the flexural curvature and the curvature due to shear, i.e. the same basic approach as Engesser's, see Equation (7.3a), with

$$\frac{\beta}{AG_{\text{eq}}} = \frac{1}{G_c A_1(1 + I_f/I_1)}. \tag{7.33}$$

Proceeding in the same way as before, the critical load is obtained in the same form as Equation (7.8) with $\beta/(AG_{\text{eq}})$ defined in (7.33) and the Euler load P_E based on the bending rigidity (7.32b).

7.5 Huang and Kardomateas Shear Correction Formulas

A shear correction formula for sandwich columns in terms of the face sheet and core geometrical and mechanical properties was presented in Huang and Kardomateas (2002). This formula can be used in either the Engesser (7.8), or the Haringx (7.13) expression. It essentially provides for proper definitions of $(EI)_{\text{eq}}$, G_{eq} and β for the sandwich section.

In particular, the equivalent bending rigidity includes both the face sheets and the core and the bending rigidity of the faces about their own separate centroidal axes is included. Referring again to Figure 7.2, the equivalent flexural rigidity of the sandwich section is

$$(EI)_{\text{eq}} = 2E_f \frac{bf^3}{12} + 2E_f bf \left(\frac{f}{2} + c\right)^2 + E_c \frac{b(2c)^3}{12}. \tag{7.34}$$

Denoting the shear stresses in the face sheet and the core by $\tau_f(x, z)$ and $\tau_c(x, z)$, respectively, we can write the shear energy in the sandwich beam as

$$\begin{aligned} U_\gamma &= b \int\int \frac{\tau^2(x, z)}{2G(z)} dz dx \\ &= 2b \left\{ \int_0^L \int_c^{c+f} \frac{\tau_f^2(x, z)}{2G_f} dz dx + b \int_0^L \int_0^c \frac{\tau_c^2(x, z)}{2G_c} dz dx \right\}. \end{aligned} \tag{7.35}$$

An "effective" or "equivalent" shear modulus for the sandwich section, G_{eq}, which includes the contribution of the face sheets, can be defined based on the compliances of the constituent layers, as follows:

$$\frac{2f + 2c}{G_{\text{eq}}} = \frac{2f}{G_f} + \frac{2c}{G_c}, \tag{7.36}$$

where G_f is the shear modulus of the face sheets and G_c the shear modulus of the core. Equation (7.36) shows that when the core is of very low modulus, the second term would dominate and G_{eq} would approach G_c.

Now, assume that the shear stress is distributed in a uniform fashion over the entire section, $A = b(2c + 2f)$, then the corresponding equivalent shear stress and strain are

$$\tau_{\text{eq}} = \frac{V(x)}{A}; \quad \gamma_{\text{eq}} = \frac{\beta V(x)}{G_{\text{eq}} A}, \tag{7.37}$$

where β is the shear correction coefficient, which takes into account the non-uniform distribution of shear stresses over the entire cross-section.

Then, the energy due to shear is

$$U_\gamma = A\int \frac{1}{2}\tau_{\text{eq}}\gamma_{\text{eq}}dx = \frac{\beta}{2G_{\text{eq}}A}\int_0^L V^2(x)dx, \tag{7.38}$$

Now the shear stresses, from simple bending theory, are distributed as follows:

- *Face sheets*

$$\tau_f(z) = \frac{V}{(EI)_{\text{eq}}}\frac{E_f}{2}\left[(f+c)^2 - z^2\right]. \tag{7.39a}$$

- *Core*:

$$\tau_c(z) = \frac{V}{(EI)_{\text{eq}}}\left[E_f f\left(\frac{f}{2}+c\right) + \frac{E_c}{2}\left(c^2 - z^2\right)\right]. \tag{7.39b}$$

Substituting into Equation (7.35) gives

$$U_\gamma = \frac{b}{(EI)_{\text{eq}}^2}\left(\frac{a_f}{G_f} + \frac{a_c}{G_c}\right)\int_0^L V^2 dx, \tag{7.40}$$

where

$$a_f = \frac{E_f^2}{4}\left[(f+c)^4 f - \frac{7}{15}(f+c)^5 - \frac{c^5}{5} + \frac{2}{3}(f+c)^2c^3\right], \tag{7.41a}$$

$$a_c = E_f^2 f^2 c\left(\frac{f}{2}+c\right)^2 + \frac{2}{15}E_c^2c^5 + \frac{2}{3}E_f E_c f\left(\frac{f}{2}+c\right)c^3. \tag{7.41b}$$

Comparing (7.38) and (7.40) gives the shear correction as

$$\frac{\beta}{AG_{\text{eq}}} = \frac{2b}{(EI)_{\text{eq}}^2}\left(\frac{a_f}{G_f} + \frac{a_c}{G_c}\right). \tag{7.42}$$

For a homogeneous section (this can be most easily seen by setting $c = 0$, $A = 2fb$), $\beta = 6/5$, which is a well-established shear correction factor for a rectangular homogeneous section.

Notice that this shear correction formula is not exclusively based on the shear modulus of the core, but includes the shear modulus of the faces and the extensional modulus of the core. Hence, it can account for sandwich constructions with stiffer cores and/or more compliant faces.

It should also be noted that a more general formula for the transverse shear correction coefficient β, which is applicable to a sandwich section with dissimilar faces can be found in Huang and Kardomateas (2002). This formula is also given in Chapter 12 in conjuction with the debond buckling problem.

The shear correction formula (7.42) can now be used by substituting this expression for β in either the Engesser critical load formula (7.8) or the Haringx one (7.13), where P_E is the Euler load based on the equivalent rigidity (7.34).

7.6 Comparison of the Global Buckling Formulas

Let us consider a sandwich column with unidirectional carbon/epoxy faces and hexagonal glass/phenolic honeycomb core. The orthotropic carbon/epoxy face moduli are (in GPa): $E_1^f = 181$, $E_2^f = E_3^f = 10.3$, $G_{23}^f = 5.96$, $G_{12}^f = G_{13}^f = 7.17$; and the face Poisson's ratios: $\nu_{12}^f = \nu_{13}^f = 0.277$, $\nu_{32}^f = 0.400$. The orthotropic honeycomb core moduli are (in GPa): $E_1^c = E_2^c = 0.032$, $E_3^c = 0.300$, $G_{23}^c = G_{13}^c = 0.048$, $G_{12}^c = 0.013$; and the core Poisson ratios are $\nu_{12}^c = \nu_{32}^c = \nu_{31}^c = 0.25$.

The total thickness is considered constant at $h = 2f + 2c = 30$ mm, the length over total thickness, $L/h = 30$, and we examine a range of face thicknesses defined by the ratio of face sheet thickness over total thickness, f/h, between 0.010 and 0.20. Figure 7.5 shows the critical load for a simply supported sandwich column, normalized with the Euler load (without transverse shear), P_{E0}. The different formulas from the literature are plotted. Notice also that we use G_{13}^c in place of G_c in these formulas, which were originally derived for isotropy.

Since it is possible that face wrinkling could dominate the failure of the column for very thin face sheets (see Chapter 9), Figure 7.5 also shows the critical wrinkling load calculated from Allen's wrinkling formula (8.63a). It is indeed noted that wrinkling would dominate for ratios f/h below 0.02.

From these results we can make the following observations:

(a) Allen's thin-face formula (7.15) and the Bazant and Cedolin (1991) formulas (7.8) and (7.33) produce similar results. In Figure 7.5, the curves from these two formulas can hardly be distinguished.
(b) Allen's thick-face formula (7.27) and the Engesser formula (7.8) with the Huang and Kardomateas shear correction (7.42) give predictions which are also practically identical and the corresponding curves can hardly be distinguished in Figure 7.5.
(c) The transverse shear effect is very large and results in a critical load being about only one third of the Euler load for face sheet thickness ratios, f/h, above 0.10.

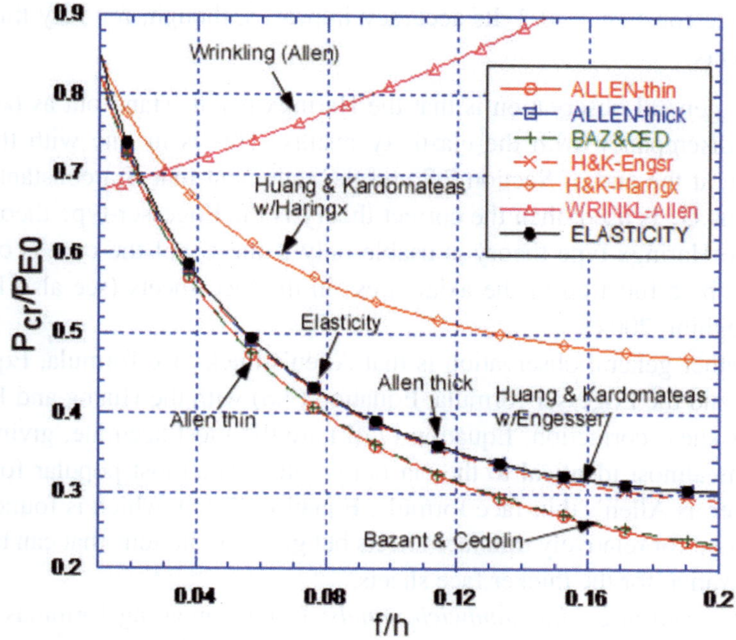

Figure 7.5 Critical loads for sandwich columns calculated from different formulas. The elasticity curve is from Kardomateas (2008b).

An exact three-dimensional elasticity solution to the problem was derived by Kardomateas (2008b). This solution can serve as a benchmark for assessing the accuracy of all these different formulas. From this solution, it was concluded that

(a) Allen's thin and thick-face formulas, the Bazant and Cedolin formula, and the Engesser formula with the Huang and Kardomateas shear correction are all conservative.
(b) Allen's thick-face formula and the Engesser formula with the Huang and Kardomateas shear correction are the most accurate, giving predictions almost identical to the elasticity value.
(c) Allen's thin-face formula and the Bazant and Cedolin formulas are accurate within about 5% of the elasticity value for f/h below 0.05, so they are very good for relatively thin face sheets; however, both give predictions that can be very conservative for the thicker face sheets (of the order of 20% below the elasticity value for $f/h = 0.2$).
(d) The Haringx formula gives predictions which are non-conservative and it is the most inaccurate, being of the order of 50% above the elasticity

value for $f/h = 0.2$. Its accuracy improves, though, for very thin face sheets.

One general observation is that the Haringx results stand out as being in much discrepancy with the elasticity results. This is in line with the discussion at the end of Section 7.2 and the statement that if a constant shear modulus G_c is used, then the correct theory is the Engesser-type theory and that the Haringx-type theory is usable only if the G_c of the core is considered to be a function of the axial stress in the face sheets (see also Bazant and Beghini, 2004).

Another general observation is that Allen's thick-face formula, Equation (7.27) and the Engesser formula, Equation (7.8) with the Huang and Kardomateas shear correction, Equation (7.42) are the most accurate, giving predictions almost identical to the elasticity value. The most popular formula, however, is Allen's thin-face formula, Equation (7.15), which is found to be very good for relatively thin face sheets but gives predictions that can be very conservative for the thicker face sheets.

Note regarding wide sandwich panels: In the foregoing formulas, when dealing with a wide panel, E_f must be replaced by $E_f/(1-\nu_f^2)$ where ν_f is the Poisson ratio of the faces, since in a wide panel, the lateral strains ε_{yy} must be zero, or else the bending could not be cylindrical and curvature would arise also in the lateral direction y. Therefore, with this modification, all of the previous formulas are also applicable to the buckling of wide sandwich panels.

7.7 First-Order Shear Deformation Analysis of Buckling of a Simply-Supported Sandwich Panel

Buckling of sandwich panels has been considered by several researchers and an excellent review of early work is presented in Plantema (1966). Solutions for plate buckling problems are also presented in the texts by Allen (1969) and Zenkert (1997). These solutions are derived using the approach of "partial deflections". In this section we will approach the buckling of a simply-supported sandwich panel using the classical first-order shear deformation approach outlined in Chapter 3.

A rectangular panel under biaxial compressive loading is considered, see Figure 7.6. The edges are loaded by uniform forces of magnitudes N_x and N_y. The initially flat symmetric sandwich plate is compressed until the flat shape deviates into a slightly bent mode shape once a critical set of loads

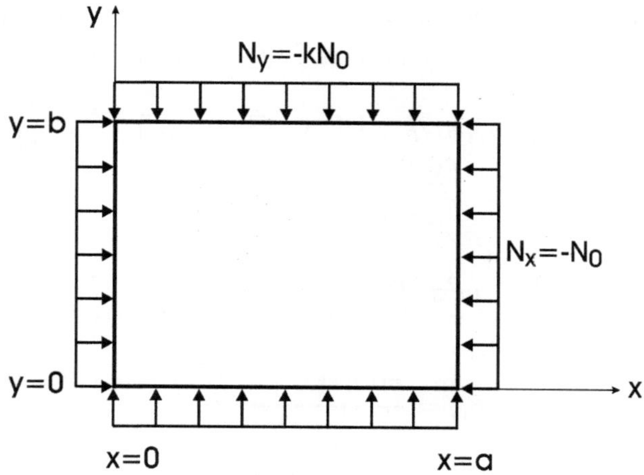

Figure 7.6 Sandwich panel under in-plane biaxial compressive loading.

(N_x and N_y) is reached. To determine the buckling loads and mode shapes the equilibrium equation (3.58e), which includes in-plane forces, is utilized with $q = 0$, and $N_{xy} = 0$,

$$\frac{\partial Q_x}{\partial x} + \frac{\partial Q_y}{\partial y} + N_x \frac{\partial^2 w}{\partial x^2} + N_y \frac{\partial^2 w}{\partial y^2} = 0. \tag{7.43}$$

Substitution of the expressions for the shear forces, Equations (3.68), yields

$$h_c G_{xz} \left(\frac{\partial \psi_x}{\partial x} + \frac{\partial^2 w}{\partial x^2} \right) + h_c G_{yz} \left(\frac{\partial \psi_y}{\partial y} + \frac{\partial^2 w}{\partial y^2} \right) + N_x \frac{\partial^2 w}{\partial x^2} + N_y \frac{\partial^2 w}{\partial y^2} = 0. \tag{7.44}$$

Simply-supported boundary conditions, see Section 3.2.3, are assumed

$$w(x, y) = M_x(x, y) = M_y(x, y) = 0. \tag{7.45}$$

These conditions apply along the edges $x = (0, a)$ and $y = (0, b)$ of the panel. The above boundary conditions are satisfied by

$$\psi_x = A_{mn} \cos \frac{m\pi x}{a} \sin \frac{n\pi y}{b}, \tag{7.46a}$$

$$\psi_y = B_{mn} \sin \frac{m\pi x}{a} \cos \frac{n\pi y}{b}, \tag{7.46b}$$

$$w = C_{mn} \sin \frac{m\pi x}{a} \cos \frac{n\pi y}{b}, \tag{7.46c}$$

where m and n are integers ($m = 1, 2, \ldots, n = 1, 2, \ldots$). Substitution of Equations (7.46) into (3.69c and d) and (7.44) yields the following matrix equation:

$$\begin{bmatrix} F_{11} F_{12} F_{13} \\ F_{12} F_{22} F_{23} \\ F_{13} F_{23} F_{33} \end{bmatrix} \begin{bmatrix} A_{mn} \\ B_{mn} \\ C_{mn} \end{bmatrix} = \begin{bmatrix} 0 \\ 0 \\ 0 \end{bmatrix}, \tag{7.47}$$

where

$$F_{11} = \frac{m^2\pi^2 D_{11}}{a^2} + \frac{n^2\pi^2 D_{66}}{b^2} + h_c G_{xz}, \tag{7.48a}$$

$$F_{12} = \frac{mn\pi^2(D_{12} + D_{66})}{ab}, \tag{7.48b}$$

$$F_{13} = \frac{m\pi h_c G_{xz}}{a}, \tag{7.48c}$$

$$F_{22} = \frac{n^2\pi^2 D_{22}}{b^2} + \frac{m^2\pi^2 D_{66}}{a^2} + h_c G_{yz}, \tag{7.48d}$$

$$F_{23} = \frac{n\pi h_c G_{yz}}{b}, \tag{7.48e}$$

$$F_{33} = \pi^2 \left[\frac{m^2 h_c G_{xz}}{a^2} + \frac{n^2 h_c G_{yz}}{b^2} + \frac{m^2 N_x}{a^2} + \frac{n^2 N_y}{b^2} \right], \tag{7.48f}$$

A non-trivial solution can be obtained by choosing N_x and N_y such that the determinant of the matrix $[F]$ in Equation (7.47) vanishes,

$$\det[F] = F_{11}(F_{22}F_{33} - F_{23}^2) - F_{12}(F_{12}F_{33} - F_{23}F_{13}) + F_{13}(F_{11}F_{23} - F_{22}F_{13}). \tag{7.49}$$

The only element of the matrix containing the in-plane loads N_x and N_y is F_{33} (Equation (7.48f)). With $\det[F] = 0$, Equation (7.49) gives the condition for a non-trivial solution in terms of F_{33}

$$F_{33} = \frac{F_{11}F_{23}^2 + F_{22}F_{13}^2 - 2F_{12}F_{13}F_{23}}{F_{11}F_{22} - F_{12}^2}. \tag{7.50}$$

In a typical problem, N_x and N_y are proportional,

$$N_x = -N_o, \tag{7.51a}$$

$$N_y = -kN_o, \tag{7.51b}$$

where N_o is the magnitude of compression load per unit length applied in the x direction. The critical buckling load is given by the set of m and n that minimizes the load N_o. The buckling mode is defined by the integers m and

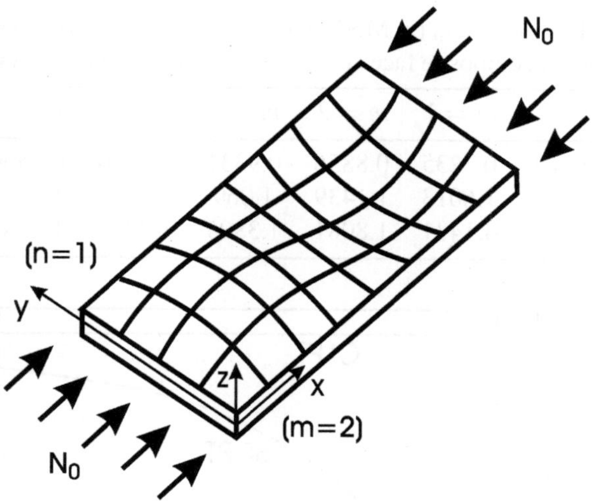

Figure 7.7 Illustration of buckling mode corresponding to $m = 2$, $n = 1$, where the panel buckles into a full sine wave in the x direction and a half sine wave in the y direction.

n defining the number of half sine waves into which the panel buckles in the x and y directions, respectively, see the expression for the panel deflection in Equation (7.48c) and the example $m = 2$, $n = 1$ shown in Figure 7.7.

We will specifically examine the buckling of a sandwich panel with isotropic core, $G_{xz} = G_{yz} = G_c$, under uniaxial compressive loading in the x direction, which in Equations (7.51) corresponds to $k = 0$. Equation (7.48f) applied to this loading yields

$$N_o = h_c\left(G_{xz} + \left(\frac{n}{m}\right)^2\left(\frac{a}{b}\right)^2 G_{yz}\right) - \frac{a^2}{\pi^2 m^2}F_{33}, \qquad (7.52)$$

with F_{33} given by Equation (7.50).

For numerical calculations, a square ($a = b = 0.5$ m) sandwich panel consisting of 2 mm thick unidirectional composite face sheets with the fibers aligned with the x axis (loading direction), and a 16 mm thick H100 PVC foam core was considered. The face and core mechanical properties are the same as those considered in Section 3.3. Calculation of the buckling load N_o for a set of values m and n was conducted based on Equation (7.52). Table 7.1 lists the results for $m = 1, 2, 3$ and $n = 1, 2, \ldots, 5$.

It is observed that the lowest buckling load (critical load) corresponds to a mode shape with one half sine wave in both the x and y directions $m = n = 1$. This is also the case for calculation of the critical load using classical laminated plate theory (CLPT) where transverse shear is not incorporated.

Table 7.1 Buckling load, N_o (in MN/m) for a square 0.5 m × 0.5 m sandwich panel with unidirectional composite face sheets and an isotropic H100 PVC foam core.

	$m = 1$	$m = 2$	$m = 3$	$m = 4$	$m = 5$
$n = 1$	0.7235	0.8856	0.9331	0.9500	0.9569
$n = 2$	1.6013	1.1439	1.069	1.0393	1.9221
$n = 3$	4.1431	1.8081	1.3805	1.2255	1.1487

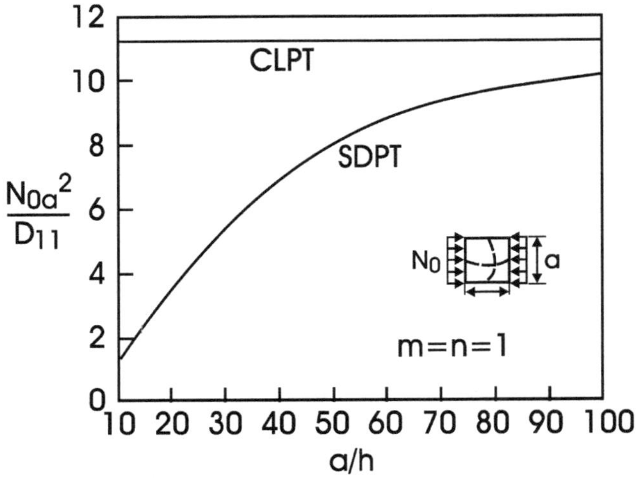

Figure 7.8 Critical buckling load for square sandwich panel vs. size calculated using shear deformation plate theory (SDPT) and classical laminated plate theory (CLPT).

For uniaxial compression of a square panel, CLPT yields (Whitney, 1987)

$$N_o = \frac{\pi^2}{a^2}\left(D_{11}m^2 + 2\left(D_{12} + 2D_{66}\right)n^2 + D_{22}\frac{n^4}{m^2}\right). \tag{7.53}$$

The smallest buckling load for any panel occurs for $n = 1$. For the current panel with D_{11} much larger than D_{12}, $D_{22}$and D_{66}, the smallest value of N_o occurs for $m = 1$. This was found to be the case for any size of the square sandwich panel considered. Figure 7.8 shows the buckling load normalized with the bending stiffness D_{11}, and panel area a^2 plotted vs. the normalized side length.

It is observed that transverse shear deformation reduces the critical load. As the panel size increases, the difference between shear deformation plate theory and classical laminated plate theory decreases, similar to the bending case discussed in Section 3.3. For small panels, shear deformation has a strong influence on the critical load and neglecting this important mode of deformation of sandwich panels will produce very unconservative estimates of buckling load. See also the results for sandwich columns presented in Sections 7.1–7.6.

7.8 Panel Compression Testing

It is general design practice of sandwich structures to determine the dimensions and supports of the panels to eliminate the possibility for buckling under service conditions. The experimental study of the buckling behavior of thin panels has been motivated by the emphasis of using structurally efficient materials in engineering applications such as naval ship structures, wind turbine blades, airplane structures, and packaging containers. In the analysis of such panels (see Section 3.2.3) idealized boundary conditions are imposed, typically "simply supported" or "clamped". These boundary conditions are introduced in order to obtain a tractable solution to the specific problem under investigation. In actual structures and experimental test fixtures, such idealized boundary conditions are often difficult to assess in a precise manner. Panel compression tests are commonly devised with the purpose of verifying an analytical or numerical finite element solution and to determine the actual mechanisms leading to the collapse of the panel such as localized buckling of the face sheets (face wrinkling) or compression failure of the face sheets. When performing testing for such purposes, it is important to assess the details on how the test fixture introduces load and how it supports the panel. In compression testing of sandwich panels into the post-buckling regime of the panel, the actual load-deformation behavior critically depends on the manner in which the edges of the panel are supported and how the load is introduced into the panel. Most experimental studies of the compressive response of thin panels have been focused on the implementation of simply supported boundary conditions, such as was discussed in the analysis of edge-loaded panels in Section 3.2.3. Simply-supported edge conditions are in this context commonly defined as being achieved by a fixture that allows all the edges to freely rotate around axes parallel to the edges and allows unconstrained movement of the edges in the plane of the panel, while restricting out-of-plane deflections. Farris and Filippov (1982), Khot and Bauld (1983), Souza et al. (1983), and Minguez (1986) have discussed several aspects of testing fixtures that provide support conditions close to the idealized boundary conditions.

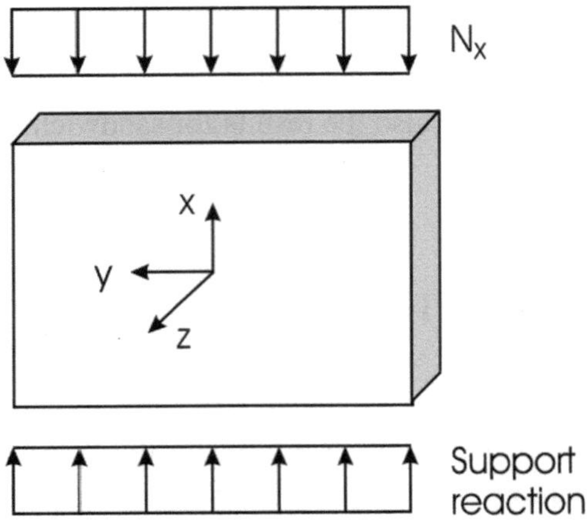

Figure 7.9 Schematic of uniaxial in-plane compressive loading of a sandwich panel.

Most experimental studies reported in the literature consider uniaxial compressive loading of panels. The testing of such panels is typically accomplished in a general-purpose vertical testing machine containing a metal fixture that allows the introduction of distributed load (line load) on the top horizontal edge while the vertical edges are unloaded, see Figure 7.9.

As the magnitude of the load is increased, the panel will deform in compression, and the upper horizontal edge will displace downwards. The vertical side supports in the test fixture must allow for such deformation. Furthermore, the edge supports should allow moment-free rotation of the edges of the panel.

Minguez (1986) designed a panel test where out-of-plane deflections of the vertical edges were constrained by attaching steel wires in regularly spaced slots machined along the edges, as shown in Figure 7.10. Each steel wire was attached to the panel using a brass collar with a set screw and a steel angle section fitting the machined slot. The ends of the wire were attached with screws to frames mounted on each side of the panel. The 80 cm long and 40 cm wide panel was supposed to buckle into one full sine wave along the loading direction and one half sine wave transverse to the loading direction. Minguez placed the wires 5 cm apart. To allow for sufficient tightening of the wires, high-strength piano wire was used.

A more common method to constrain out-of-plane deflection of the unloaded edges, is to use knife-edge supports (Figure 7.11). As will be discussed later, both the wire support and the knife-edge supports appear to approximate simply supported edge conditions, as judged from buckling mode

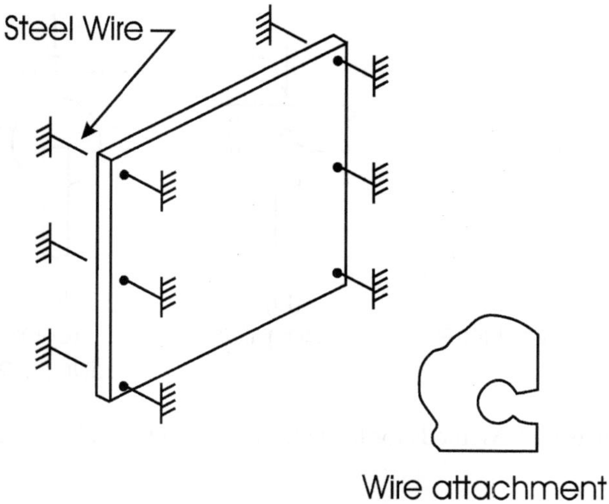

Figure 7.10 Edge support using regularly spaced steel wires.

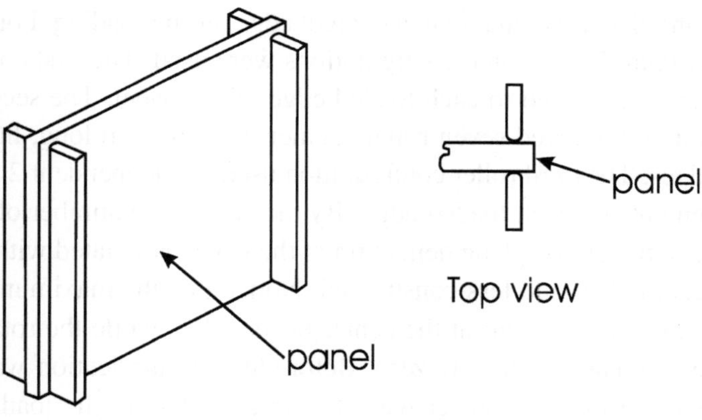

Figure 7.11 Knife-edge supports to constrain lateral deflections of the unloaded edges.

shapes and the magnitude of the measured buckling load. A disadvantage with the wire supports, however, is that the cut-outs for the steel wire attachments will weaken the panel. After buckling, the load distribution is no longer uniform and the compressive load becomes concentrated to the edge regions. If the panels are loaded to collapse, the cut-outs may reduce the ultimate load.

Minguez (1986) considered several other options for load introduction, see Figure 7.12. Each of the configurations shown in Figure 7.12 were employed for the directly loaded top and bottom edges. The flat plate configura-

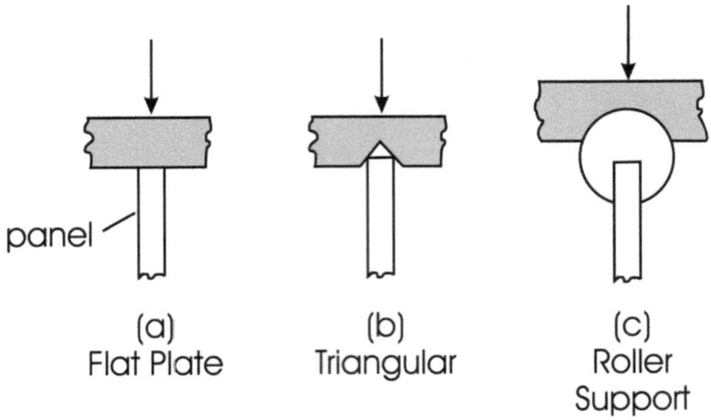

Figure 7.12 Methods for load introduction into sandwich panel.

tion, Figure 7.12a consisted of a flat stiff plate in direct contact with the panel edge. The triangular slot configuration, Figure 7.12b, used a loading platen with a triangular slot to maintain a straight edge during loading. For the roller support (Figure 7.12c) three configurations were used. The first consisted of two single rollers fitted to each loaded edge of the panel. The second roller configuration employed seven roller segments, each 5 cm long, fitted to the loaded edges. The third roller configuration used 13 independent 2.5 cm long roller segments on each loaded edge. By increasing the number of independent rollers, the out-of-plane deflection of the panel associated with buckling would become less and less constrained. Notice that the maximum edge rotation is expected to occur at the center of the edge, while the rotation near the corners should be close to zero. Before testing, lubrication was applied to the roller surface in contact with the circular slots in the loading plates to reduce friction. Nordstrand (2003) used a similar slotted roller arrangement as Minguez (1986), Figure 7.13, although the segments were shorter, approximately 1.7 cm, and the rollers were resting on needle bearings. For a panel size of 40×40 (cm) 23 segments were used on each horizontal edge.

The compression test fixture designed by Nordstrand (2003), moreover, employed an aluminum frame consisting of U-shaped extruded beams to provide rigid support to the panel. Knife-edge supports (Figure 7.11), were used to constrain out-of-plane deflections of the unloaded vertical edges. The upper loading beam was connected to the moving cross-head of the testing machine using a pin connection so that the panel is loaded uniformly before buckling and symmetrically after buckling, in the post-buckling regime. To maintain a load path along the undeformed reference plane of the panel, the upper loading beam was guided by two pairs of roller bearings in contact

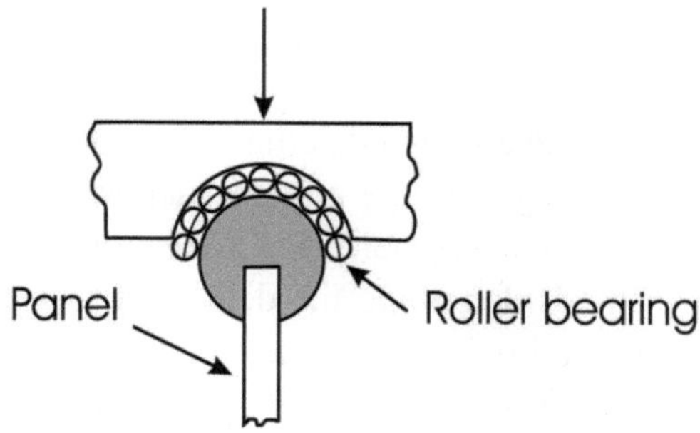

Figure 7.13 Rollers resting in circular lots supported by needle roller bearings.

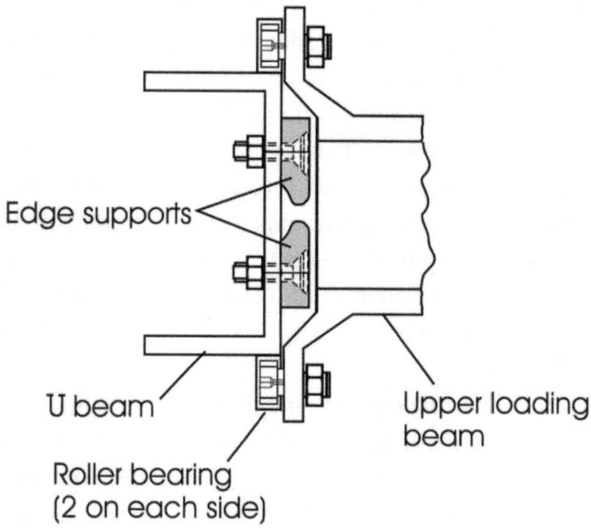

Figure 7.14 The upper loading beam is guided by two pairs of roller bearings in contact with the vertical U-beams of the compression test fixture.

with the outer surfaces of the vertical U-shaped beams, see the top view in Figure 7.14.

To examine whether a fixture is able to provide the desired loading and support conditions for a test panel, analytical and numerical predictions of the critical load and the associated buckling mode shape are compared to those determined experimentally. Such an approach seems very straightforward and it is for perfectly flat and defect-free panels. Actual panels, however, tend to deviate from the ideal flat form due to process-induced asymmetric residual stresses or other reasons. Hence, the panels tend to be slightly bent or warped in the unloaded state and when external loads are applied the panel deforms further without any obvious indication of a bifurcation behavior.

Experimental studies have shown that the out-of-place deflection is much more indicative of buckling than the in-plane deformations. Consequently, several methods to monitor the out-of-plane deflection have been developed. The most simple methods determine the deflection at a point, such as the crest of a buckle, or several points, using deflectometers or non-contact laser interferometry. More sophisticated methods enable measurement of the full displacement field for the deflected panel. Such methods are the shadow-moiré method (Sciammarella, 1982), and the more recent digital image correlation technique described by Helm et al. (1996).

Once the load vs. out-of-plane deflection response has been measured, there are several methods available to determine the buckling load from the measured data. A commonly applied method is the Southwell graphical procedure outlined in Appendix C. This method was developed by Southwell (1932) for the evaluation of the buckling load for slightly bowed simply-supported columns. This method amounts to plotting the column deflection, w, vs. deflection divided by the load (w/P). The slope of the line represents P_{cr}, see e.g. the article by Souza et al. (1983). Such a method is adequate for structural members such as columns that display "neutral" post-buckling response (see Figure 7.15).

Neutral post-buckling response means that the load remains constant after buckling as long as the material is loaded within the elastic regime. A perfect column would buckle at a load, $P = P_{cr}$, which would remain constant up to very large deflections. Perfect here means that the column is initially straight and that the load acts along the specimen centroidal axis. Figure 7.15 shows the load, P, vs. additional out-of-plane deflections, w, for the perfect column. w_o represents the amplitude of the initial imperfection. Hence, the total out-of-plane deflection is $w_T = w + w_o$. w is the deflection one would measure after zeroing the displacement gage before load application. During compressive loading of an imperfect column, the column would already deflect at small applied loads and the load would asymptotically approach the buckling load at large deflections.

It can be readily observed that there are substantial difficulties in accurate determination of P_{cr} from experimental data for imperfect columns. The Southwell method (Appendix C) has proven to be an excellent method to determine the buckling load for columns. Panels loaded past the critical load, on the other hand, display a stable post-buckling response (see Figure 7.15), meaning that the panel can support loads substantially greater than the critical load. A perfect panel ($w_o = 0$) would not display any out-of-plane dis-

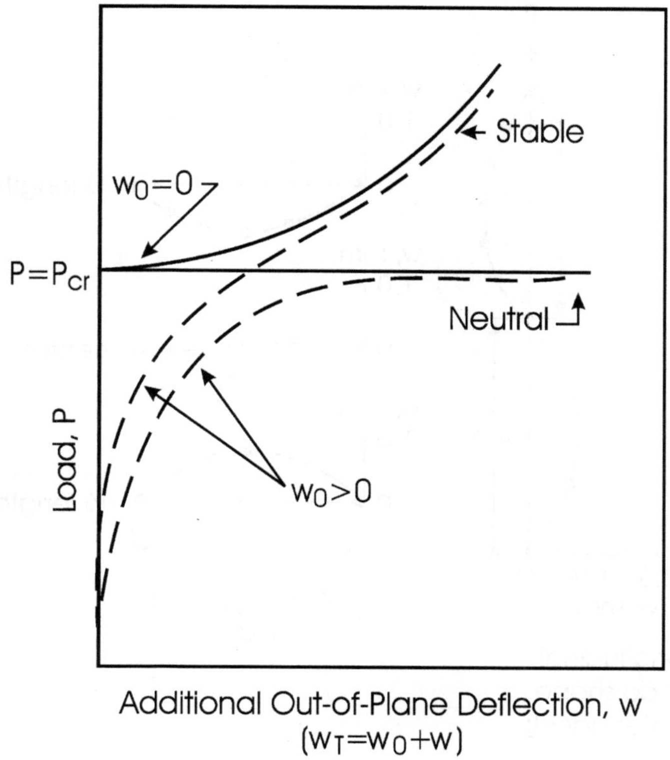

Figure 7.15 Illustration of neutral and stable post-buckling response.

placement until the critical load, P_{cr}, is reached. After such a panel buckles, the load will steadily increase until the material yields. Notice that a sandwich panel may fail by local buckling (wrinkling or intercell buckling, see Chapters 1 and 8).

As may be observed in Figure 7.15, extraction of the buckling load (P_{cr}) from the experimentally measured load vs. out-of-plane displacement record for an imperfect panel is not straightforward. Minguez (1986) applied the Southwell method to extract P_{cr} from the measured load vs. out-of-plane displacement (P–w) response of an aluminum panel under the various edge boundary conditions shown in Figure 7.12. In addition to the determination of P_{cr}, Minguez (1986) also examined the buckling mode shapes along and transverse to the panel. Figure 7.16 shows an example of mode shapes along and transverse to the loading direction for a panel loaded using the triangular slotted configurations shown in Figure 7.12b. The results show that all load introduction configurations produced a mode shape in agreement with

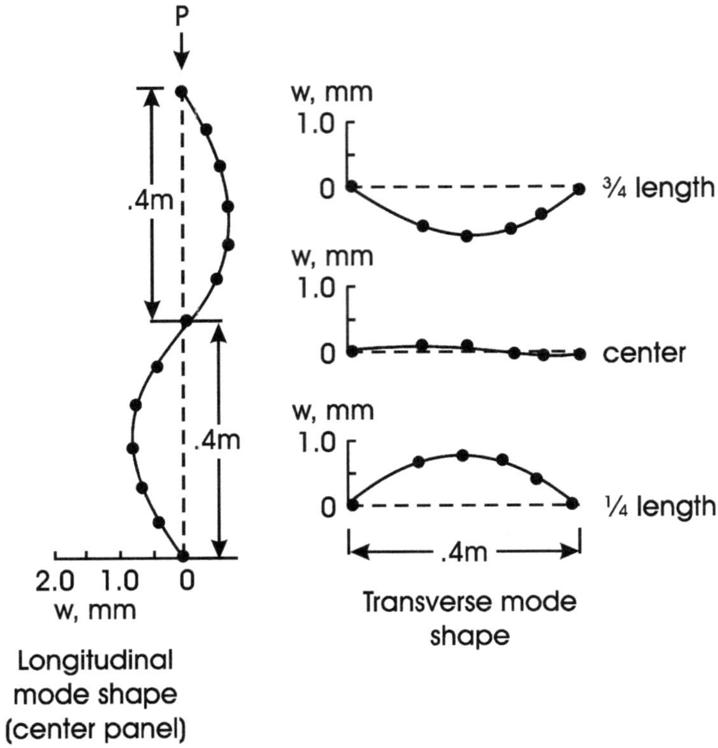

Figure 7.16 Buckling mode shape for a rectangular aluminum panel loaded using a triangular slot configuration. Data from Minguez (1986).

predictions (full sine wave along loading direction and half sine wave transversely).

Quantitatively, however, Minguez (1986) found that the various methods of load introduction produced substantially different amplitudes of deflection. It was found that the method of using 13 segmented rollers on each loaded edge (Figure 7.12c) produced the largest deflections at any given load above P_{cr}. This arrangement allows each section of the load-carrying edges to accommodate the buckling shape (Figure 7.16) with the maximum slope (rotation) at the center. Southwell plots were constructed for each load introduction configuration (Figure 7.12), see the example of a Southwell plot shown in Figure 7.17 for the triangular slotted configuration.

The buckling load, P_{cr}, determined from the slope of the fitted line in Figure 7.17, is $P_{cr} = 2.88$ kN. Table 7.2 summarizes buckling loads determined from Southwell plots for the various load introduction configurations. Table 7.2 also lists the buckling loads normalized by the theoretical critical

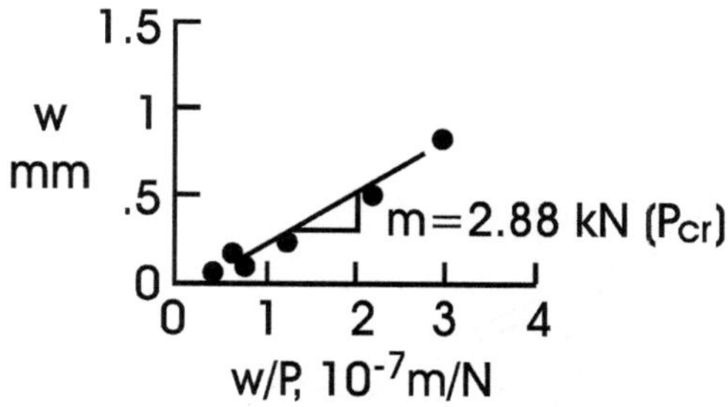

Figure 7.17 Southwell plot for panel with triangular slotted load introduction.

Table 7.2 Buckling loads for various load introduction configurations (Figure 7.7).

Load introduction	P_{cr} (exp), kN	$\frac{P_{cr}(\text{exp})}{P_{cr}(\text{SS})}$
Flat	N.A.	–
Triangular slots	2.88	1.11*
Rigid rollers	2.80	1.08*
13 roller segments	2.71	1.05*
Simply supported*	2.59	1.00*

* P_{cr}(SS) calculated from plate theory assuming simply supported (SS) boundary conditions.

load calculated for the panel assuming simply-supported edges (Minguez, 1986). For all load introduction configurations, the experimental buckling load exceeds the critical load by 5 to 11%, depending on the actual configuration. This result indicates that simply-supported conditions were not fully achieved for any of the configurations examined, although the one with 13 roller segments on each horizontal edge, provides boundary conditions close to simply-supported.

7.8.1 Experimental Determination of the Buckling Load of Panels

The method to extract the buckling load from the test results by Minguez (1986) was criticized by Chau (1987), because the Southwell method strictly applies only to columns, not to panels. To enable accurate experimental evaluation of the buckling load for panels, Spencer and Walker (1975) used load and deflection data in the pre- and post-buckling regimes of isotropic homogeneous panels in connection with a generalized Donnell (1938) equation,

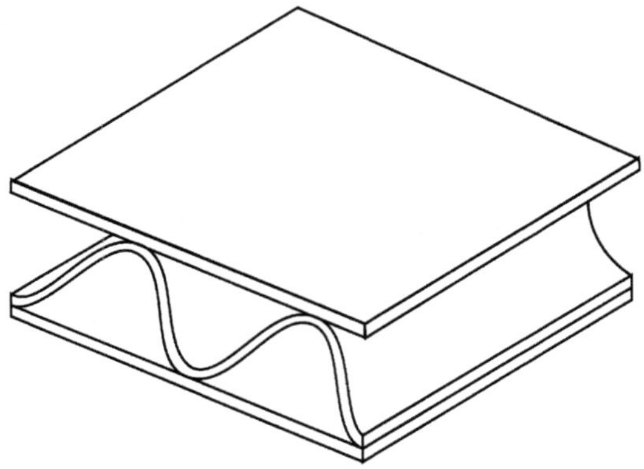

Figure 7.18 Element of a corrugated core panel.

$$\frac{P}{P_{\text{cr}}} = \frac{1}{w + w_o} + \frac{w + 2w_o}{(Ah)^2}, \tag{7.54}$$

where w_o and w are the initial out-of-plane imperfection, and w is the additional deflection. A is constant and h is the panel thickness. Equation (7.54) was fitted to the experimentally measured load (P) vs. out-of-plane displacement (w) data with w_o, A and P_{cr} as undetermined parameters to extract P_{cr}.

Nordstrand (2003) analyzed and tested orthotropic sandwich panels in uniaxial compression, as shown schematically in Figure 7.9. Geometric nonlinear analysis using classical plate theory without transverse shear deformation was developed by extending the post-buckling analysis of Rhodes and Harvey (1977) to an orthotropic panels with initial imperfection. This analysis yields an equation for the load (P) as a function of the out-of-plane displacement w, which contains the critical load as a parameter,

$$P = P_{\text{cr}}\left(1 - \frac{w_o}{w}\right) + \psi(w^2 - w_o^2), \tag{7.55}$$

where ψ is a post-buckling parameter. Consequently, this formula can be employed for experimental evaluation of the buckling load from measured load and out-of-plane deflection data.

Compression testing was done on 4 mm thick, 0.4 × 0.4 (m) corrugated core sandwich panels with an areal weight of 556 g/m^2. The corrugation wave length was 7.26 mm. Figure 7.18 shows the structure of corrugated core sandwich, a very common sandwich for packaging applications. Table 7.3 lists bending and shear stiffnesses of the sandwich panels.

Only reasonably flat panels with an imperfection of less than 2 mm (half thickness) were tested. The compression testing utilized a fixture described

Table 7.3 Bending and shear stiffnesses of corrugated core sandwich panels. h_c = core thickness = 3.51 mm.

Stiffness*	Value
D_{11}, Nm	14.6
D_{12}, Nm	2.71
D_{22}, Nm	5.43
D_{66}, Nm	3.34
$h_c G_{yz}$, kN/m	39.2
$h_c G_{xz}$, kN/m	5.6

*The in-plane principal directions refer to a coordinate system with the 1 axis perpendicular and the 2 axis parallel to the corrugations.

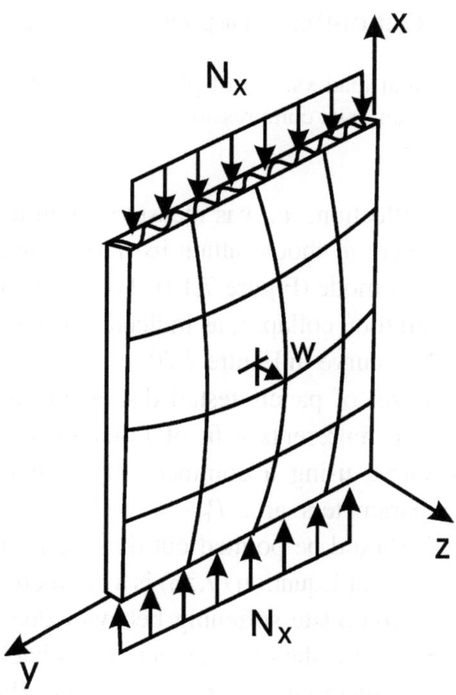

Figure 7.19 Uniaxial compression loading of corrugated core sandwich panel.

in connection with Figures 7.13 and 7.14. The panels were loaded uniaxially along the corrugations, i.e., the material direction "2" was along the loading axis as shown in Figure 7.19.

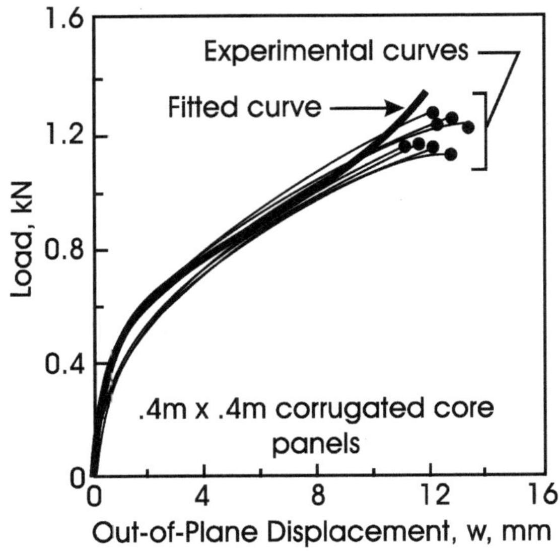

Figure 7.20 Experimental load vs. out-of-plane deflection curves for corrugated core sandwich loaded in uniaxial compression.

The out-of-plane deflection, w, was measured with a displacement gage at the panel center, where w should attain its maximum for buckling of the panel in its fundamental mode (Figure 7.19). The panels were loaded in displacement control until total collapse, as indicated by the circle at the end of each experimental P–w curve in Figure 7.20.

It is noted that the set of panels tested displayed quite a consistent response. The bold curve represents a fit of Equation (7.56) to the average experimental P–w curve using a commercially available software; SAS (2003). The fitting parameters are: $P_{cr} = 814$ N, $w_o = 0.8$ mm and $\psi = 3.55$ MN/m^2. It should be pointed out that the post-buckling analysis underlying the derivation of Equation (7.56) is a geometric nonlinear-elastic, and is not able to accommodate softening behavior due to the plasticity of the constituent materials. It is therefore essential that buckling occurs within the elastic regime of the material prior to localized buckling or yield.

It should be pointed out that the loading shown in Figure 7.18 refers to a panel loaded parallel to the corrugations and that the bending stiffnesses listed in Table 7.3 refer to a principal coordinate system (1-2) with the 1 axis perpendicular and the 2 axis parallel to the corrugations. To accommodate the loading configuration shown in Figure 7.19 the stiffnesses were transformed (Table 7.4). The first-order shear analysis presented in Section 7.7 was used to determine the critical buckling load for the sandwich panel with

Table 7.4 Transformed bending and shear stiffness for corrugated core sandwich panel (Figure 7.19).

Stiffness	Value
D_{11}, Nm	5.43
D_{12}, Nm	2.71
D_{22}, Nm	14.6
D_{66}, Nm	3.34
$h_c G_{yz}$, kN/m	5.6
$h_c G_{xz}$, kN/m	39.2

the data listed in Table 7.4. Calculations revealed that the panel should buckle in the fundamental mode ($m = n = 1$) at a load, $P_{cr} = 820$ N, which agrees very favorable with the critical load determined using the nonlinear regression analysis above, $P_{cr} = 814$ N.

7.8.2 Analysis of Collapse Load

As indicated in Figure 7.15, a distinctive feature of nearly flat slender sandwich panels is their ability to support loads significantly larger than the buckling load. Thus, in several situations the load design allowable load of such panels is governed by the collapse load rather than just the critical load. For the specific corrugated sandwich panel discussed earlier, the results in Figure 7.20, indicate that the collapse load exceeds the buckling load (814 N) by almost 50%.

The analysis of the collapse of sandwich panels is complicated by the fact that the strain in the middle plane of the panel due to buckling cannot be neglected once the panel is loaded above the buckling load. When the deflection, w, becomes comparable to the panel thickness, second-order terms in the expression for the components of strain in the middle plane of the plate must be taken into account, which substantially complicates the analysis of the buckling response (Timoshenko, 1936). As shown in Figure 7.15, the response of the panel is highly nonlinear and it becomes a formidable task to determine the distribution of load in the post-buckled panel. Such analysis shows that N_x becomes non-uniform and most of the load is supported by the regions of the panel near the unloaded edges, see Figure 7.21.

Because of the difficulties in analyzing the distribution of load in post-buckled panels, semi-empirical and simplified analytical approaches have

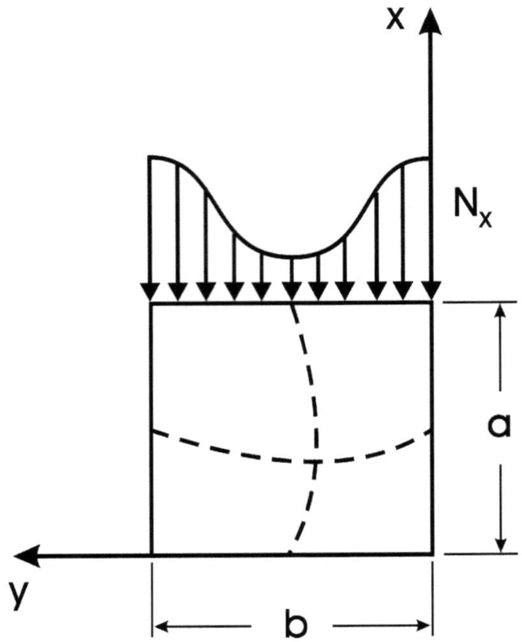

Figure 7.21 Distribution of compressive load in a panel loaded in uniaxial compression beyond the critical buckling load.

been developed. One of the early approaches to determining the post-buckling strength is attributed to Cox (1933), which was later modified by Norris (1942) for use with orthotropic materials such as plywood, and further modified for corrugated core sandwich panels by McKee et al. (1963). According to this approach, the compressive strength of the panel is assumed to follow a power function given by

$$\frac{P_{\text{col}}}{P_{\text{cr}}} = c\left(\frac{X_c}{P_{\text{cr}}}\right)^b, \tag{7.56}$$

where c and b are empirical constants, X_c is the uniaxial compressive strength of the sandwich in the direction of loading, and P_{cr} is the critical buckling load of the panel per unit width. To establish the parameter values for a given sandwich (X_c = constant), the size of the panel may be varied which results in variations of P_{col} and P_{cr}. To establish their numerical values, a logarithmic form of Equation (7.56) is used:

$$\log\left(P_{\text{col}}/P_{\text{cr}}\right) = \log c + b\log(X_c/P_{\text{cr}}). \tag{7.57}$$

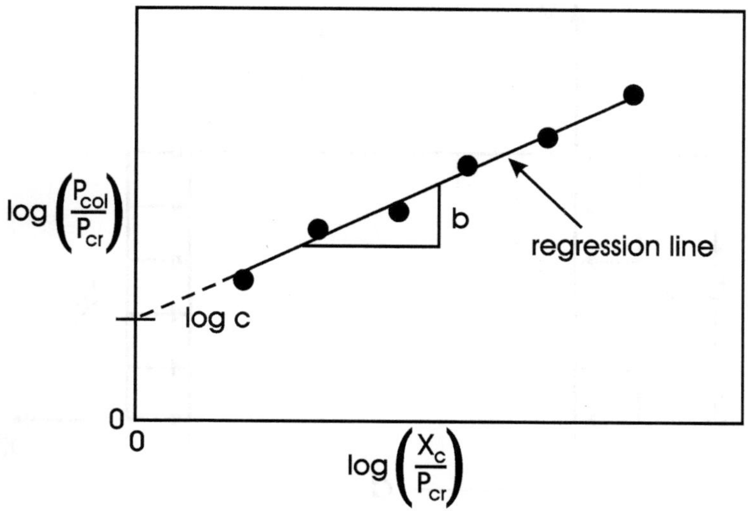

Figure 7.22 Determination of parameters c and b in the McKee et al. equation.

The critical buckling load is determined from analytical methods (see Section 7.7) or extracted from the experimental load vs. out-of-plane deflection curve, such as the one shown in Figure 7.20, using the methods described earlier in this section. The collapse load is also readily obtained from the experimental P–w curve. Once the data set has been established, a log-log graph may be constructed (Figure 7.22).

It is observed in Figure 7.22 that the slope of the linear regression line is b and the intercept with the vertical axis is log c. This method for strength determination has gained much acceptance within the corrugated board industry. Properly calibrated, the McKee et al. (1963) method produces reliable predictions of the collapse load of corrugated core packages. One shortcoming, however, is that the semi-empirical foundation requires experimental testing of several panels before new predictions can be made, and even if empirical data exits, there is always uncertainty about the accuracy of the predictions of sandwich panels that are different from those employed in calibration.

An approximate closed-form approach to predict the collapse load of panels loaded in uniaxial compression by a rigid frame into the post-buckling regime has been proposed by Timoshenko (1936). His analysis should also be applicable to relatively slender sandwich panels able to support loads greater than the critical buckling load. Since the panels are slender, it is here assumed that transverse shear deformation can be neglected. The critical

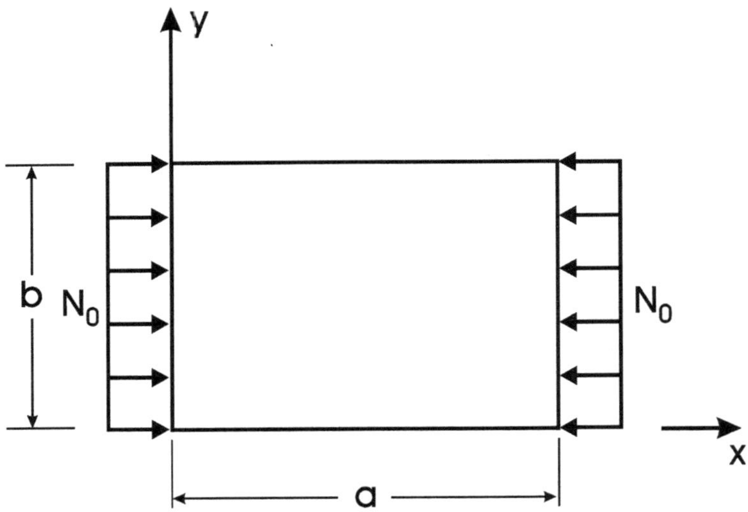

Figure 7.23 Uniaxial compressive loading of simply supported sandwich panel.

load, N_o of such a simply-supported panel loaded in uniaxial compression (Figure 7.23) is given by (Jones, 1999)

$$\frac{N_o b^2}{\pi^2 D_{11}} = m^2 \left(\frac{b}{a}\right)^2 + \frac{2\left(D_{12} + 2D_{66}\right)}{D_{11}} + \frac{D_{22}}{D_{11}} \left(\frac{a}{b}\right)^2 \frac{1}{m^2}, \tag{7.58}$$

where m is the number of half sine waves into which the panel buckles in the direction of loading, i.e. the x direction. Recall that the panel buckles into one half sine wave transverse to the direction of loading. Analysis of sandwich panels under a more general loading configuration (see Section 7.7) shows that the number of buckling half-waves (n) in the y direction must in general be considered.

The number of half-waves the panel buckles into in the x direction, m, depends on ratios of the material stiffnesses and the panel aspect ratio (length/width ratio $= a/b$). Figure 7.24 shows the buckling load, N_o, normalized by the bending stiffness, D_{11}, and square of the width plotted vs. the panel aspect ratio for the following set of bending stiffness ratios $D_{22}/D_{11} = (D_{12} + 2D_{66})/D_{11} = 0.1$.

It is observed that the buckling load corresponding to a particular mode shape (defined by the parameter m) undergoes a minimum at a certain panel aspect ratio and that the minimum gets more and more shallow as the aspect ratio increases. It is further noted that a square panel ($a = b$) would buckle into one half-wave ($m = 1$) while the minimum buckling load for long and

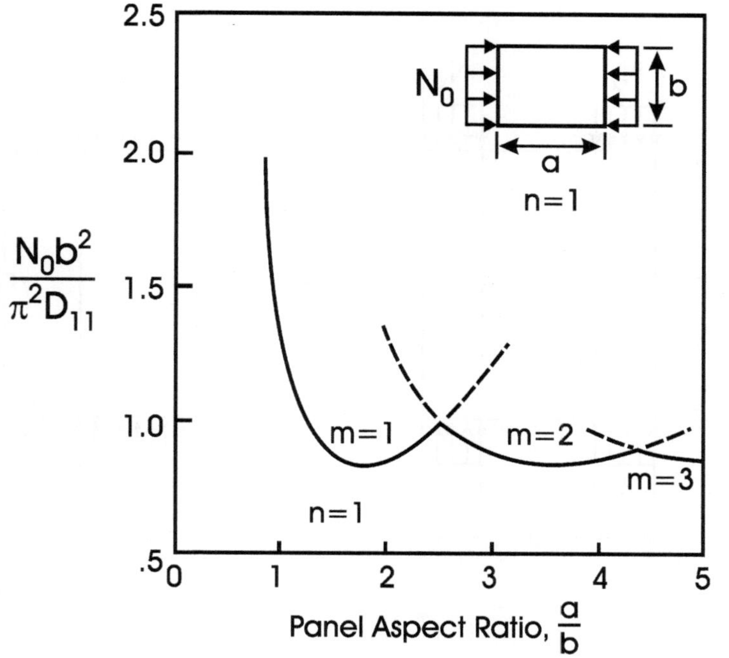

Figure 7.24 Buckling load of orthotropic panel loaded in uniaxial compression. $D_{22}/D_{11} = (D_{12} + 2D_{66})/D_{11} = 0.1$.

narrow panels (with a high aspect ratio, a/b) corresponds to $m > 1$, i.e., several half-waves.

Moreover, the minimum buckling load for each curve (with fixed value of m) does not depend on m. Detailed analysis based on Equation (7.58) reveals that each minimum occurs at an aspect ratio given by

$$a/b = m\sqrt[4]{\frac{D_{11}}{D_{22}}}. \tag{7.59}$$

The minimum buckling load, independent of the value of m is

$$\left(\frac{N_o b^2}{\pi^2 D_{11}}\right)_{\min} = 2\left[\sqrt{\frac{D_{22}}{D_{11}}} + \frac{D_{12} + 2D_{66}}{D_{11}}\right]. \tag{7.60}$$

Timoshenko (1936) presents an analysis of the post-buckling strength for a panel loaded in uniaxial compression based the distribution of load according to an approximation by von Kármán et al. (1932). As shown in Figure 7.21, most of the compressive after buckling is supported by the edge regions of

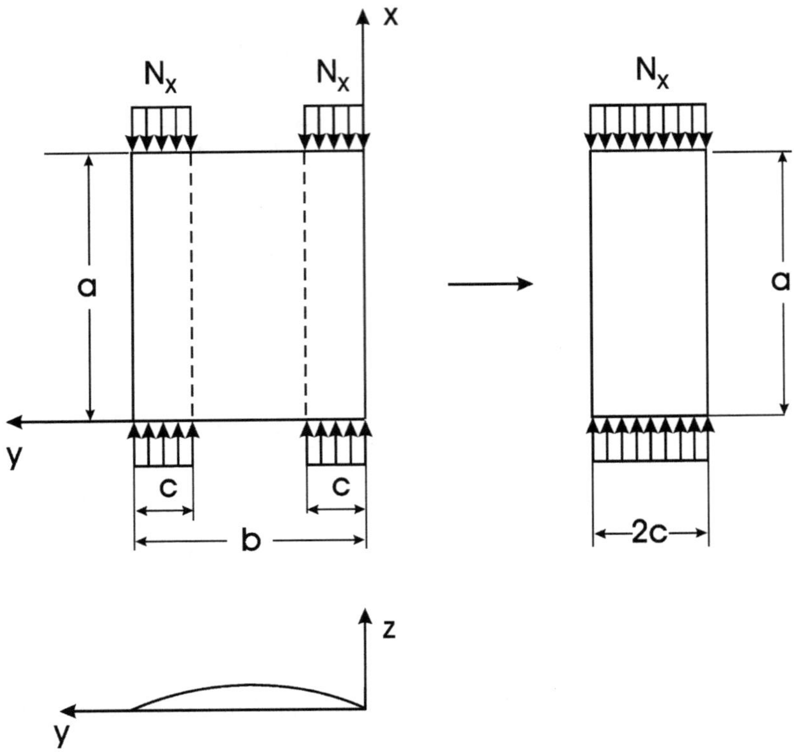

Figure 7.25 Timoshenko model of load distribution in post-buckled panel.

the panel. Timoshenko (1936) proposed that the load distribution in the post-buckling regime may be approximated by assuming a uniform load over each edge region of the panel, see Figure 7.25. The width of each such region is denoted by c. The middle region of the panel is completely disregarded, and the non-uniformly loaded panel of width, b, may be represented by a uniformly loaded panel of width, $2c$, Figure 7.25.

For a long narrow panel, the aspect ratio is much larger than 1, and such a panel is expected to buckle into several half-waves along the loading direction corresponding to a value of m greater than 1. The critical load for such a panel should approximately be equal to the minimum value, $(N_o)_{\min}$, as provided by Equation (7.60)

$$N_o = \frac{\pi^2}{2c^2}\left[\sqrt{D_{11}D_{22}} + D_{12} + 2D_{66}\right]. \tag{7.61}$$

Collapse of the panel is assumed to occur when the critical buckling load reaches the uniaxial compression failure load per unit width of the sandwich,

X_c. With this assumption ($N_o = X_c$) Equation (7.61) yields the width of the load-bearing region

$$c = \pi \left[\frac{\left(\sqrt{D_{11}D_{22}} + D_{12} + 2D_{66}\right)}{2X_c} \right]^{1/2}. \tag{7.62}$$

Hence, the collapse load of the panel is given by

$$P_c = 2cX_c = 2\pi \left[\frac{X_c}{2} \left(\sqrt{D_{11}D_{22}} + D_{12} + 2D_{66}\right) \right]^{1/2}. \tag{7.63}$$

Notice that the buckling factor within the parenthesis remains invariant if the directions 1 and 2 of the panel are interchanged. The only factor changing if the material directions 1 and 2 are interchanged is the compression strength, X_c.

Grangard and Rudstrom (1970) followed the Timoshenko analysis and derived an equation similar to (7.63) for the prediction of the collapse load of paperboard packages, but omitted the D_{12} stiffness which leads to under-prediction of the collapse load. To the knowledge of the authors, this analysis has not been applied to sandwich panels.

For the corrugated board panel examined by Nordstrand (2003), the bending stiffnesses D_{ij} are listed in Table 7.4. The uniaxial compression failure load per unit width of the sandwich, X_c is (Westerlind and Carlsson, 1992)

$$X_c = X_1 + X_2 + \alpha X_w, \tag{7.64}$$

where X_1, X_2, and X_w are the compressive strengths failure load per unit width of the two faces (1 and 2), and web (w), and α is the "take-up factor", i.e. length of web per unit width of the sandwich. For the core considered, $\alpha = 1.43$ (Nordstrand, 2003). With the strengths and thicknesses of the faces and web provided by Nordstrand (2003), Equation (7.64) yields $X_c = 4.16$ kN/m. Substitution of the strength X_c and the stiffnesses D_{ij} (Table 7.4) in Equation (7.63) yields a collapse load $P_c = 1{,}226$ N. This value may be compared to the experimental average from Figure 7.20, $P_c = 1{,}200$ N. This good agreement seems to support the simplified analysis, although more general acceptance would demand a much larger experimental data base.

An alternative analysis, which does not involve the empirical calibration of the McKee et al. (1963) approach, or the simplifying assumptions in the Timoshenko (1936) model, is to perform geometric nonlinear analysis to determine the stress distribution in the most highly stressed face sheet, which would be the face on the concave side of the post-buckled panel subject to stresses due to axial compression and bending. Nordstrand (2003) conducted geometric nonlinear finite element analysis to determine the state of stress in the face sheets and used this in combination with a widely used biaxial failure criterion for orthotropic materials, viz. the "Tsai–Wu criterion" (Tsai and Wu, 1971). This analysis provided a collapse load of 1,270 N, in good agreement with the measured average collapse load, Figure 7.20, $P_{col} = 1{,}200$ N. Based on this favorable agreement Nordstrand (2003) concluded that local buckling or wrinkling of the face were not failure mechanisms governing the collapse load. However, once the panel reached the collapse load it was observed to fail in a face wrinkling mode leading to the collapse of the web core and collapse of the panel.

Chapter 8
Wrinkling and Local Instabilities

Compression loaded faces of sandwich members are sometimes subject to local instability phenomena, the most prominent being the *wrinkling or rippling* and the *intracell buckling or dimpling*. This chapter presents the mechanics associated with these phenomena and the classical formulas that predict the conditions for inducing these forms of local instability.

8.1 Wrinkling

Wrinkling refers to a particular form of local instability of the compression faces of a sandwich panel in which the wavelength of the buckled form is of the same order as the thickness of the core. This short-wavelength instability of the faces can occur at lower load levels than the ordinary global, "Euler", buckling of the structure, which is characterized by a half-wavelength of the order of the compressed length of the sandwich panel.

As shown in Figure 8.1, in the case of the global, Euler, buckling, the core may exhibit a substantial shearing deformation; in the case of local wrinkling it acts like an elastic foundation and the buckling deformation is mainly confined to the layers adjacent to the face sheets. Wrinkling of a symmetric configuration can occur in a symmetric mode or an antisymmetric one (Figure 8.1).

Wrinkling will be considered for wide sandwich panels or sandwich columns. Thus, referring to Figure 8.2, the panel is so wide that lines along the y axis can be taken as uncurved. Therefore, a unit width can be treated as a Euler column.

The classical wrinkling formulas of Hoff and Mautner (1945), Plantema (1966), and Allen (1969) will be presented in the following.

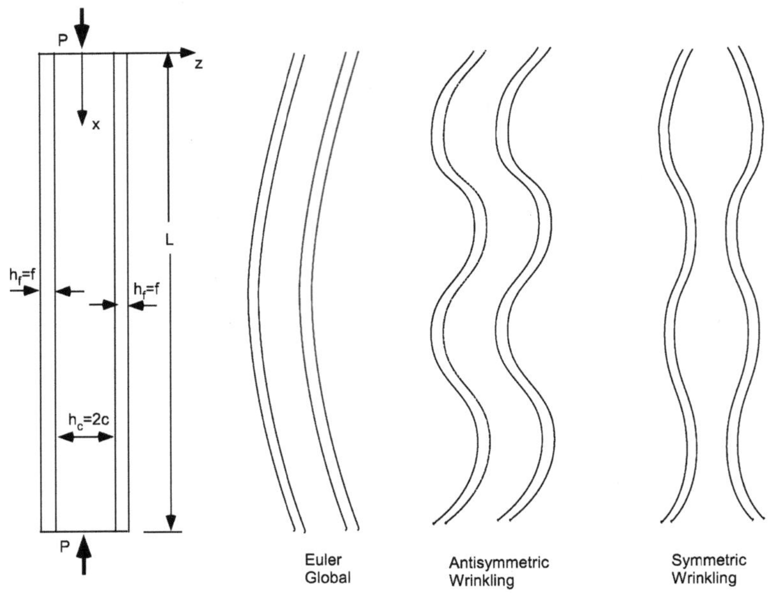

Figure 8.1 Buckling modes.

8.1.1 Hoff and Mautner's Formulas

I. Symmetric Wrinkling

The wrinkled state involves the bending of the faces of thickness $h_f = f$ and the elongation and compression of the core of thickness. $h_c = 2c$. A short wave on the surface can hardly have any effect upon the material in the middle of the core when c is large. It is assumed that the displacements occur only in marginal zones of depth d (Figure 8.3).

Assuming that the face undergoes a sinusoidal displacement and that the wave damps out linearly (linear decay) through the thickness, the transverse displacement, w of the top marginal zone is given with respect to the local coordinate system (x, z) in Figure 8.3, as follows:

$$w = \frac{Bz}{d} \sin \frac{\pi x}{a}. \tag{8.1}$$

where the origin of the z coordinate is at the boundary between the affected and unaffected core regions.

The critical load is now calculated from the requirement that the work done by the compressive force be equal to the strain energy of bending stored in the face material plus the strain energy of extension and shear stored in the core. Because of the symmetry, it is sufficient to calculate the work and the

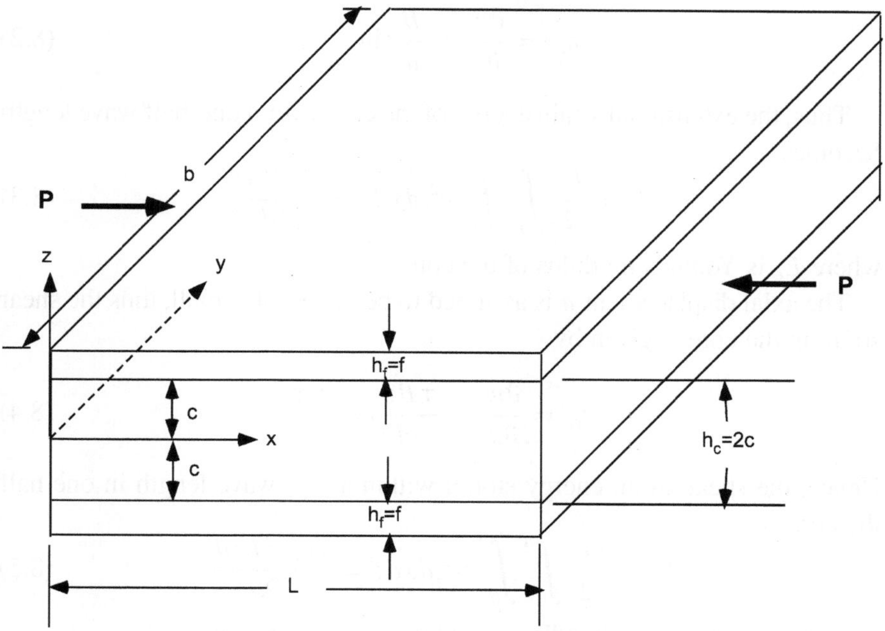

Figure 8.2 Definition of the geometry of a sandwich wide panel/beam under axial compression.

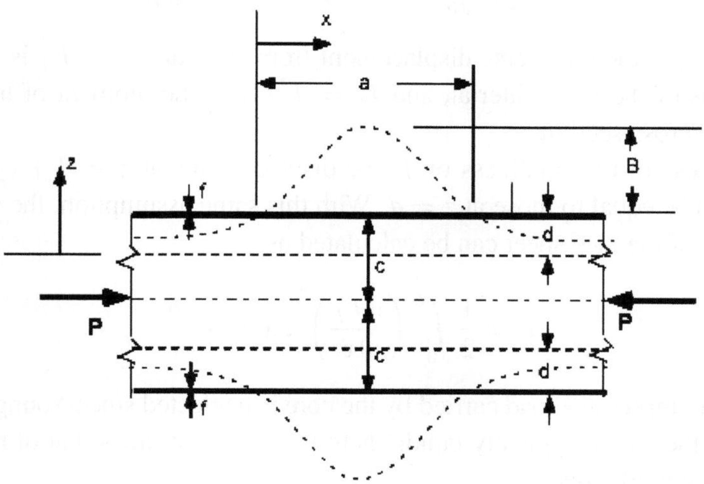

Figure 8.3 Hoff and Mautner's model for symmetric wrinkling.

strain energy for one-half of the cross-section. The width of the sandwich perpendicular to the plane of the drawing is taken as unity.

The normal (z directional) strain in the core is

$$\varepsilon_{zz} = \frac{\partial w}{\partial z} = \frac{B}{d}\sin\frac{\pi x}{a}. \tag{8.2}$$

Thus, the extensional strain energy of the core within one-half wave length becomes

$$U_e = \frac{E_c}{2}\int_0^a\int_0^d \varepsilon_{zz}^2 dxdz = \frac{E_c B^2 a}{4d}, \tag{8.3}$$

where E_c is Young's modulus of the core.

The axial displacement u is assumed to be negligibly small, thus the shear strain in the core is given by

$$\gamma_{xz} = \frac{\partial w}{\partial x} = \frac{\pi B z}{ad}\cos\frac{\pi x}{a}. \tag{8.4}$$

Hence, the shear strain energy stored within a half wave length in one half the core is

$$U_s = \frac{G_c}{2}\int_0^a\int_0^d \gamma_{xz}^2 dxdz = \frac{G_c \pi^2 B^2 d}{12a}. \tag{8.5}$$

The strain energy of bending stored in one face sheet is

$$U_f = \frac{E_f I_f}{2}\int_0^a \left(\frac{\partial^2 w_f}{\partial x^2}\right)^2 dx = \frac{\pi^4 E_f B^2 f^3}{48a^3}, \tag{8.6}$$

where w_f is the transverse displacement from (8.1) at $z = d$, E_f is Young's modulus of the face material, and $I_f = f^3/12$ is the moment of inertia of the face cross-section.

Because of the smallness of f, the displacements at $z = d + (f/2)$ are taken to be equal to those at $z = d$. With this same assumption, the shortening ΔL of the face sheet can be calculated as

$$\Delta L = \frac{1}{2}\int_0^a \left(\frac{\partial w_f}{\partial x}\right)^2 dx = \frac{\pi^2 B^2}{4a}. \tag{8.7}$$

The compressive load carried by the core is neglected since Young's modulus of the face is typically hundreds to thousands of times that of the core. Thus, the work done is

$$W = (\sigma_{\text{cr}}^f f)\Delta L = \sigma_{\text{cr}}^f \frac{\pi^2 B^2 f}{4a}. \tag{8.8}$$

The equation

$$W = U_e + U_s + U_f \tag{8.9}$$

can be solved for σ_{cr}^f after substitution of the expressions thus developed:

$$\sigma_{\mathrm{cr}}^{f} = \frac{E_c a^2}{\pi^2 f d} + \frac{G_c d}{3f} + \frac{\pi^2 E_f f^2}{12a^2}. \tag{8.10}$$

The critical stress in this equation depends upon the parameters d and a. The actual values of these parameters are those which make the critical stress a minimum. Consequently, $\partial\sigma_{\mathrm{cr}}^{f}/\partial d$ and $\partial\sigma_{\mathrm{cr}}^{f}/\partial a$ must vanish:

$$\partial\sigma_{\mathrm{cr}}^{f}/\partial d = -\frac{E_c a^2}{\pi^2 f d^2} + \frac{G_c}{3f} = 0, \tag{8.11a}$$

$$\partial\sigma_{\mathrm{cr}}^{f}/\partial a = \frac{2E_c a}{\pi^2 f d} - \frac{\pi^2 E_f f^2}{6a^3} = 0, \tag{8.11b}$$

Simultaneous solution of the two equations gives

$$d/f = 0.91\sqrt[3]{\frac{E_f E_c}{G_c^2}}, \tag{8.12a}$$

$$a/f = 1.65\sqrt[6]{\frac{E_f^2}{E_c G_c}}, \tag{8.12b}$$

Substitution in the equation for the critical stress yields

$$\sigma_{\mathrm{cr}}^{f} = 0.91\sqrt[3]{E_f E_c G_c}\,. \tag{8.13}$$

This formula is correct only if d is smaller than or equal to c, as may be seen from Figure 8.3. Thus, from (8.12a), the condition for validity is

$$0.91 f\sqrt[3]{\frac{E_f E_c}{G_c^2}} \leq c, \tag{8.14a}$$

that is the core to face thickness ratio,

$$\frac{2c}{f} \geq 1.82\sqrt[3]{\frac{E_f E_c}{G_c^2}}. \tag{8.14b}$$

When this inequality does not hold, then the marginal zone depth is equal to half the core thickness, so equation (8.1) is replaced by

$$w = \frac{Bz}{c}\sin\frac{\pi x}{a}. \tag{8.15}$$

The expression for the critical stress then becomes

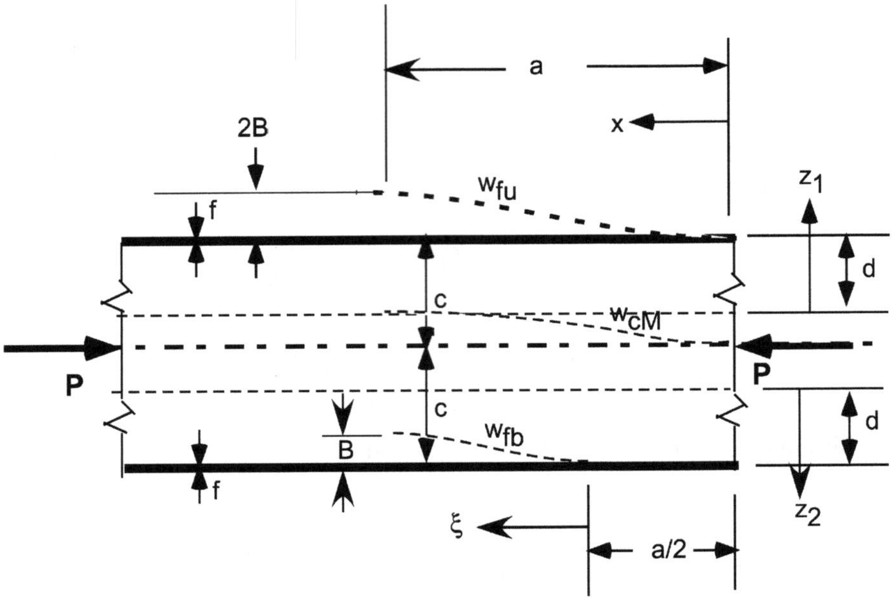

Figure 8.4 Hoff and Mautner's assumed shape for anti-symmetric wrinkling.

$$\sigma_{cr}^f = \frac{E_c a^2}{\pi^2 f c} + \frac{G_c c}{3f} + \frac{\pi^2 E_f f^2}{12a^2}. \tag{8.16}$$

The derivative of σ_{cr}^f with respect to a must again vanish. This condition yields

$$a/f = 1.42\sqrt[4]{\frac{E_f}{E_c}}\sqrt[4]{\frac{2c}{f}}. \tag{8.17}$$

Substitution in Equation (8.16) gives

$$\sigma_{cr}^f = 0.577\sqrt{E_f E_c}\sqrt{\frac{f}{c}} + 0.333 G_c \left(\frac{c}{f}\right). \tag{8.18}$$

II. Anti-Symmetric Wrinkling
In this case the deflected shape is assumed by Hoff and Mautner as indicated by the dashed line in Figure 8.4. Again marginal zones of depth d are assumed to which all the extensions are restricted, although shear deformation occurs throughout the entire core.

For $0 \le x \le a/2$ (i.e. the right segment of the figure), the displacement of the upper face is assumed to be

$$w_{fu} = B\left(1 - \cos\frac{\pi x}{a}\right), \tag{8.19}$$

whereas the displacement of the bottom face is

$$w_{fb} = 0. \tag{8.20}$$

Consequently, the displacement w_{cM} of the median line of the core is assumed to be

$$w_{cM} = \frac{B}{2}\left(1 - \cos\frac{\pi x}{a}\right). \tag{8.21}$$

There are two marginal zones, one for each face sheet. The displacement of any point in the upper marginal zone is assumed to be

$$w_{cu} = \frac{B}{2}\left(1 - \cos\frac{\pi x}{a}\right) + \frac{B}{2}\frac{z_1}{d}\left(1 - \cos\frac{\pi x}{a}\right), \quad \text{for } 0 \le z_1 \le d. \tag{8.22a}$$

In the bottom marginal zone

$$w_{cb} = -\frac{B}{2}\left(1 - \cos\frac{\pi x}{a}\right) + \frac{B}{2}\frac{z_2}{d}\left(1 - \cos\frac{\pi x}{a}\right), \quad \text{for } 0 \le z_2 \le d. \tag{8.22b}$$

In the middle portion of the core, the displacement depends on z and is assumed to be as in (8.21), i.e.

$$w_{cm} = w_{cM} = \frac{B}{2}\left(1 - \cos\frac{\pi x}{a}\right), \tag{8.22c}$$

The extensional strain in the upper marginal zone is

$$\varepsilon_{cu} = \frac{\partial w_{cu}}{\partial z_1} = \frac{B}{2d}\left(1 - \cos\frac{\pi x}{a}\right). \tag{8.23a}$$

In the lower marginal zone,

$$\varepsilon_{cb} = \frac{\partial w_{cb}}{\partial z_2} = \frac{B}{2d}\left(1 - \cos\frac{\pi x}{a}\right) = \varepsilon_{cu}. \tag{8.23b}$$

In the middle portion, the extensional strain is zero.

Consequently, the extensional strain energy is

$$U_\varepsilon = 2\frac{E_c}{2}\int_0^{a/2}\int_0^d \varepsilon_{cu}^2 dx dz_1 = E_c\frac{B^2 a(3\pi - 8)}{16\pi d}. \tag{8.24}$$

The shear strain in the upper marginal zone is (axial displacement again assumed to be negligibly small):

$$\gamma_{cu} = \frac{\partial w_{cu}}{\partial x} = \frac{B}{2}\left(1 + \frac{z_1}{d}\right)\frac{\pi}{a}\sin\frac{\pi x}{a}. \tag{8.25a}$$

In the bottom marginal zone

$$\gamma_{cb} = \frac{\partial w_{cb}}{\partial x} = -\frac{B}{2}\left(1 - \frac{z_2}{d}\right)\frac{\pi}{a}\sin\frac{\pi x}{a}. \tag{8.25b}$$

In the middle portion of the core

$$\gamma_{cm} = \frac{\partial w_{cm}}{\partial x} = \frac{B}{2}\frac{\pi}{a}\sin\frac{\pi x}{a}. \tag{8.25c}$$

The shear strain energy can be now calculated from

$$U_\gamma = \frac{G_c}{2}\int_0^{a/2}\left[\int_0^d \gamma_{cu}^2 dz_1 + \int_0^d \gamma_{cb}^2 dz_2 + \gamma_{cm}^2(2c - 2d)\right]dx. \tag{8.26a}$$

Integration gives

$$U_\gamma = \frac{\pi^2 B^2 G_c}{48a}(3c + d). \tag{8.26b}$$

In the middle segment of the figure, i.e. for

$$\frac{a}{2} \leq x \leq a, \quad 0 \leq \xi \leq \frac{a}{2},$$

the displacements of the faces and the median line of the core, respectively, are assumed to be

$$w_{fu} = B\left(1 + \sin\frac{\pi\xi}{a}\right), \tag{8.27a}$$

$$w_{fb} = B\left(1 - \cos\frac{\pi\xi}{a}\right). \tag{8.27b}$$

$$w_{cM} = B\left(1 - \frac{1}{2}\cos\frac{\pi\xi}{a} + \frac{1}{2}\sin\frac{\pi\xi}{a}\right). \tag{8.27c}$$

The displacements of points in the upper marginal zone, the middle portion and the bottom marginal zone, respectively, are

$$w_{cu} = B\left(1 - \frac{1}{2}\cos\frac{\pi\xi}{a} + \frac{1}{2}\sin\frac{\pi\xi}{a}\right) + \frac{Bz_1}{d}\left(\frac{1}{2}\sin\frac{\pi\xi}{a} + \frac{1}{2}\cos\frac{\pi\xi}{a}\right), \tag{8.28a}$$

$$w_{cm} = w_{cM} = B\left(1 - \frac{1}{2}\cos\frac{\pi\xi}{a} + \frac{1}{2}\sin\frac{\pi\xi}{a}\right), \tag{8.28b}$$

$$w_{cb} = B\left(-1 + \frac{1}{2}\cos\frac{\pi\xi}{a} - \frac{1}{2}\sin\frac{\pi\xi}{a}\right) + \frac{Bz_2}{d}\left(\frac{1}{2}\sin\frac{\pi\xi}{a} + \frac{1}{2}\cos\frac{\pi\xi}{a}\right). \tag{8.28c}$$

The corresponding extensional strains are

$$\varepsilon_{cu} = \frac{\partial w_{cu}}{\partial z_1} = \frac{B}{2d}\left(\sin\frac{\pi\xi}{a} + \cos\frac{\pi\xi}{a}\right) = \varepsilon_{cb} = \frac{\partial w_{cb}}{\partial z_2}; \quad \varepsilon_{cm} = 0. \quad (8.29)$$

The shear strains are

$$\gamma_{cu} = \frac{\partial w_{cu}}{\partial \xi} = \frac{B\pi}{2a}\left(\sin\frac{\pi\xi}{a} + \cos\frac{\pi\xi}{a}\right) + \frac{B\pi}{2a}\frac{z_1}{d}\left(\cos\frac{\pi\xi}{a} - \sin\frac{\pi\xi}{a}\right). \quad (8.30a)$$

$$\gamma_{cm} = \frac{\partial w_{cm}}{\partial \xi} = \frac{B\pi}{2a}\left(\sin\frac{\pi\xi}{a} + \cos\frac{\pi\xi}{a}\right). \quad (8.30b)$$

$$\gamma_{cb} = \frac{\partial w_{cb}}{\partial \xi} = -\frac{B\pi}{2a}\left(\sin\frac{\pi\xi}{a} + \cos\frac{\pi\xi}{a}\right) + \frac{B\pi}{2a}\frac{z_2}{d}\left(\cos\frac{\pi\xi}{a} - \sin\frac{\pi\xi}{a}\right). \quad (8.30c)$$

The strain energy can be calculated as before (Equations (8.24) and (8.26a) with ξ in place of x). Integration yields

$$U_\varepsilon = E_c\frac{B^2a(\pi+2)}{8d\pi}, \quad (8.31a)$$

$$U_\gamma = G_c\frac{B^2\pi}{8a}\left[(\pi+2)c + (\pi-2)\frac{d}{3}\right]. \quad (8.31b)$$

Because of the point symmetry of the distorted shape, the strain energy in the left half shown in Figure 8.4 is the same as in the right half. The total strain energy U_c stored in the core is, therefore, the sum of the strain energies given in Equations (8.31) plus twice those given in Equations (8.24) and (8.26b), i.e.,

$$U_c = E_c\frac{B^2a(2\pi-3)}{4\pi d} + G_c\frac{B^2\pi}{12a}\left[3c(\pi+1) + (\pi-1)d\right]. \quad (8.32)$$

The strain energy of bending stored in the two faces, U_f, is found by substituting w_f into $(E_fI_f/2)\int w_f''^2 dx$ where w_f is obtained from (8.27a) and (8.27b) plus twice the contributions from (8.19) and (8.20), i.e.,

$$U_f = E_f\frac{B^2\pi^4 f^3}{24a^3}. \quad (8.33)$$

Similarly, the work W done by the external forces is found from $(\sigma_{cr}^f A_f/2)\int w_f'^2 dx$ where A_f is the cross-sectional area of each face sheet, and w_f is given by Equations (8.27a) and (8.27b) plus twice the contributions from (8.19) and (8.20), i.e.,

$$W = \sigma_{cr}^f\frac{B^2\pi^2 f}{2a}. \quad (8.34)$$

Substitution in the equation

$$W = U_f + U_c$$

and solution for σ_{cr}^f yields

$$\sigma_{cr}^f = \beta_1\left(\frac{f}{a}\right)^2 + \beta_2\left(\frac{a^2}{df}\right) + \beta_3\left(\frac{c}{f}\right) + \beta_4\left(\frac{d}{f}\right), \tag{8.35a}$$

where

$$\beta_1 = \frac{\pi^2 E_f}{12}; \quad \beta_2 = \frac{(2\pi - 3)E_c}{2\pi^3}; \quad \beta_3 = \frac{(\pi + 1)G_c}{2\pi}; \quad \beta_4 = \frac{(\pi - 1)G_c}{6\pi}. \tag{8.35b}$$

Minimization with respect to a and d, yields

$$\frac{\partial \sigma_{cr}^f}{\partial a} = -\frac{\beta_1 f^2}{a^4} + \frac{\beta_2}{df} = 0; \quad \frac{\partial \sigma_{cr}^f}{\partial d} = -\frac{\beta_2 a^2}{d^2 f} + \frac{\beta_4}{f} = 0. \tag{8.36}$$

Solution of the two equations yields

$$a/f = \left(\frac{\beta_1^2}{\beta_2\beta_4}\right)^{1/6} = 2.19\sqrt[6]{\frac{E_f^2}{E_c G_c}}, \tag{8.37a}$$

$$d/f = \left(\frac{\beta_1\beta_2}{\beta_4^2}\right)^{1/3} = 1.50\sqrt[3]{\frac{E_f E_c}{G_c^2}}. \tag{8.37b}$$

Substitution into (8.35a) results in the expression

$$\sigma_{cr}^f = 3\,(\beta_1\beta_2\beta_4)^{1/3} + \beta_3(c/f) = 0.51\sqrt[3]{E_f E_c G_c} + 0.66G_c(c/f). \tag{8.38}$$

These formulas are valid only if d proves to be smaller than or equal to c, hence if

$$c/f \leq 1.50\sqrt[3]{\frac{E_f E_c}{G_c^2}}. \tag{8.39}$$

When this inequality is not satisfied, d must be replaced by c in the expressions assumed for the deflected shapes. If the calculations are carried out on the basis of this assumption, the following expression is obtained for the critical stress (by replacing d with c in (8.35a)):

$$\sigma_{cr}^f = \frac{\beta_1 f^2}{a^2} + \frac{\beta_2}{cf}a^2 + (\beta_3 + \beta_4)\frac{c}{f}. \tag{8.40a}$$

The critical stress is a minimum when

$$\partial\sigma_{\mathrm{cr}}^{f}/\partial a = -\frac{\beta_1 f^2}{a^4} + \frac{\beta_2}{cf} = 0. \tag{8.40b}$$

Consequently,

$$a^4 = \frac{\beta_1 f^3 c}{\beta_2}$$

and

$$a/f = 1.98\sqrt[4]{E_f/E_c}\sqrt[4]{c/f}. \tag{8.41}$$

Substitution into Equation (8.40a) gives

$$\begin{aligned}\sigma_{\mathrm{cr}} &= 2\sqrt{\beta_1\beta_2 f/c} + (\beta_3 + \beta_4)c/f \\ &= 0.417\sqrt{E_f E_c}\sqrt{f/c} + 0.773G_c(c/f).\end{aligned} \tag{8.42}$$

Summary: Hoff and Mautner's wrinkling formulas are (i) for symmetric wrinkling, the critical stress is from Equation (8.13) provided the inequality (8.14b) is fulfilled; if not then the critical wrinkling stress is from Equation (8.18); (ii) for anti-symmetric wrinkling, the critical stress is from Equation (8.38) provided the inequality (8.39) is fulfilled; if not then the critical wrinkling stress is obtained from Equation (8.42). Calculations made by Hoff and Mautner (1945) for a sandwich consisting of papreg faces (a paper/plastic laminate) and a cellular cellulose acetate core showed that the anti-symmetric wrinkling would dominate (i.e. the critical stress corresponding to anti-symmetry would be less than that for symmetry) if the core thickness of face thickness ratio, $2c/f$, is smaller than 20.6. This is, of course, only valid for this particular sandwich material system and is not a general conclusion, but it gives some idea of the level of the parameters influencing the occurrence of symmetric vs. anti-symmetric wrinkling. Nevertheless, Equation (8.13) for symmetric wrinkling is the most popular wrinkling formula and the one mostly used, albeit with a factor of 0.5 instead of 0.91 for safety (e.g. Zenkert, 1997).

8.1.2 Plantema's Formula

The Plantema (1966) analysis assumes that the points of the core undergo vertical displacements with an exponential decay. With z defined from the upper face sheet (Figure 8.5) and n denoting the number of half

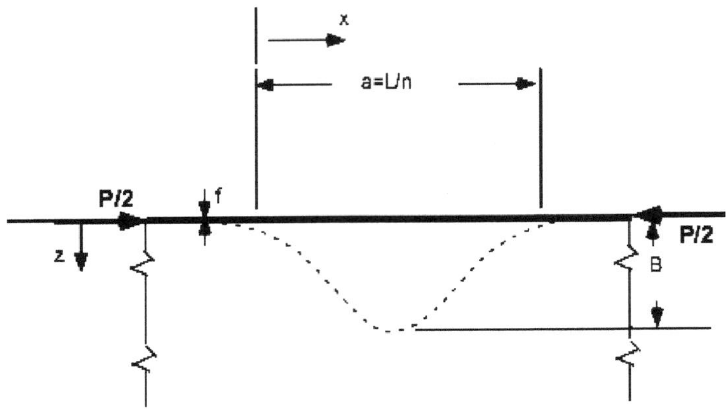

Figure 8.5 Plantema's wrinkling model.

waves over the length L of the panel, the transverse displacement expression (8.1) becomes

$$w = Be^{-kz}\sin\frac{n\pi x}{L}. \tag{8.43}$$

Strictly speaking, this equation implies that the core is infinitely thick. It is also assumed that vertical lines remain vertical during wrinkling ($\partial u/\partial z = 0$). Notice that no marginal depth is assumed in this model.

The transverse normal stress σ_{zz} and the shear stress τ_{xz} in the core are

$$\sigma_{zz} = E_c\frac{\partial w}{\partial z} \quad \text{and} \quad \tau_{xz} = G_c\frac{\partial w}{\partial x}. \tag{8.44}$$

The total strain energy per unit width consists of the bending strain energy in the face and the strain energy in the core associated with the stresses σ_{zz} and τ_{xz} computed from

$$\begin{aligned} U &= \frac{E_f I_f}{2}\int_0^L\left(\frac{\partial^2 w_f}{\partial x^2}\right)^2 dx + \frac{1}{2E_c}\int_0^L\int_0^\infty \sigma_{zz}^2 dxdz \\ &\quad + \frac{1}{2G_c}\int_0^L\int_0^\infty \tau_{xz}^2 dxdz \\ &= B^2\left(\frac{\pi^4 n^4}{4L^3}E_f I_f + \frac{kL}{8}E_c + \frac{\pi^2 n^2}{8kL}G_c\right), \end{aligned} \tag{8.45}$$

where $I_f = f^3/12$ is the moment of inertia of the face cross-section. For a plate, the first term in the above equation, $E_f I_f$ should be replaced by $E_f I_f/(1-\nu_f^2)$ where ν_f is the Poisson ratio of the face.

In the same manner as for Hoff and Mautner, the stress $\sigma_{xx,c}$ in the core is neglected in comparison with the stress $\sigma_{xx,f}$ in the faces, and the work done by the external load, $P/2$ per unit width, becomes

$$W = \frac{P}{4}\int_0^L \left(\frac{\partial w_f}{\partial x}\right)^2 dx = \frac{\pi^2 n^2 B^2}{8L} P. \tag{8.46}$$

The total potential is $U - W$ and it contains the parameters B, n and k. Minimizing this total potential with respect to B, n, and k, and assuming n to be a continuous variable, we obtain

$$a = \frac{L}{n} = 1.26\pi \left[\frac{(E_f I_f)^2}{E_c G_c}\right]^{1/6}; \quad k = \left[\frac{G_c^2}{2(E_f I_f) E_c}\right]^{1/3};$$
$$P = 3[2(E_f I_f) E_c G_c]^{1/3}. \tag{8.47}$$

Again, for a plate, $E_f I_f$ should be replaced by $E_f I_f/(1 - \nu_f^2)$.

Strictly speaking, these equations are valid only for a plate length $L = \infty$, but in practice, n will be so large that they may be used for finite values of L. The ratio $a = L/n$ is the half wavelength of the wrinkles.

For a sandwich plate of unit width, substituting $I_f = f^3/(1 - \nu_f^2)$, with $\nu_f = 0.3$, yields

$$a = 1.78 f \left(\frac{E_f^2}{E_c G_c}\right)^{1/6}; \quad k = \frac{1.76}{f}\left(\frac{G_c^2}{E_f E_c}\right)^{1/3};$$
$$P = 1.70 f (E_f E_c G_c)^{1/3}. \tag{8.48}$$

It should be noted that the wrinkling stress, $P/(2f)$, is independent of the face thickness, f.

8.1.3 Allen's Formula

Allen's (1969) formula is based on simplifying the face wrinkling problem to that of an infinitely long strut (face sheet) attached to an elastic medium (core), which extends to infinity on one side of the strut. To this extent, the geometry is the same as that in Plantema's model (Figure 8.5). The strut (representing the face sheet of a sandwich with a core of infinite thickness) is assumed to be of rectangular section, of thickness f and width b in the y direction.

The differential equation of the face is

$$D_f \frac{d^4 w_f}{dx^4} + (P/2)\frac{d^2 w_f}{dx^2} = b\sigma_{zz}^0, \tag{8.49}$$

where D_f is the flexural rigidity of the face, P is the axial force, w_f is the displacement of the face in the z direction, and $b\sigma_{zz}^0$ is the transverse load on the face (σ_{zz}^0 is the normal stress between face and core).

Suppose that the face buckles into sinusoidal waves with half-wavelength a, as in Section 8.2, i.e.

$$w_f = B \sin\frac{\pi x}{a}. \tag{8.50}$$

An Airy stress function approach is used to derive the normal stress in the core, σ_{zz}. In particular, the stress function

$$\phi(x, z) = A(1 - dz)e^{-\pi z/a} \sin\frac{\pi x}{a}, \tag{8.51}$$

satisfies the bi-harmonic

$$\frac{\partial^4 \phi}{\partial x^4} + 2\frac{\partial^4 \phi}{\partial x^2 \partial z^2} + \frac{\partial^4 \phi}{\partial z^4} = 0$$

and results in stresses

$$\sigma_{zz} = \frac{\partial^2 \phi}{\partial x^2} = -A(1 - dz)\frac{\pi^2}{a^2}e^{-\pi z/a} \sin\frac{\pi x}{a}, \tag{8.52a}$$

$$\sigma_{xx} = \frac{\partial^2 \phi}{\partial z^2} = \left[(1 - dz)\frac{\pi}{a} + 2d\right] A\frac{\pi}{a}e^{-\pi z/a} \sin\frac{\pi x}{a}. \tag{8.52b}$$

At the face/core interface, $z = 0$

$$\sigma_{zz}^0 = \left.\frac{\partial^2 \phi}{\partial x^2}\right|_{z=0} = -A\frac{\pi^2}{a^2} \sin\frac{\pi x}{a}, \tag{8.53a}$$

$$\sigma_{xx}^0 = \left.\frac{\partial^2 \phi}{\partial z^2}\right|_{z=0} = A\frac{\pi}{a}\left(2d + \frac{\pi}{a}\right) \sin\frac{\pi x}{a}. \tag{8.53b}$$

It is assumed that the face is attached to the surface of the core and is permitted to deform in the z direction only; there are no x displacements. Consequently, the axial strain at the face/core interface, e_{xx}^0, is zero, which gives

$$e_{xx}^0 = \left.\frac{\partial u}{\partial x}\right|_{z=0} = \frac{1}{E_c}\left(\sigma_{xx} - \nu_c \sigma_{zz}\right)|_{z=0}$$

$$= \frac{A}{E_c}\frac{\pi}{a}\left[2d + \frac{\pi}{a}(1 + \nu_c)\right] \sin\frac{\pi x}{a} = 0, \tag{8.54}$$

where E_c and ν_c are the modulus of elasticity and the Poisson ratio of the core.

Therefore, the constant d must take the value

$$d = -\frac{\pi}{2a}(1+\nu_c). \tag{8.55}$$

The displacement w is found by integrating the normal strain equation

$$\frac{\partial w}{\partial z} = e_{zz} = \frac{1}{E_c}\left(\sigma_{zz} - \nu_c \sigma_{xx}\right), \tag{8.56a}$$

resulting in

$$w = \left\{\left[(1+\nu_c)\frac{\pi}{a} + (\nu_c - 1)d\right]\frac{a}{\pi} - d(1+\nu_c)z\right\}\frac{A}{E_c}\frac{\pi}{a}e^{-\pi z/a}\sin\frac{\pi x}{a}. \tag{8.56b}$$

Substituting the expression for d, (8.55), and evaluating w at the face/core interface, $z = 0$, gives

$$w_f = w|_{z=0} = \frac{\pi}{2a}(1+\nu_c)(3-\nu_c)\frac{A}{E_c}\sin\frac{\pi x}{a}. \tag{8.57a}$$

Combining with (8.50) gives A in terms of B:

$$A = \frac{BE_c}{(1+\nu_c)(3-\nu_c)}\frac{2a}{\pi}, \tag{8.57b}$$

which can be substituted in the expression for σ^0_{zz} in (8.53a) which gives

$$\sigma^0_{zz} = -\frac{B}{a}\frac{2\pi E_c}{(3-\nu_c)(1+\nu_c)}\sin\frac{\pi x}{a}. \tag{8.58}$$

Substituting for w_f and σ^0_{zz} from Equations (8.50) and (8.58) into (8.49) yields an expression in which $(B \sin \pi x/a)$ cancels, leaving the result

$$D_f\frac{\pi^4}{a^4} - (P/2)\frac{\pi^2}{a^2} = -\frac{2\pi E_c b}{(3-\nu_c)(1+\nu_c)a}. \tag{8.59}$$

This equation defines the critical value of P for a given a.

It is convenient to write

$$D_f = \frac{E_f b f^3}{12}; \quad P/2 = \sigma_f b f, \tag{8.60}$$

where E_f is the modulus of elasticity of the face and σ_f is the compressive stress in the face. Then from (8.59) (and after cancelling b), we obtain

$$\sigma_f = \frac{\pi^2 E_f}{12}\left(\frac{f}{a}\right)^2 + \frac{2E_c}{\pi(3-\nu_c)(1+\nu_c)}\left(\frac{a}{f}\right), \tag{8.61}$$

i.e., a function of a/f.

There is one last step in order to determine the critical wrinkling stress. We have to minimize σ_f from (8.61) with respect to a/f:

$$\frac{d\sigma_f}{d(a/f)} = 0. \tag{8.62}$$

This gives the following values for the critical stress and the half-wavelength at which it occurs:

$$\sigma_f^{\mathrm{cr}} = B_1 E_f^{1/3} E_c^{2/3}; \quad \text{where} \quad B_1 = 3\left[12(3-\nu_c)^2(1+\nu_c)^2\right]^{-1/3}. \tag{8.63a}$$

$$\left(\frac{a}{f}\right)_{\mathrm{cr}} = C\left(\frac{E_f}{E_c}\right)^{1/3}; \quad \text{where} \quad C = \pi\left[(3-\nu_c)(1+\nu_c)/12\right]^{1/3}. \tag{8.63b}$$

8.1.4 The Winkler Elastic Foundation Approach

This approach assumes that the core supports the faces as an elastic foundation, i.e. an array of continuously distributed linear springs, as shown in Figure 8.6. In the anti-symmetrical case, it is seen that the springs remain unloaded even after wrinkling and, furthermore, the mode of deformation in the core is shear rather than tension/compression, which the set of springs is unable to model and, hence, no solution can be derived in this case.

In the symmetrical case, on the other hand, the model becomes more realistic since the mode of deformation in the core is both tension/compression and shear. We can refer again to Figure 8.5 and we suppose that the elastic foundation modulus is λ, which is the force needed to displace a unit area of the face through a unit distance in the z direction. Then the corresponding normal stress at the face/core interface (tension positive) is

$$\sigma_{zz}^0 = -\lambda w_f. \tag{8.64}$$

Substituting Equation (8.64) into the differential equation of the face, Equation (8.49), leads to

$$\frac{d^4 w_f}{dx^4} + \frac{(P/2)}{D_f}\frac{d^2 w_f}{dx^2} + \frac{b\lambda}{D_f} w_f = 0. \tag{8.65}$$

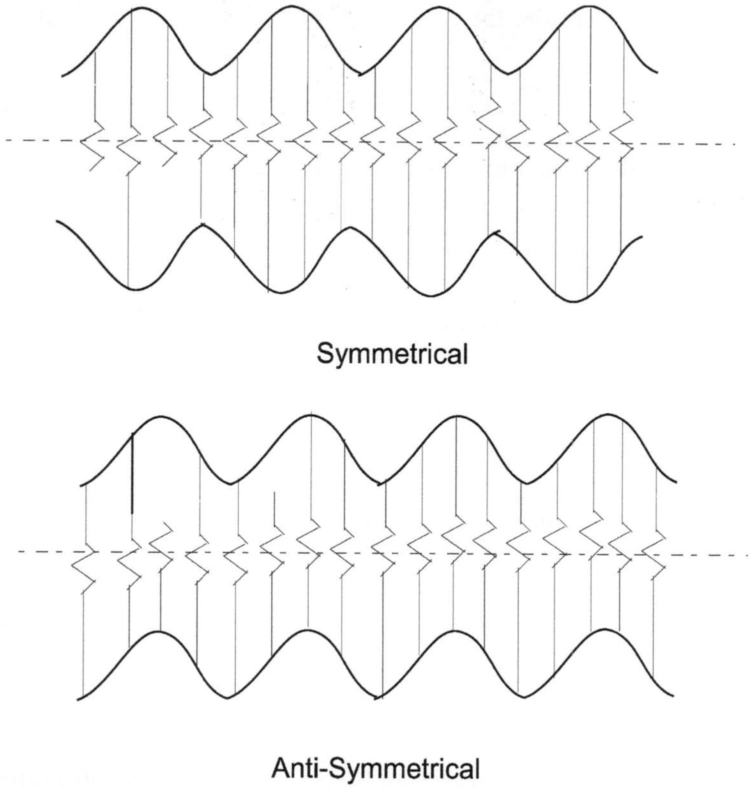

Figure 8.6 Winkler elastic foundation model.

The face will assume the same displacement as in Equation (8.50), which, when substituted into Equation (8.65), yields an expression for P:

$$P/2 = D_f \left(\frac{\pi}{a}\right)^2 + b\lambda \left(\frac{a}{\pi}\right)^2. \tag{8.66}$$

To find the critical load we must further minimize P with respect to the unknown wavelength, a, by setting $dP/da = 0$, which gives

$$a^4 = \frac{\pi^4 D_f}{b\lambda} \tag{8.67}$$

and the corresponding critical load from Equation (8.66) is

$$P_{\mathrm{cr}}/2 = 2\sqrt{D_f b\lambda}. \tag{8.68}$$

An expression for λ in terms of the properties of the core can be found as follows: since the springs are assumed to be linear, the core stress must be independent of z, therefore the core strain is also independent of z and equal to

$$\varepsilon_{zz}^{c}(x,z) = -\frac{\sigma_{zz}^{0}}{E_c} = -\frac{\lambda w_f(x)}{E_c}. \tag{8.69}$$

Integrating with respect to z gives the core displacement as

$$w_c(x,z) = -\frac{\lambda w_f(x)}{E_c} z + e(x). \tag{8.70}$$

Next we use the fact that w_c must be zero for $z = c$ (Figure 8.6), which gives:

$$e(x) = \frac{\lambda w_f(x)}{E_c} c. \tag{8.71}$$

Therefore,

$$w_c(x,z) = -\frac{\lambda w_f(x)}{E_c}(z - c). \tag{8.72}$$

At $z = 0$,

$$w_c(x,z)|_{z=0} = w_f, \tag{8.73}$$

which gives

$$\lambda = \frac{E_c}{c}. \tag{8.74}$$

This provides the critical load from Equation (8.68), after substituting the expression for D_f from (8.60), as

$$P_{\mathrm{cr}}/2 = 2bf\sqrt{\frac{E_f E_c f}{12c}}, \tag{8.75}$$

or the critical stress in the face

$$\sigma_{\mathrm{cr}}^{f} = \frac{P_{\mathrm{cr}}/2}{bf} = 2\sqrt{\frac{E_f E_c f}{12c}}. \tag{8.76}$$

8.1.5 Example and Comparison of the Wrinkling Formulas

Let us now perform a comparison of the predictions from these different formulas. We consider a sandwich with isotropic face and core materials with $E_f/E_c = 1{,}000$ and $\nu_c = 0$ (this case has been historically emphasized in the early sandwich literature). Also, the plate length to total thickness ratio, $L/h = 5$. We will examine a range of face thicknesses defined by the

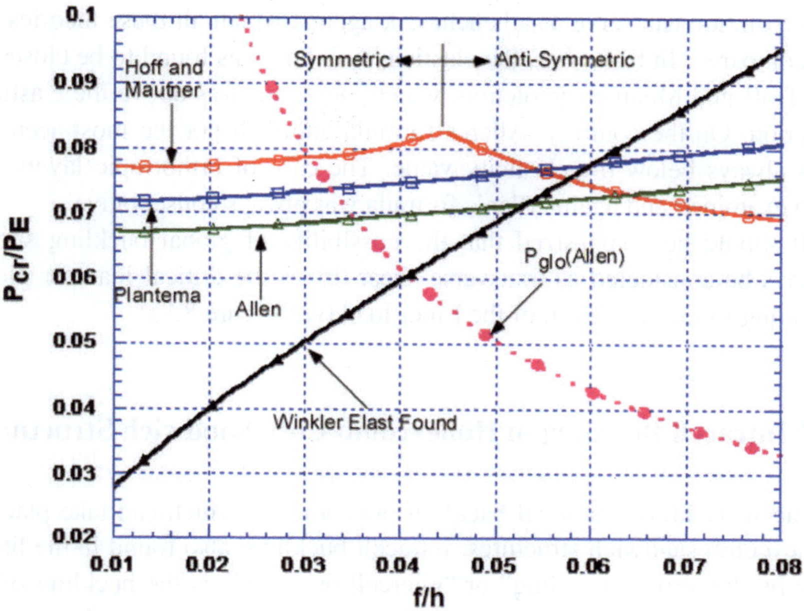

Figure 8.7 Wrinkling load calculated from different formulas for a range of thickness ratios, f/h.

ratio of face sheet thickness over total thickness, f/h, between 0.01 and 0.08. Figure 8.7 shows the critical wrinkling load, normalized with the Euler load for a simply supported configuration (without transverse shear), $P_E = \pi^2(EI)_{eq}/L^2$. The figure also shows the critical global buckling load, based on Allen's formula, which includes transverse shear (see Chapter 7).

We can observe the following: (a) for this configuration, wrinkling definitely dominates for ratios f/h below about 0.03. Above this level, global buckling would dominate, although there is a "gray" area between 0.03 and about 0.04, in which global buckling dominance would depend on the wrinkling formula used; (b) the Winkler elastic foundation formula seems to be outside the "cluster" of the other formulas; (c) for the range of f/h where wrinkling dominates, all formulas with the exception of the Winkler formula, are close to each other; among these, the Allen formula is most conservative and Hoff and Mautner's the least; (d) based on Hoff and Mautner's approach, there is a switch from symmetric to anti-symmetric wrinkling at a ratio f/h of about 0.044. This is very evident on the curve from the marked discontinuity in the slope of the curve at this point.

An elasticity solution of the wrinkling problem was presented by Kardomateas (2005) and the same configuration as the above was examined.

This solution can serve as a benchmark against which all these theories can be compared. In that work, the elasticity solution was found to be closest to the Hoff and Mautner prediction, which, however, was above the elasticity solution. On the contrary, Allen's formula although not the most accurate, was always below the elasticity value. The case of orthotropic layers was also examined and again Allen's formula was always conservative.

It should be emphasized that the possibility of global buckling should always be considered as transverse shear drives the critical load for global buckling to just a fraction of the Euler load (see Figure 8.7).

8.2 Intracell Buckling in Honeycomb Core Sandwich Structures

Intracell buckling is a local buckling phenomenon which can take place in honeycomb sandwich structures. Intracell buckling, also found in the literature by the terms "dimpling" or "intercell buckling", is the buckling of the face sheet within an individual honeycomb cell. Although the honeycomb structure will retain part of its load-carrying capacity after intracell buckling has occurred, the buckled face sheet may have undesirable influences, for example, it can adversely affect the aerodynamic properties of the structure; thus, it is important to know the load at which this intracell buckling occurs.

A contrast can be made with the previously treated phenomenon of wrinkling, which is local face sheet buckling over a row of several honeycomb cells (i.e., the buckling wavelength is greater than the honeycomb cell size). Wrinkling would appear as a sharp trough on the face sheet and would be accompanied either by the crushing of the honeycomb core or the separation of the core and face sheet at this sharp trough. When wrinkling occurs, for all practical purposes, the honeycomb structure has lost its load carrying capacity.

The oldest intracell buckling formula is the one suggested by Norris and Krommer (1950), as derived empirically from tests. An early theoretical investigation was also done by Weikel and Kobayashi (1959).

8.2.1 The Norris Formula

We shall consider two cases of a sandwich panel with square, honeycomb cells loaded in uniaxial compression: (a) the honeycomb cells are oriented such that this load is parallel to the diagonal of the cell (Figure 8.8) and

(b) the honeycomb cells are oriented such that this load is parallel to a cell wall (Figure 8.9). In both cases, we assume that the honeycomb cell is square in shape, a being the length of each side. We also assume that, in both cases, the face sheet is simply supported along its four edges at the cell and that the thickness of the honeycomb core is large compared to the cell size.

Within these assumptions, let us approximate the face sheet deflection by a one-term double trigonometric series:

$$w = A \sin \frac{\pi x}{a} \sin \frac{\pi y}{a}. \tag{8.77}$$

In connection with this assumed displacement profile (8.77), we generally assume a deflection of the form $\sin(m\pi x/a)\sin(n\pi y/b)$, where a and b are the sides of the rectangle. However, for the simple square geometry considered here, the minimum load is obtained for $m = n = 1$.

Denoting by D_f the flexural rigidity of the face sheet and by ν_f Poisson's ratio of the face sheet (considered isotropic), the strain energy due to bending of the face sheet is

$$U_B = \frac{D_f}{2} \int_0^a \int_0^a \left[w_{,xx}^2 + w_{,yy}^2 + 2\nu_f w_{,xx} w_{,yy} + 2(1 - \nu_f) w_{,xy}^2 \right] dx dy. \tag{8.78}$$

The work done by the external force is

$$W = \int_0^a \int_0^a \left(N_{xx} \frac{w_{,x}^2}{2} + N_{yy} \frac{w_{,y}^2}{2} + N_{xy} w_{,x} w_{,y} \right) dx dy. \tag{8.79}$$

For both loading configurations, the strain energy of bending is

$$U_B = \frac{D_f \pi^4}{2a^2} A^2. \tag{8.80}$$

The first loading configuration yields (Figure 8.8): $N_{xx} = N_{yy} = N/\sqrt{2}$ and $N_{xy} = -N/\sqrt{2}$, therefore the work done by the external force is

$$W = \frac{N \pi^2 \sqrt{2}}{8} A^2. \tag{8.81}$$

Setting the work done by the external forces equal to the strain energy of bending, $U_B = W$, and substituting the expression for the bending rigidity of the face, $D_f = E_f f^3 / [12(1 - \nu_f^2)]$, gives the critical load

$$N_{\text{cr}} = \frac{E_f}{1 - \nu^2} f \left(\frac{f}{a} \right)^2 \frac{\pi^2}{3\sqrt{2}} \simeq 2.3 \frac{E_f}{1 - \nu^2} f \left(\frac{f}{a} \right)^2. \tag{8.82}$$

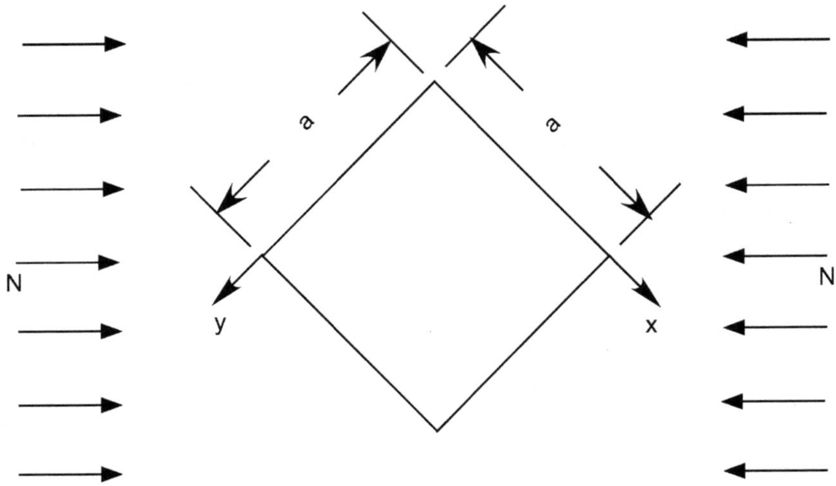

Figure 8.8 Intracell buckling model with the load parallel to the diagonal of the honeycomb cell.

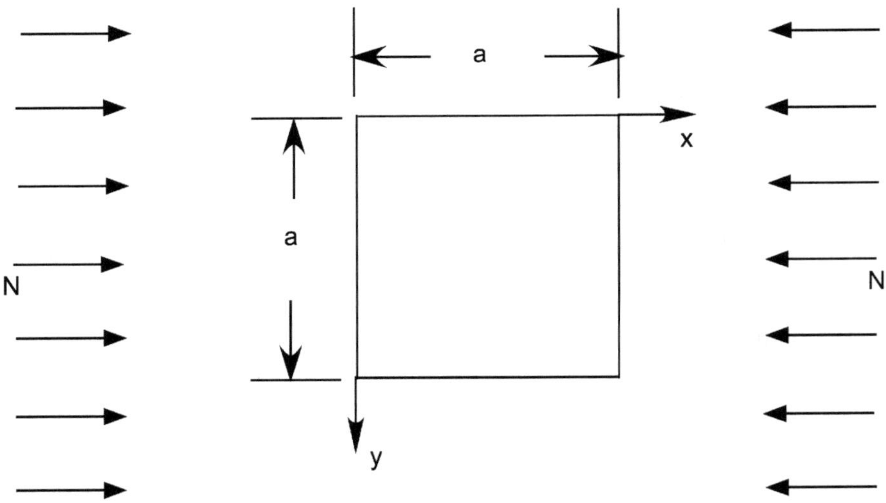

Figure 8.9 Intracell buckling model with the load parallel to a cell wall of the honeycomb core.

The second loading configuration yields (Figure 8.9): $N_{xx} = N$, $N_{yy} = N_{xy} = 0$, therefore

$$W = \frac{N\pi^2}{8}A^2. \tag{8.83}$$

Again, setting the work done by the external forces equal to the strain energy of bending, and substituting the expression for the bending rigidity of the face, gives the critical load

$$N_{\rm cr} = \frac{E_f}{1-\nu^2} f \left(\frac{f}{a}\right)^2 \frac{\pi^2}{3} \simeq 3.3 \frac{E_f}{1-\nu^2} f \left(\frac{f}{a}\right)^2. \tag{8.84}$$

Norris provided the formula in terms of an empirical factor K_d, as

$$N_{\rm cr} = K_d \frac{E_f}{1-\nu^2} f \left(\frac{f}{s}\right)^2, \tag{8.85}$$

where s is the diameter of the circle that can be inscribed in the honeycomb cell and K_d was specified to be about $K_d \simeq 2$. Obviously, for square cells, $s = a$.

From the two derived formulas (8.82) and (8.84), it can be seen that the orientation of the cells is important, i.e. it makes a difference whether the uniaxial compression is parallel to the diagonal of the cells (factor of 2.3 in place of K_d) or parallel to a side of the cells (factor of 3.3 in place of K_d).

Weikel and Kobayashi (1959) examined the conditions that would make intracell buckling dominate over wrinkling and they concluded that intracell buckling would dominate over wrinkling for a value of $\lambda a^4/D_f > 100$, where λ is the apparent core stiffness (when viewed as an elastic foundation, see Equation (8.74)). In other words, the predominant mode of buckling failure will be wrinkling for a relatively soft core, but with increased core stiffness, λ, intracell buckling becomes the predominant mode of local buckling failure.

8.2.2 The Fokker Dimpling Formula

The foregoing formulas can be easily revised for the case of orthotropic faces. Let us consider the case of the honeycomb cells oriented such that the applied compressive load is parallel to a cell wall (Figure 8.9). If $x \equiv 1$, $y \equiv 2$, then the strain energy of bending for an orthotropic plate is

$$U_B = \frac{1}{2}\int_0^a \int_0^a \left[D_{11} w_{,xx}^2 + D_{22} w_{,yy}^2 + 2D_{12} w_{,xx} w_{,yy} + 4D_{66} w_{,xy}^2\right] dxdy, \tag{8.86a}$$

where

$$D_{11} = \frac{E_1 f^3}{12(1-\nu_{12}\nu_{21})}; \quad D_{22} = \frac{E_2 f^3}{12(1-\nu_{12}\nu_{21})}, \tag{8.86b}$$

$$D_{12} = D_{11}\nu_{21}; \quad D_{66} = \frac{G_{12}f^3}{12}. \tag{8.86c}$$

Therefore, by using the displacement profile (8.77), the strain energy of bending is

$$U_B = (D_{11} + D_{22} + 2D_{12} + 4D_{66})\frac{\pi^4}{8a^2}A^2. \tag{8.87}$$

The work done by the external load is the same as in the isotropic face case and is given by (8.83). Equating the strain energy of bending with the work done gives the critical load as

$$N_{\text{cr}} = [D_{11} + 2(D_{12} + 2D_{66}) + D_{22}]\frac{\pi^2}{a^2}. \tag{8.88}$$

This is known as the "Fokker Dimpling Formula", as it was first suggested in a study conducted by the Fokker Aircraft BV for the European Space Agency (Blaas et al., 1984).

For the case of the honeycomb cells oriented such that the applied compressive load is parallel to the diagonal of the cell (Figure 8.8), we have to transform the elastic constants to the x and y system oriented at an angle $\phi = 45°$ to the principal elastic axes of the face (Lekhnitskii, 1968):

$$\begin{aligned} D'_{11} &= D_{11}\cos^4\phi + 2(D_{12} + 2D_{66})\sin^2\phi\cos^2\phi + D_{22}\sin^4\phi \\ &= \frac{1}{4}(D_{11} + D_{22} + 2D_{12} + 4D_{66}), \end{aligned} \tag{8.89a}$$

$$D'_{22} = D_{11}\sin^4\phi + 2(D_{12} + 2D_{66})\sin^2\phi\cos^2\phi + D_{22}\cos^4\phi = D'_{11}, \tag{8.89b}$$

$$\begin{aligned} D'_{12} &= D_{12} + [D_{11} + D_{22} - 2(D_{12} + 2D_{66})]\sin^2\phi\cos^2\phi \\ &= \frac{1}{2}D_{12} + \frac{1}{4}(D_{11} + D_{22} - 4D_{66}), \end{aligned} \tag{8.89c}$$

$$\begin{aligned} D'_{66} &= D_{66} + [D_{11} + D_{22} - 2(D_{12} + 2D_{66})]\sin^2\phi\cos^2\phi \\ &= \frac{1}{4}(D_{11} + D_{22} - 2D_{12}), \end{aligned} \tag{8.89d}$$

$$\begin{aligned} D'_{16} &= \frac{1}{2}[D_{22}\sin^2\phi - D_{11}\cos^2\phi + (D_{12} + 2D_{66})\cos 2\phi]\sin 2\phi \\ &= \frac{1}{4}(D_{22} - D_{11}), \end{aligned} \tag{8.89e}$$

$$D'_{26} = \frac{1}{2}[D_{22}\cos^2\phi - D_{11}\sin^2\phi - (D_{12} + 2D_{66})\cos 2\phi]\sin 2\phi = D'_{16}. \tag{8.89f}$$

In this case, the plate behaves as being anisotropic rather than orthotropic. The strain energy of bending is

$$U_B = \frac{1}{2}\int_0^a \int_0^a [D'_{11}w_{,xx}^2 + D'_{22}w_{,yy}^2 + 2D'_{12}w_{,xx}w_{,yy} + 4D'_{66}w_{,xy}^2$$
$$+ 4D'_{16}w_{,xx}w_{,xy} + 4D'_{26}w_{,yy}w_{,xy}]dxdy. \tag{8.90}$$

With the displacement profile (8.77), the strain energy of bending becomes

$$U_B = (D'_{11} + D'_{22} + 2D'_{12} + 4D'_{66})\frac{\pi^4}{8a^2}A^2. \tag{8.91}$$

Equating with the work done by the external load, Equation (8.81), gives the critical load as

$$N_{\text{cr}} = (D'_{11} + D'_{22} + 2D'_{12} + 4D'_{66})\frac{\pi^2}{a^2\sqrt{2}}. \tag{8.92}$$

Substituting the expressions of the transformed elastic constants from (8.89a–d) we obtain

$$N_{\text{cr}} = 2(D_{11} + D_{22})\frac{\pi^2}{a^2\sqrt{2}}. \tag{8.93}$$

Equation (8.93) is a newly derived formula, given for the first time herein, that can be used with orthotropic faces when the honeycomb cells are oriented such that the applied compressive load is parallel to the diagonal of the cell (Figure 8.8).

Chapter 9
Fracture Mechanics Analysis of Face/Core Debonds

The superior performance of light-weight sandwich structures requires that the face sheets be successfully bonded to the core. Lack of bonding, or inadequate bonding, will compromise the transfer of shear stress between the face and core, and if debonding occurs over a large area, the debond is likely to grow further. It is also obvious that the face/core adhesion may vary in a large panel with composite face sheets due to inadequate wet-out of the face fabrics resulting in "islands" of poor face/core bonding. Service loads are also known to be a potential source for face/core debonding, in particular wave-slamming loads on the bottom of a ship hull or hard object impact loads transverse to the surface of a sandwich structure.

The face/core interface in a sandwich panel may be toughened by the addition of a thin low fiber content fabric layer such as a continuous fiber mat (CFM) or a chopped strand mat (CSM). Such a mat will provide a transition zone between the high modulus face sheets and the low modulus core and may improve the face/core adhesion. The preparation of the core bonding surfaces is also very important, as discussed by Pfund (2005). Ideally, the critical link for the separation of the face and core should not be the actual interface between face and core. If the adhesive layer is strong and tough, debonding tends to occur in the face sheet in the form of delamination or inside the core, see the scheme in Figure 9.1.

The morphology of the face/core interface is thus an important factor. The actual morphology depends of the core material and the manner in which the face sheets are attached to the core. For example, if fiber composite faces are used, fabric preforms may be injected with resin, forming the face sheets and the face/core bond simultaneously creating an integral, co-cured sandwich structure (Advani and Sozer, 2002). Alternatively, preformed composite or metal faces may be attached to the core using adhesive bonding procedures.

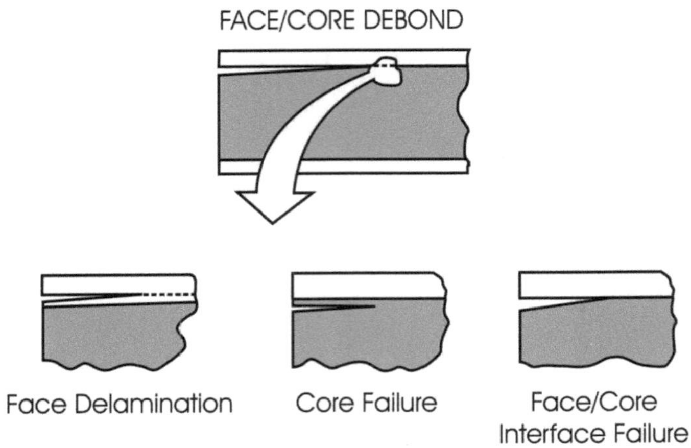

Figure 9.1 Crack growth scenarios in foam core sandwich.

The face/core debond is commonly represented as a sharp discontinuity, i.e. a crack, and the condition when this crack will propagate are analyzed using fracture mechanics principles. This chapter aims to review the fracture mechanics of face/core interface cracks and analysis of the crack path in sandwich-test specimens. For sandwich specimens that fail after developing large fiber bridging zones behind the crack tip, linear elastic fracture mechanics becomes inadequate and cohesive zone models may better represent the fracture process.

9.1 Linear Elastic Fracture Mechanics Concepts

Fracture mechanics is a science developed to analyze the tendency for a pre-existing crack in a structure to grow as a result of applied external loads. The presence of a crack reduces the strength of the structure and, if the crack is sufficiently long, the structure will fall below the designed limit load. The assessment of the defect criticality for the structural performance is a major objective of fracture mechanics. Another common situation where fracture mechanics has been found extremely useful is in the prediction of the maximum crack size that can be allowed in a structure.

The field of fracture mechanics traditionally involves applied mechanics and materials science. Applied mechanics relates external loads applied to a flawed, or cracked, structural component to crack-tip stress fields and elastic deformations of the material in the vicinity of the crack tip. Materials

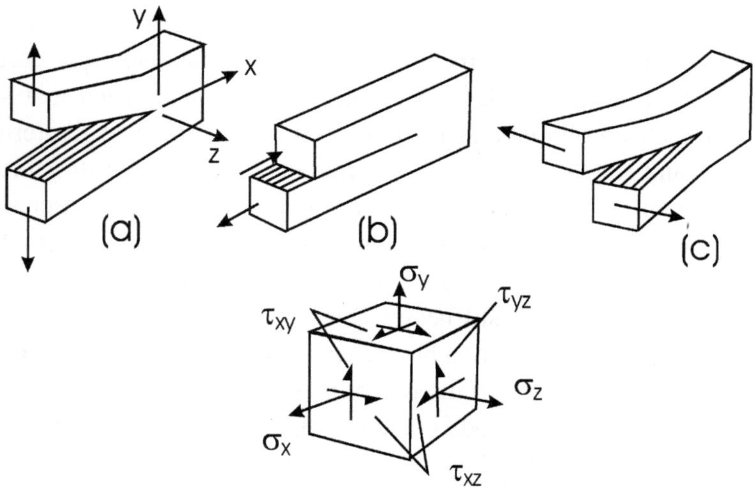

Figure 9.2 Modes of crack loading and definition of stress components: (a) mode I (opening), (b) mode II (forward shear), and (c) mode III (tearing).

science, on the other hand, views fracture mechanics as a means of characterizing the fracture resistance of a material and this understanding will aid in the development of improved processing strategies and materials design.

Most published work on fracture mechanics concerns fracture of isotropic materials such as metals and polymers (Ewalds and Wanhill, 1989; Kinloch and Young, 1983) although there are more recent compilations on fracture mechanics of composites (e.g., Friedrich, 1989; Armanios, 1996). The review article by Hutchinson and Suo (1992) is an excellent source for fracture mechanics of orthotropic and layered dissimilar materials.

Fracture mechanics analysis is most commonly concerned about the opening mode of fracture, i.e. mode I, associated with opening displacements of the crack faces and tensile stresses near the crack tip (see Figure 9.2a). In general, analysis of the forward shear (mode II) and tearing modes (mode III), illustrated in Figures 9.2b and c, is required. Figure 9.2 also shows a commonly used xyz coordinate system and the associated stress element with σ_y, τ_{xy} and τ_{yz}, being the dominant mode I, mode II, and mode III stresses. As discussed by Hutchinson and Suo (1992), an interface crack between two dissimilar materials is inherently loaded in mixed mode because of the mismatch in material properties across the crack interface. This means that both normal and shear stresses act on the interface in front of the crack tip.

We will first consider a true interface crack between a homogeneous, isotropic or orthotropic face sheet and a homogeneous, isotropic core, i.e., the case shown at the far right in Figure 9.1. It is recognized that such a crack constitutes a "bimaterial interface crack" according to the terminology of Suo and Hutchinson (1990). For foam-cored sandwich structures, the combination of stiff faces and a soft core is associated with a tremendous stiffness mismatch. Face-to-core modulus ratios in the range from 100 to 1,000 are common. A bimaterial crack displays some peculiar behavior due to the mismatch in elastic stiffnesses across the crack plane. Mixed mode loading is inherent at the tip of a bimaterial interface crack even if the global load is pure mode I. The asymmetries of moduli and Poisson's ratios across the interface cause mixed mode loading at the crack tip because tensile and shear stresses must appear along the interface to maintain continuity in displacements between the two materials. The stresses near the crack tip experience the typical square root singularity of homogeneous fracture and, in addition, oscillatory behavior (Williams, 1959; England, 1965; Erdogan, 1965; Rice and Sih, 1965). Hutchinson (1990) proposed the following complex representation of the interfacial normal and shear stresses (σ_y and τ_{xy}), see Figure 9.3, ahead of the crack tip ($\theta = 0$):

$$\sigma_y + i\tau_{xy} = \frac{(K_1 + iK_2)x^{i\varepsilon}}{\sqrt{2\pi x}}, \tag{9.1}$$

where x is the distance from the tip along the interface, $i = \sqrt{-1}$, and $K = K_1 + iK_2$ is called "the complex stress intensity factor" for an interfacial crack (Hutchinson, 1990). Notice here that K_1 and K_2 do not represent the opening and sliding modes as K_I and K_{II} do in classical fracture mechanics and no simple physical interpretation of K_1 and K_2 exists. The parameter ε is called the "oscillatory index" (Hutchinson, 1990), defined in terms of one of Dundurs' (1969) elastic mismatch parameters, β,

$$\varepsilon = \frac{1}{2\pi} \ln\left(\frac{1-\beta}{1-\beta}\right), \tag{9.2a}$$

$$\beta = \frac{1}{2}\frac{G_1(1-2\nu_2) - G_2(1-2\nu_1)}{G_1(1-\nu_2) + G_2(1-\nu_1)}, \tag{9.2b}$$

where subscripts 1 and 2 on the shear modulus G and Poisson's ratio ν represent the (isotropic) materials above and below the crack plane. The parameter β vanishes if both materials are identical ($G_1 = G_2$), and when both materials are incompressible ($\nu_1 = \nu_2 = 1/2$).

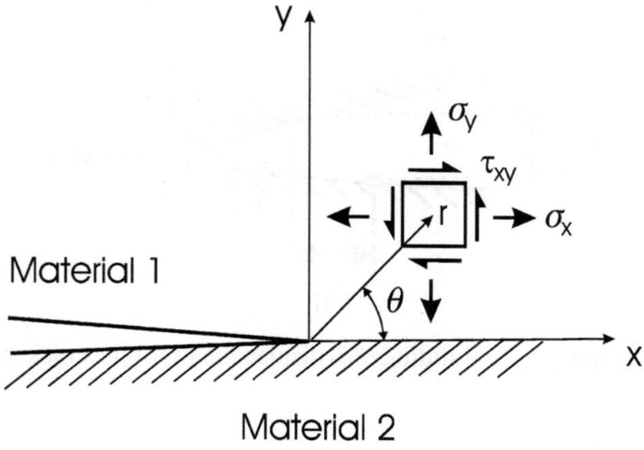

Figure 9.3 Stress element near a bimaterial crack tip.

With $x^{i\varepsilon} = \cos(\varepsilon \ln x) + i \sin(\varepsilon \ln x)$, it is noted that $x^{i\varepsilon}$ in Equation (9.1) is a function that changes sign infinitely often, thus oscillating in a violent manner as the distance x tends to zero. Also, the near tip displacements of the crack face behind the crack tip display oscillatory behavior.

He and Hutchinson (1989) argued that the oscillation is a mathematical artifact of the elasticity solution and showed that the oscillatory region is usually exceedingly small. They proposed a consistent, pragmatic approach to suppress the role of the oscillations by assuming $\varepsilon = 0 = \beta$ both in the evaluation of fracture toughness and in the subsequent application of the toughness data for prediction of crack propagation. For this case, the mode I and mode II stress intensity factors, K_{I} and K_{II}, assume their classical, conventional relation to the opening (σ_y) and shear stresses (τ_{xy}) at the interface ahead of the crack tip ($\theta = 0$).

Suo (1990) examined a crack between two orthotropic solids (or one orthotropic and one isotropic) and determined the near tip stress and displacement fields. For the case where $\varepsilon = \beta = 0$, the following expressions for the stresses and crack flank displacements were derived:

$$\left(\frac{H_{22}}{H_{11}}\right)^{1/2} \sigma_y + i\tau_{xy} = \frac{K_{\text{I}} + iK_{\text{II}}}{\sqrt{2\pi x}}, \tag{9.3a}$$

$$\left(\frac{H_{11}}{H_{22}}\right)^{1/2} \delta_{\text{I}} + i\delta_{\text{II}} = \frac{2H_{11}(K_{\text{I}} + iK_{\text{II}})\sqrt{x}}{\sqrt{2\pi}}, \tag{9.3b}$$

where the stress components are defined in the element shown in Figure 9.3. δ_{I} and δ_{II} are the opening (y directional) and sliding (x directional) displace-

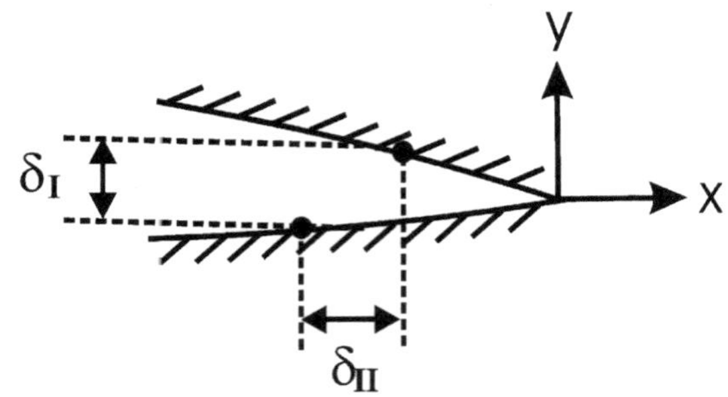

Figure 9.4 Opening (δ_I) and sliding (δ_{II}) displacements of the crack surfaces behind the crack tip.

ments of a point located at a distance, x, behind the crack tip before the cracked structure is loaded, see Figure 9.4.

The parameters H_{11} and H_{22} are defined in terms of the orthotropic compliances (b_{ij}) of the materials above and below the crack plane, Figure 9.3,

$$H_{11} = \sum_{i=1}^{2} \left[2n\lambda^{1/4}(b_{11}b_{22})^{1/2}\right]_i, \tag{9.4a}$$

$$H_{22} = \sum_{i=1}^{2} \left[2n\lambda^{-1/4}(b_{11}b_{22})^{1/2}\right]_i, \tag{9.4b}$$

where $i = 1$ for the material above the crack, and $i = 2$ for the material below the crack, and

$$n_i = \left[\frac{1}{2}(1 + \rho_i)\right]^{1/2}. \tag{9.5}$$

λ and ρ are defined below in terms of b_{ij}, where the subscripts i, j refer to the material description and assume values of 1, 2, and 6,

$$\lambda = \frac{b_{11}}{b_{22}}, \tag{9.6a}$$

$$\rho = \frac{b_{12} + b_{66}/2}{\sqrt{b_{11}b_{22}}}, \tag{9.6b}$$

$$b_{ij} = \begin{cases} s_{ij} & \text{(plane stress)}, \\ s_{ij} - \dfrac{s_{i3}\,s_{j3}}{s_{33}} & \text{(plane strain)}. \end{cases} \tag{9.6c}$$

s_{ij} ($i, j = 1, 2, 6$) are the compliance elements defined by, e.g., Hyer (1998), for a specially orthotropic material with the 1 direction along the x axis, 2 direction along the y axis, and 3 direction along the z axis.

$$s_{11} = \frac{1}{E_1}, \quad s_{12} = -\frac{\upsilon_{12}}{E_1}, \quad s_{13} = -\frac{\upsilon_{13}}{E_1},$$
$$s_{22} = \frac{1}{E_2}, \quad s_{23} = -\frac{\upsilon_{23}}{E_2}, \quad s_{33} = \frac{1}{E_3},$$
$$s_{44} = \frac{1}{G_{23}}, \quad s_{55} = \frac{1}{G_{13}}, \quad s_{66} = \frac{1}{G_{12}}. \tag{9.7}$$

Once the stress intensity factors are determined, the energy release rate is obtained according to Suo (1990),

$$G = \frac{H_{11}}{4}[K_{\mathrm{I}}^2 + K_{\mathrm{II}}^2]. \tag{9.8}$$

If both materials above and below the crack plane are isotropic, Equation (9.3a) simplifies to (9.1), and (9.3b) simplifies to

$$\delta_{\mathrm{I}} + i\delta_{\mathrm{II}} = 4\sqrt{\frac{x}{2\pi}}\left(\frac{1}{\bar{E}_1} + \frac{1}{\bar{E}_2}\right)(K_{\mathrm{I}} + iK_{\mathrm{II}}), \tag{9.9}$$

where $\bar{E} = E/(1-\nu^2)$ for plane strain, where ν is Poisson's ratio, and $\bar{E} = E$ for plane stress. Subscripts 1 and 2 denote the materials above and below the crack plane. For this case the energy release rate, Equation (9.8), becomes

$$G = \frac{1}{2}\left(\frac{1}{\bar{E}_1} + \frac{1}{\bar{E}_2}\right)(K_{\mathrm{I}}^2 + K_{\mathrm{II}}^2). \tag{9.10}$$

If both materials above and below the crack plane are the same, and if they are isotropic, the crack face displacements become

$$\delta + i\delta_{\mathrm{II}} = 8\sqrt{\frac{x}{2\pi}}\frac{(K_{\mathrm{I}} + K_{\mathrm{II}})}{\bar{E}} \tag{9.11}$$

and the energy release rate becomes

$$G = \frac{K_{\mathrm{I}}^2 + K_{\mathrm{II}}^2}{\bar{E}}. \tag{9.12}$$

It should be pointed out that the area of bimaterial fracture mechanics is still emerging. Several approaches, such as the method of distributed dislocations to represent a crack-like discontinuity, have emerged since the pioneering works of Eshelby et al. (1953) and Stroh (1958). Ting (1986) extended this approach to anisotropic materials and Kardomateas and co-workers applied it to bimaterial cracks between dissimilar anisotropic materials (see Huang and Kardomateas, 2001; Liu et al., 2004; Li and Kardomateas, 2006). This last paper also considers crack branching (crack kinking), which is an important failure mode of sandwich structures, see Section 9.2.

9.1.1 Finite Element Crack Closure Method

Several studies of debonding and delamination of composite materials express the crack loading in terms of the energy release rate, G. Since the fracture toughness, G_c, typically depends on the fracture mode, it is common to separate G into "components" corresponding to the opening and shearing modes of fracture, i.e.,

$$G = G_{\mathrm{I}} + G_{\mathrm{II}} + G_{\mathrm{III}}. \tag{9.13}$$

Theoretically, the mode separations is based on Irwin's contention that if the crack extends by a small amount, Δa, the energy absorbed in the process is equal to the work required to close the crack to its original length (Irwin, 1958). This equality is expressed in the following crack closure integrals

$$G_{\mathrm{I}} = \lim_{\Delta a \to 0} \frac{1}{2\Delta a} \int_0^{\Delta a} \sigma_y(\Delta a - r, 0)\bar{v}(r, \pi)dr, \tag{9.14a}$$

$$G_{\mathrm{II}} = \lim_{\Delta a \to 0} \frac{1}{2\Delta a} \int_0^{\Delta a} \tau_{xy}(\Delta a - r, 0)\bar{u}(r, \pi)dr, \tag{9.14b}$$

$$G_{\mathrm{III}} = \lim_{\Delta a \to 0} \frac{1}{2\Delta a} \int_0^{\Delta a} \tau_{yz}(\Delta a - r, 0)\bar{w}(r, \pi)dr, \tag{9.14c}$$

where r is the radial distance from the crack tip, σ_y, τ_{xy}, and τ_{yz} are the normal and shear stresses ahead of the crack tip, and $\bar{v}$, $\bar{u}$ and $\bar{w}$ are the relative opening and sliding displacements between points on the crack faces behind the crack tip.

The mode separation assumes that the crack will extend in a planar fashion, which it may or may not. For mixed mode loading, the crack may tend to deflect (kink) and propagate in a different plane than the original reference plane. This will be discussed in the next section. Furthermore, as pointed out by several authors, if the crack is at a bimaterial interface, the oscillations of stresses and displacements near the crack tip discussed early in this chapter will make the separation of G into components (Equations (9.13) and (9.14)) problematic (Raju et al., 1988).

The crack closure integrals in Equations (9.14) have been implemented into a finite element computation by Rybicki and Kanninen (1977). This method is called "the finite element crack closure method" and has been widely applied to fracture problems, see the review by Kruger (2004). Berggreen (2004) developed a method to extract the stress intensity factors from a finite element solution of the near crack-tip displacements called the

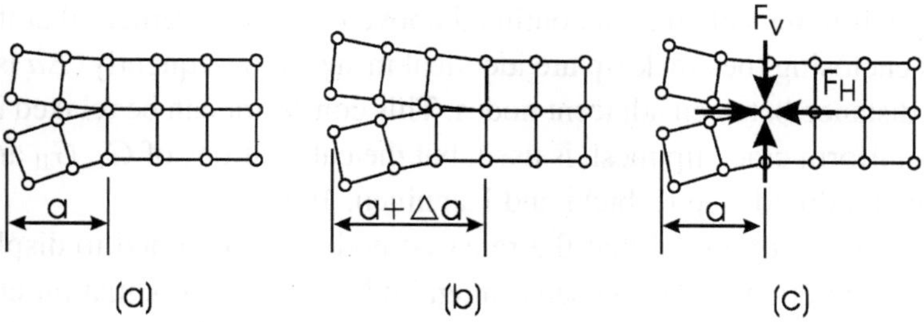

Figure 9.5 The crack-closure technique in finite elements. F_H and F_V are horizontal and vertical forces applied to close the crack. (a) Original configuration, (b) configuration after release of crack-tip nodes, (c) the released nodes are brought back to their initial position by application of nodal forces.

"crack surface displacement extrapolation method". This method has been applied to several bimaterial crack problems.

Figure 9.5 illustrates the crack-closure method applied to a finite-element mesh of a mixed mode I and mode II two-dimensional problem. A certain crack extension Δa may be introduced in the finite-element mesh by releasing duplicate nodes at the crack tip. The resulting deformations are the relative opening and sliding crack-tip deformations. By applying forces to the released nodes in two orthogonal directions, it is possible to close the crack tip to its original state. The products of crack-tip nodal displacements and forces enable G_{I} and G_{II} to be evaluated numerically.

For the plane situation illustrated in Figure 9.5, G_{I} and G_{II} become

$$G_{\mathrm{I}} = \frac{F_V \Delta v}{2b\Delta a}, \tag{9.15a}$$

$$G_{\mathrm{II}} = \frac{F_H \Delta u}{2b\Delta a} \tag{9.15b}$$

where b is the width, F_H and F_V are the horizontal and vertical crack-closure forces, and Δu and Δv are the horizontal and vertical increments of displacements required to bring the released nodes to their original positions. Notice that the far-field load applied is held constant in the steps illustrated in Figure 9.5. For mode III, G_{III} is similarly obtained as

$$G_{\mathrm{III}} = \frac{F_T \Delta w}{2b\Delta a} \tag{9.16}$$

where F_T is the magnitude of the crack-closure forces applied parallel to the crack front and Δw is the z directional incremental displacement required to close the crack.

Implicit in the calculations outlined above is the requirement that the elements enclosing the crack-tip are identical in size. Consequently, Δa is equal to the distance between adjacent nodes. This constraint can be relaxed in case a non-uniform crack tip mesh is used, but the calculations of G_{I}, G_{II} and G_{III} must be modified (see Rybicki and Kanninen, 1977).

It should be observed that the released nodes are assumed to displace in opposite directions with the same magnitude. This is the situation encountered in symmetric crack geometries and specimens. For asymmetric crack geometries and cracks between dissimilar materials, the displacements of the released nodes may no longer be of equal magnitude and the procedure will have to be modified. The basic principle, as for the homogeneous and symmetric case discussed above, is the calculation of the work required to bring the nodes back to their original position, before they were released. For further details on the crack closure method, see, e.g. Kruger (2004).

9.2 Crack Kinking Analysis

If a sandwich structure that contains a face/core debond is loaded, the debond may propagate under certain conditions governed by the crack driving force and the fracture resistance of the material around the tip of the debond. It has been observed that face/core debonds in foam-cored sandwich specimens and panels may propagate at the face/core interface or it may deflect, "kink", away from the interface and propagate inside the foam core, see Figure 9.6.

Erdogan and Sih (1963) examined crack propagation in a homogeneous, isotropic, brittle plastic sheet. Specifically, a plate containing a central crack of length $2a$ loaded in biaxial tension was analyzed using the Griffith theory assuming the crack will grow in a direction where the energy release per unit crack extension is maximum. They found that the crack growth initiates at the tip and extends in a plane perpendicular to the direction of maximum tangential stress, σ_θ as confirmed by experiments. The near tip stress element in polar coordinates (r, θ) is shown in Figure 9.7.

Following the analysis of Erdogan and Sih (1963), Prasad and Carlsson (1994a) considered a crack of length $2a$ in a homogeneous, isotropic material loaded under plane strain in mixed mode I and mode II. The stresses near the right crack tip, illustrated in Figure 9.7, can be expressed as

$$\sigma_r = \frac{1}{\sqrt{2\pi r}} \cos\frac{\theta}{2}\left(K_I\left(1+\sin^2\frac{\theta}{2}\right) + \frac{3}{2}K_{II}\sin\theta - 2K_{II}\tan\frac{\theta}{2}\right), \tag{9.17a}$$

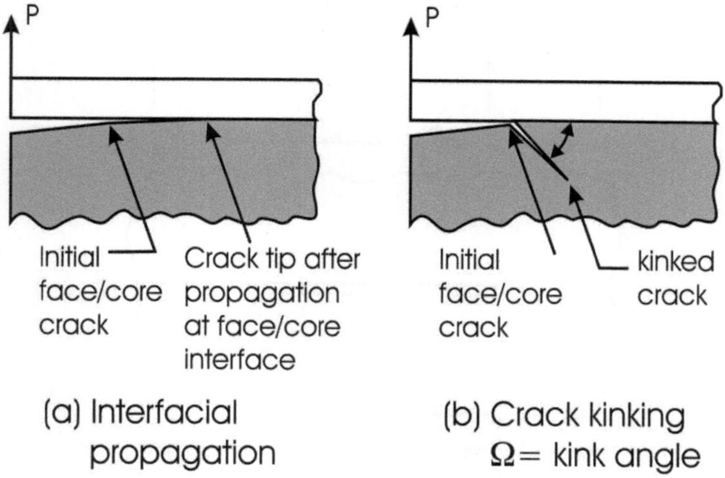

Figure 9.6 Growth scenarios of a face/core debond in foam-cored sandwich.

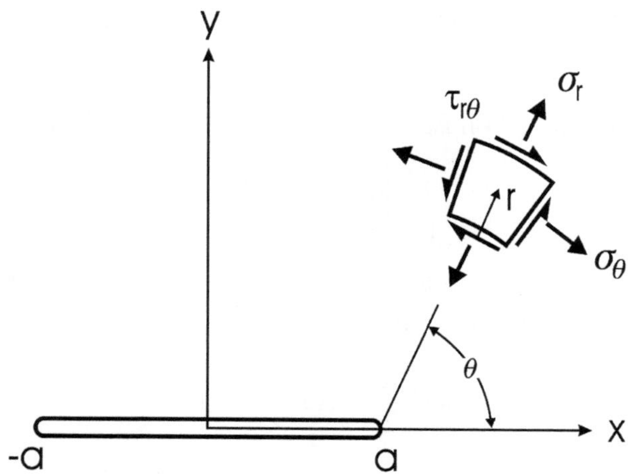

Figure 9.7 Near tip stresses in a cracked sheet.

$$\sigma_\theta = \frac{1}{\sqrt{2\pi r}} \cos\frac{\theta}{2}\left(K_I \cos^2\frac{\theta}{2} - \frac{3}{2}K_{II}\sin\theta\right), \tag{9.17b}$$

$$\tau_{r\theta} = \frac{1}{2\sqrt{2\pi r}} \cos\frac{\theta}{2}\left(K_I \sin\theta + K_{II}\left(3\cos\theta - 1\right)\right). \tag{9.17c}$$

To determine the kink angle, Ω, Figure 9.8, for a general mixed mode loading case, one may determine the direction where the stress σ_θ is maximum. This direction coincides with the direction, $\theta = \Omega$, where the shear stress $\tau_{r\theta}$ vanishes. An explicit expression for Ω was derived by Prasad and

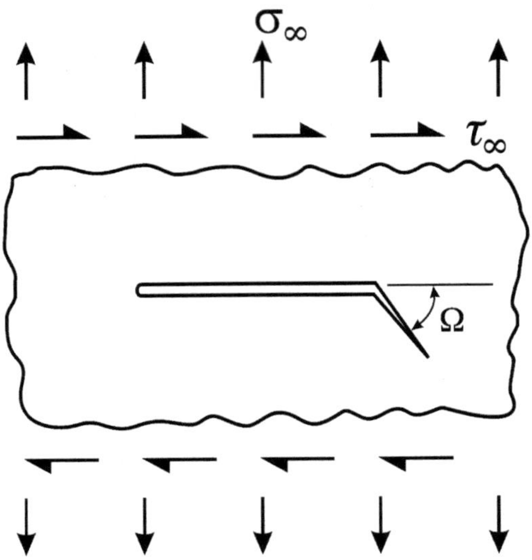

Figure 9.8 Definition of kink angle, Ω, for a crack under mixed mode loading.

Carlsson (1994a) based on Equation (9.17c) with $\tau_{r\theta} = 0$ for $\theta = \Omega$, which yields

$$\frac{\sin\Omega}{3\cos\Omega - 1} = \frac{K_{\mathrm{II}}}{K_{\mathrm{I}}}. \tag{9.18}$$

To solve for the angle Ω in terms of the mode ratio $K_{\mathrm{II}}/K_{\mathrm{I}}$, the following trigonometric identities are employed:

$$\sin^2\frac{\Omega}{2} + \cos^2\frac{\Omega}{2} = 1, \tag{9.19a}$$

$$\cos\Omega = \cos^2\frac{\Omega}{2} - \sin^2\frac{\Omega}{2}, \tag{9.19b}$$

$$\sin\Omega = 2\sin\frac{\Omega}{2} - \cos\frac{\Omega}{2}. \tag{9.19c}$$

Substitution of Equations (9.19) in (9.18) yields

$$\frac{\sin\frac{\Omega}{2}\cos\frac{\Omega}{2}}{\cos^2\frac{\Omega}{2} - 2\sin^2\frac{\Omega}{2}} = \frac{K_{\mathrm{II}}}{K_{\mathrm{I}}}. \tag{9.20}$$

After factoring out $\cos^2(\Omega/2)$, this equation simplifies to

$$\frac{\tan\frac{\Omega}{2}}{1 - 2\tan^2\frac{\Omega}{2}} = \frac{K_{\mathrm{II}}}{K_{\mathrm{I}}}. \tag{9.21}$$

Substitution of $x = \tan(\Omega/2)$ into Equation (9.21) yields

$$2K_{\mathrm{II}}x^2 + K_{\mathrm{I}}x = K_{\mathrm{II}}. \tag{9.22}$$

Solving the quadratic yields

$$\tan\frac{\Omega}{2} = \frac{-1 \pm \sqrt{1 + 8(K_{\mathrm{II}}/K_{\mathrm{I}})^2}}{4(K_{\mathrm{II}}/K_{\mathrm{I}})}. \tag{9.23}$$

Hence, the kink angle may be expressed as

$$\Omega = 2\tan^{-1}\left[\frac{-1 \pm \sqrt{1 + 8(K_{\mathrm{II}}/K_{\mathrm{I}})^2}}{4K_{\mathrm{II}}/K_{\mathrm{I}}}\right]. \tag{9.24}$$

Pure mode I, $K_{\mathrm{II}} = 0$, yields a kink angle, $\Omega = 0$, i.e., self-similar propagation, while pure mode II yields $\Omega = \pm 70.5°$, where the $+$ sign refers to kinking down as shown in Figure 9.8 ($K_{\mathrm{II}} > 0$) and the $-$ sign refers to kinking up ($K_{\mathrm{II}} < 0$).

This analysis thus provides the direction of kinking if it were to occur in a homogeneous and isotropic material. The initiation of the kinked crack growth occurs under mixed mode conditions and the fracture toughness, K_c, may depend upon the amount of mode II present, i.e., the mode ratio $K_{\mathrm{II}}/K_{\mathrm{I}}$. Hence, for a given material, it is necessary to experimentally measure the fracture toughness at various mode mixes. Experimentally it has been found that the fracture toughness increases with increased shear loading (increasing mode ratio, $K_{\mathrm{II}}/K_{\mathrm{I}}$) (Liechti and Chai, 1992).

For a face/core crack, the various growth scenarios illustrated in Figure 9.1 point to one possibility that the initial debond lies between two widely dissimilar material, i.e. between a stiff face and a compliant (soft) core. Crack kinking for such a case has been examined by He and Hutchinson (1989) who considered a crack at the interface between two dissimilar isotropic materials, #1 above the interface, and #2 below the interface. A straight crack segment denoted "kinked crack" in Figure 9.6b, of short length compared to the "parent", interface crack (Figure 9.6a) was specifically analyzed. The stress field prior to kinking is thus the singularity field of an interface crack which corresponds to a strain energy release rate, G, available for further propagation along the interface or kinking into the core. The analysis provides the energy release rate of the kinked crack, G^k, using an extensive numerical solution of the governing integral equations. The energy release rate and kink angle results are presented graphically for any given combinations of materials above and below the interface specified in terms of an elastic mismatch parameter α (Dundurs, 1969),

$$\alpha = \frac{\overline{E_1} - \overline{E_2}}{\overline{E_1} + \overline{E_2}}, \tag{9.25}$$

where $\overline{E} = E$ for plane stress, and $\overline{E} = E/(1 - \upsilon^2)$ for plane strain. Subscripts 1 and 2 on Young's modulus represents the materials above and below the crack plane.

He and Hutchinson (1989) examined the energy release rate ratio G^k/G and found that this ratio increases with increasing compliance of the material into which the crack kinks. They suggested that the crack would remain at the interface only if the compliant material is tough and the interface is relatively brittle. Furthermore, as discussed by Hutchinson and Suo (1992), the kink angle that maximizes, G^k, is also the angle where $K_{\mathrm{II}} = 0$, i.e., the kinked crack propagates under pure mode I as for the homogeneous case discussed earlier. The considerations discussed above were expressed in an energy inequality governing kinking,

$$\frac{G^k_{\max}}{G} > \frac{G_{IC}}{G_C}, \tag{9.26}$$

where $G^k_{\max}$ is the maximum energy release rate for the kinked crack with respect to kink angle, Ω and G is the energy release rate of the interface crack. G_{IC} and G_C are the mode I fracture toughness of the core, and the interface fracture toughness, respectively.

Predictions of the crack propagation path in a sandwich panel according to the above analysis thus requires experimental data on the core fracture toughness, G_{IC}, and the interface toughness, G_C, at the mixed mode loading $(K_{\mathrm{I}}, K_{\mathrm{II}})$ acting on the tip of the interface crack. Measurement of G_{IC} for the core is quite straightforward, but the measurement of the interface toughness, G_C, is more difficult, unless G_c is low and the crack propagates at the interface. If the interface is much tougher than the core, however, experimental determination of G_C becomes very difficult since the crack would not select such a high resistance path. Without a value of G_C the analysis could still be used in a semi-quantitative sense providing the direction Ω of the kinked crack and the driving force for kinking indicated by the energy release ratio $G^k_{\max}/G$.

9.2.1 Crack Path in Foam-Cored DCB Specimens

Understanding of the factors that govern the crack path in foam-cored sandwich structures is essential for the design of such structures. Furthermore,

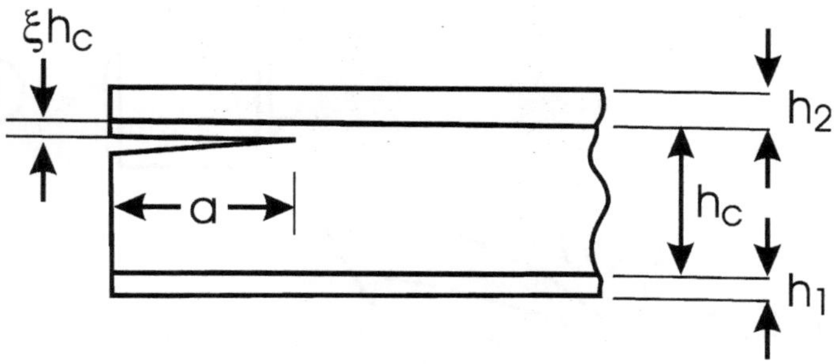

Figure 9.9 Precrack in a foam core DCB specimen. ξ is a parameter specifying the location of the precrack below the upper face in the core ($0 \leq \xi \leq 1$).

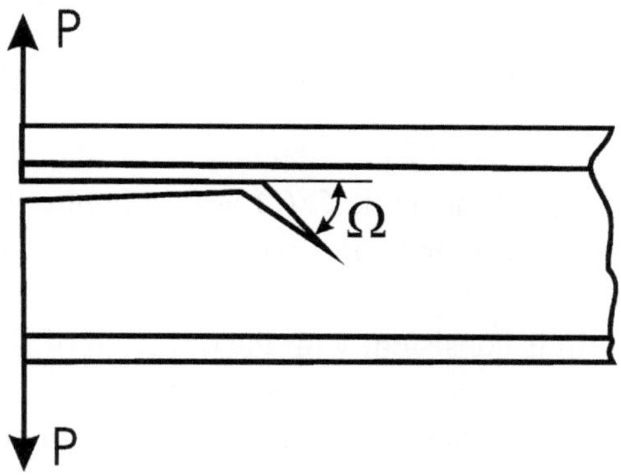

Figure 9.10 Crack kinking in foam-cored sandwich DCB specimen. Positive kink angle Ω is defined as shown ("down").

measurement of face/core debond toughness requires that the crack remains at the face/core interface. It has been found that kinking of the crack into the core is a common occurrence in DCB testing of foam-cored sandwich specimens (see Prasad and Carlsson, 1994b). Kinking would disqualify the test data for interface toughness.

When a foam core sandwich DCB specimen is tested, it is common to slightly extend the precrack beyond the tip of the artificial film insert at the face/core interface and this may tend to bring the precrack into the core below the face/core interface, as shown in Figure 9.9. If the DCB specimen is prepared from a sandwich panel without a starter film, the precrack needs to be cut into the core, near the face/core interface, which also tends to produce a precrack such as the one illustrated in Figure 9.9.

As discussed in Section 9.2, a crack in a foam-cored DCB specimen may not propagate in a self-similar manner, but kink up or down in a certain direction quantified by the kink angle Ω (Figure 9.10). The direction of kinking

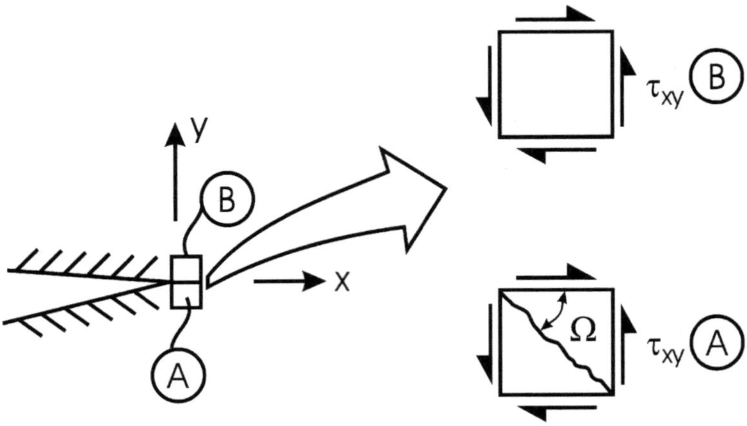

Figure 9.11 Positive shear stress and kink angle.

is indicated by the sign of the angle Ω ($\Omega > 0$ in Figure 9.10). The positive sign of the kink angle in Figure 9.10 is consistent with a positive shear stress ahead of the crack tip, see Figure 9.11. Furthermore, a positive shear stress ahead of the crack tip is consistent with the direction of sliding of the crack faces behind the crack tip, as shown in Figure 9.4.

Analysis of the kinking angle and critical load required to initiate kinking of a crack (Section 9.2) requires the stress intensity factors and energy release rate at the front of the parent crack. Such analysis, however, requires substantial detail since the definition of stress intensity factors is based on stresses and displacements in a very small region (singular domain) near the crack tip.

As a simpler alternative to detailed fracture mechanics is the beam theory analysis presented by Carlsson et al. (2006). If the influence of the small singular field domain near the crack tip is neglected, it is possible to determine the relative sliding displacement, shown in Figure 9.4, from the bending strains

$$\delta_{ii} = r(\varepsilon_x^L - \varepsilon_x^U), \tag{9.27}$$

where ε_x represents the near crack tip bending strains in the crack flanks, superscripts L and U denote the lower (L) and upper (U) sub-beams (legs), and r is the distance from the crack tip. According to this analysis, kinking would be promoted by a bending strain mismatch between the lower and upper legs of the DCB sandwich specimen, i.e. $\delta_{\mathrm{II}} \neq 0$ in Equation (9.26). $\delta_{\mathrm{II}} > 0$ would promote kinking downwards and $\delta_{\mathrm{II}} < 0$ would promote kinking upwards. Zero difference would promote self-similar growth (if the initial crack is within the core).

To quantify the strain mismatch, a DCB specimen with the generic crack configuration shown in Figure 9.9 is considered. The crack is located at distance ξh_c from the upper face/core (F/C) interface and a distance $(1-\xi)h_c$ from the lower F/C interface, where h_c is the core thickness and the parameter ξ is bounded by $0 \leq \xi \leq 1$. Consequently, $\xi = 0$ corresponds to a crack at the upper F/C interface, and $\xi = 1$ represents a crack at the lower F/C interface. The face sheets may have different moduli (E_{f1} and E_{f2}) and thicknesses (h_1 and h_2). We will here assume that each face sheet can be considered as homogeneous isotropic or orthotropic and the core is isotropic. The bending of each leg of the DCB is modeled using a laminated beam formulation where the extensional, coupling, and bending stiffness matrices of classical laminated plate theory, i.e. $[A]$, $[B]$ and $[D]$ as defined in Equations (3.18) and (3.19), are replaced by three scalar stiffnesses, A, B and D given by

$$A = \sum_{k=1}^{N} (E_x)_k (z_k - z_{k-1}), \tag{9.28a}$$

$$B = \frac{1}{2} \sum_{k=1}^{N} (E_x)_k (z_k^3 - z_{k-1}^3), \tag{9.28b}$$

$$D = \frac{1}{3} \sum_{k=1}^{N} (E_x)_k (z_k^3 - z_{k-1}^3), \tag{9.28c}$$

where the ply coordinates z_k are defined for a general laminate in Figure 3.2.

Application of this analysis to the bending of the lower and upper legs of the DCB specimen shown in Figure 9.9, each subject to a load P, yields the strains ε_x^L and ε_x^U as required for Equation (9.27).

$$\frac{\varepsilon_x^L b}{Pa} = \frac{h_L/2 - B_L/A_L}{D_L - B_L^2/A_L} \quad \text{(lower)}, \tag{9.29a}$$

$$\frac{\varepsilon_x^U b}{Pa} = \frac{h_U/2 + B_U/A_U}{D_U - B_U^2/A_U} \quad \text{(upper)}, \tag{9.29b}$$

where h_L and h_U are the thicknesses of the lower and upper legs,

$$h_L = h_1 + (1-\xi)h_c, \tag{9.30a}$$

$$h_U = h_2 + \xi h_c. \tag{9.30b}$$

The extensional, coupling, and bending stiffness of the lower and upper legs are obtained from Equations (9.28),

$$A_L = E_{f1}h_1 + E_c(1-\xi)h_c, \tag{9.31a}$$

$$A_U = E_{f2}h_2 + E_c\xi h_c, \tag{9.31b}$$

$$B_L = \frac{h_1 h_c(1-\xi)}{2}(E_c - E_{f1}), \tag{9.31c}$$

$$B_U = \frac{\xi h_2 h_c(1-\xi)}{2}(E_{f2} - E_c), \tag{9.31d}$$

$$D_L = \frac{1}{12}[E_{f1}(h_1^3 + 3h_1h_c^2(1-\xi)^2) + E_c((1-\xi)^3h_c^3 + 3h_1^2h_c(1-\xi))], \tag{9.31e}$$

$$D_U = \frac{1}{12}[E_c(\xi^3h_c^3 + 3\xi h_c h_c^2) + E_{f2}(h_2^3 + 3\xi^2h_c^2h_c)], \tag{9.31f}$$

where E_{f1} and E_{f2} are the effective Young's moduli of the lower and upper face sheets, respectively, and E_c is the core modulus.

An example (Carlsson et al., 2006) is a symmetric DCB specimen consisting of 2.4 mm thick glass/vinylester faces and a 16 mm thick H100 PVC foam core. The face and core moduli are $E_{f1} = E_{f2} = 27.6$ GPa and $E_c = 105$ MPa. The crack tip strains in the lower and upper legs of the DCB specimen were calculated from Equations (9.29) as a function of the crack position in the core ($0 \leq \xi \leq 1$). Figure 9.12 shows the crack tip strains in the upper and lower legs vs. ξ. The graph shows that the strains are equal ($\delta_{II} = 0$) at three locations of the crack, i.e., $\xi = 0.08$, 0.5, and 0.92. Consequently, a crack located at these locations is expected to propagate in a self-similar manner. Further, according to this analysis, a crack located at the top interface ($\xi = 0$), would kink down ($\varepsilon_x^L > \varepsilon_x^U$) until it reaches an equilibrium location where the strains are equal ($\xi = 0.08$). This is a stable location since Figure 9.13 indicates that any small perturbation of the crack location (up or down) would tend to return the crack to the $\xi = 0.08$ location. Similarly, a crack at the lower F/C interface ($\xi = 1$) would kink up ($\varepsilon_x^L > \varepsilon_x^U$) until a stable position is reached at $\xi = 0.92$. The actual distance from the upper and lower interface to the stable locations is only about 1.3 mm for this specimen, and a crack propagating at this location would be considered as "near interface". A crack positioned in the upper half region of the core between $\xi = 0.08$ and 0.5 would kink up and approach the $\xi = 0.08$ location while a crack located in the lower half would kink down. A center crack ($\xi = 0.5$) would potentially remain at the center, but Figure 9.12 indicates that this configuration is not stable since any small disturbance (up or down) would promote kinking up or down towards the stable $\xi = 0.08$ or 0.92 locations.

Several other cases were examined by Carlsson et al. (2006).

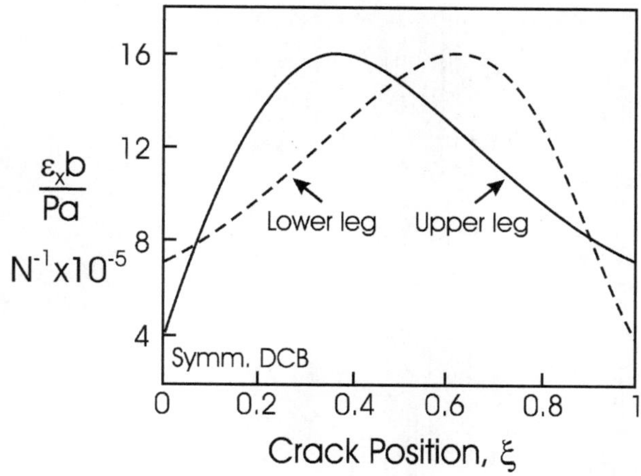

Figure 9.12 Crack tip strains in lower and upper legs of a symmetric DCB specimen.

9.3 Cohesive Zone Models of Face/Core Interface Failure

Producers of sandwich structures sometimes add a mat of continuous fibers called a "continuous filament mat" (CFM), or a "chopped strand mat" (CSM), between the face and core. The purpose of such a design is to increase the resin content between face and core to avoid "dry spots" and improve the resistance to face/core debonding by the increased resin content. The CFM and CSM may also add an additional toughening mechanism provided by fiber bridging the crack surfaces, see the photograph in Figure 9.13 showing face/core separation in a sandwich beam with a CSM added between face and core evidencing large-scale fiber bridging.

In cases where bridging zones develop, the actual material separation displacements across the crack faces can be quite substantial and the large-scale fracture process zone makes linear elastic fracture mechanics concepts difficult to satisfy. In many cases, the global response of a cracked sandwich structure, or fracture test specimen, remains linear, while the material separation process near the crack tip behaves nonlinear. This situation is conveniently represented by a cohesive zone model, where the tensile strength and the work of fracture become connected using a specific traction/separation (T/S) law (Argon, 2000). Figure 9.14 shows an idealized T/S law. As shown, the crack will start to open when the crack tip stress exceeds the cohesive strength σ_0. Once the crack opens up, the cohesion of the material elements

Figure 9.13 Fiber bridging observed during face/core separation in a sandwich beam with a 450 g/m^2 CSM. (Courtesy of Lundsgaard et al., 2008)

behind the crack tip will not be totally lost, as in linear elastic fracture mechanics, until the crack opening exceeds the critical value u_n^c.

As discussed by Nairn (2009) and shown in Figure 9.15, a cohesive zone is associated with two crack tips, viz. the notch root where the traction is zero and the actual crack tip where the material separation is initiated (see Figure 9.15). At the beginning of the fracture process, the bridging zone develops by extension of the crack tip while the notch root remains fixed. As will be discussed below, this process is associated with a rising resistance curve (R curve). After further loading, if a steady-state is reached, the crack tip and notch root propagate at the same rate, corresponding to a constant length of the bridging zone and self-similar growth.

Figure 9.15 also illustrates the separation of the crack flanks at the notch root, u_n^* where the traction vanishes. The fracture process in a test specimen or structure occurs in a "process zone" defining a layer of total thickness, Δ. The process zone is comprised of parts of the face and core materials above and below the interface where fiber pull-out, plastic deformation, and other complex, nonlinear material separation processes occur, see Figure 9.15. In modeling of cohesive failure of fracture test specimens, the local response of the material separation law typically demands a numerical finite element solution with special cohesive elements where the T/S law is specified and

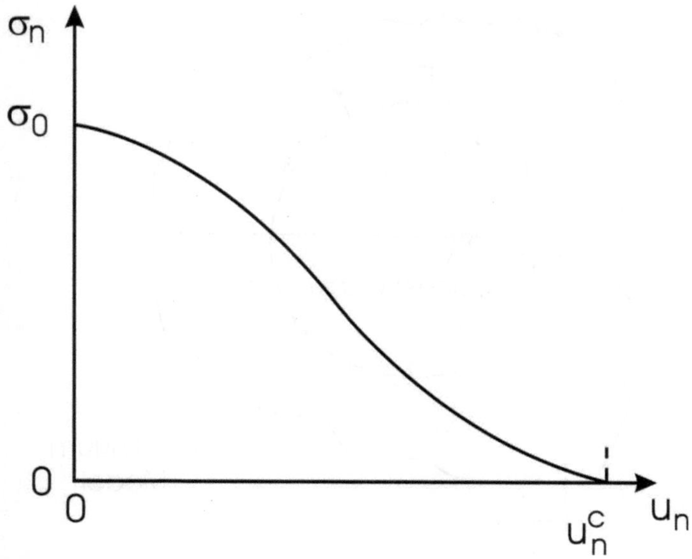

Figure 9.14 Schematic representation of a traction/separation law representing the failure process.

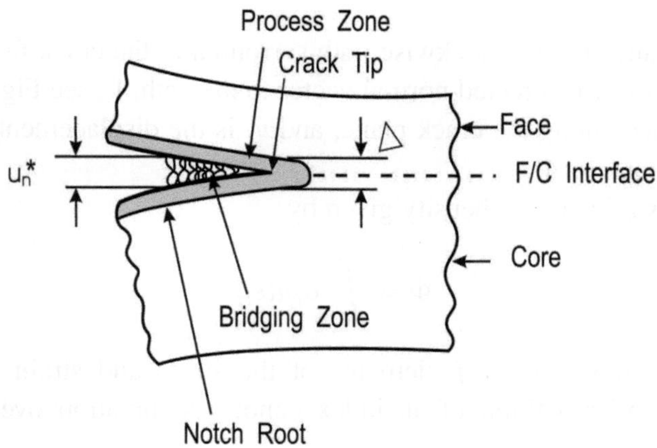

Figure 9.15 Process zone around the crack tip.

assigned along the anticipated crack propagation path, which, for a sandwich structure, is commonly along the face/core interface (Ostergaard, 2007). The layer thickness, Δ, of the process zone is typically neglected.

An important tool in cohesive zone modeling is the *J* integral proposed by Rice (1968).

$$J = \int_{\Gamma} \left(\Phi \, dx_2 - \sigma_{ij} n_j \frac{\partial u_i}{\partial x_i} ds \right), \tag{9.32}$$

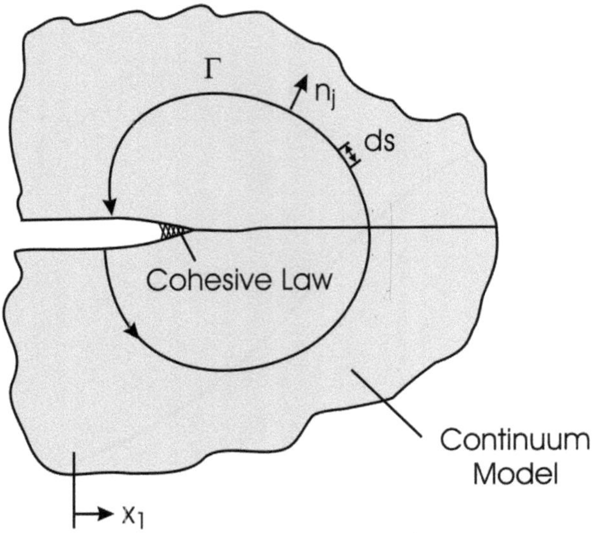

Figure 9.16 Path Γ for calculation of the J integral with normal n_j. x_1 is a coordinate along the crack propagation path and x_2 is perpendicular to the crack plane.

where Γ is any counter-clockwise path surrounding the crack faces, with n_j being the outwards directed normal vector to the path, Γ, see Figure 9.16. x_1 is a coordinate along the crack plane, and u_i is the displacement vector and ds is an element of Γ.

Φ is the strain energy density given by

$$\Phi = \int_0^{\varepsilon} \sigma_{ij} d\varepsilon_{ij}, \tag{9.33}$$

where σ_{ij} and ε_{ij} are i, j elements of the stress and strain tensors, respectively, and repetition of an index denotes summation over the range $i, j = 1, 2, 3$.

For the pure normal opening loading considered here, a cohesive law (Figure 9.14) can be expressed as

$$\sigma_n = \sigma_n(u_n). \tag{9.34}$$

When the stress on the interface reaches its ultimate value, σ_0, the fracture process is initiated. Before that, all deformation of the specimen is accommodated by the continuum. Figure 9.17 schematically illustrates the distribution of normal stress acting in the cohesive zone.

Evaluation of the J integral around the cohesive zone (Li and Ward, 1987; Bao and Suo, 1992) provides the relation between the cohesive law and the

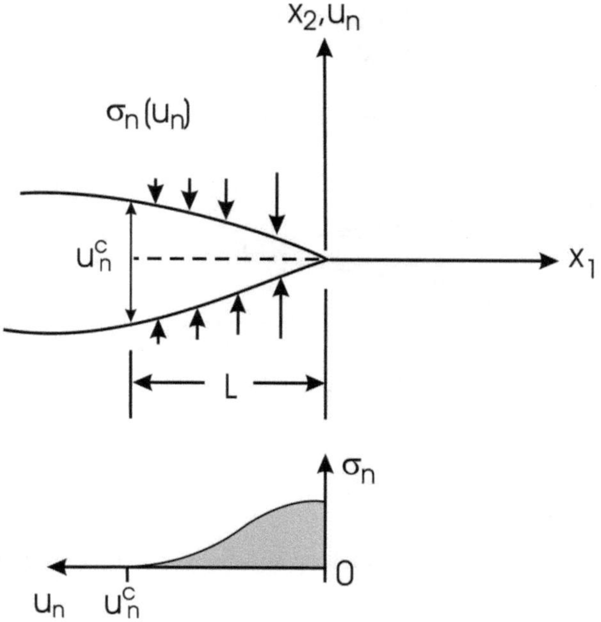

Figure 9.17 Normal stress, σ_n, in the cohesive zone.

fracture resistance:

$$J_r = \int_0^{u_n^*} \sigma_n du_n + J_0, \tag{9.35}$$

where u_n^* is the maximum separation distance encountered at any specific crack length (Figure 9.18). As shown in Figure 9.18, at the beginning of the fracture test no fibers are bridging the crack and the crack will advance when $J_R = J_0$. Upon further loading and extension of the crack tip (Figure 9.18), the first term in Equation (9.35) will start to contribute as a result of development of the bridging zone.

As the crack propagates further, the bridging zone will become more and more developed and extend behind the crack tip corresponding to a rising R curve, i.e., a plot of J_R vs. Δa, where Δa is the extension of the crack tip from its initial value (a_0), see Figure 9.19. Once the maximum end opening, u_n, reaches its critical value, u_n^c, the bridging zone is fully developed, corresponding to a steady-state fracture toughness, $J_R = J_{ss}$. Nairn (2009) has argued that the energy stored in the bridging fibers may be recovered upon unloading the specimen and that a true R curve determination prior to steady-state should subtract this contribution to the J integral. However, this should not be an issue for continuous loading considered here.

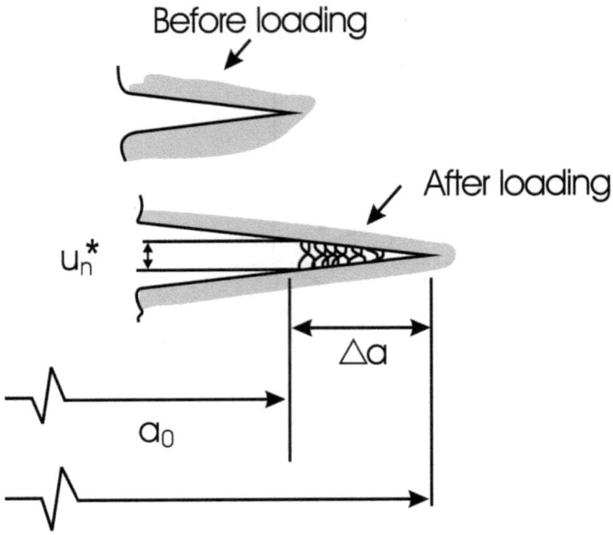

Figure 9.18 Extension of crack tip and development of bridging zone.

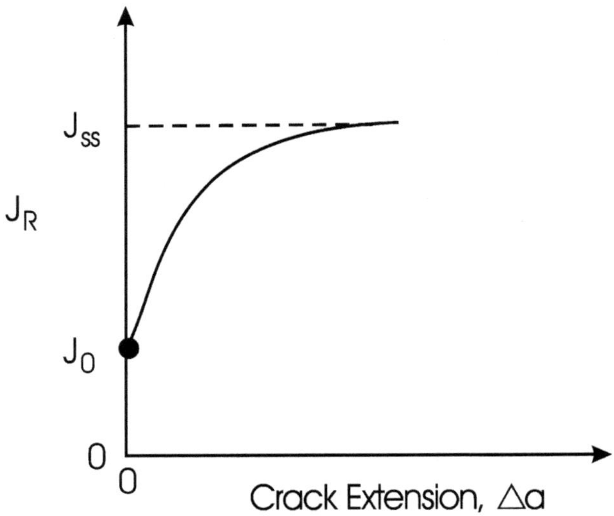

Figure 9.19 Schematic R curve showing crack growth resistance, J_R, vs. crack extension in a material where fiber bridging develops.

As discussed by Ostlund (1995) and Sorensen and Jacobsen (1998), the bridging fibers represent distributed discrete forces, here smeared into a continuum mechanics cohesive stress, σ_n. Such an approach is assumed appropriate if there is a sufficient number of bridging fibers per unit area.

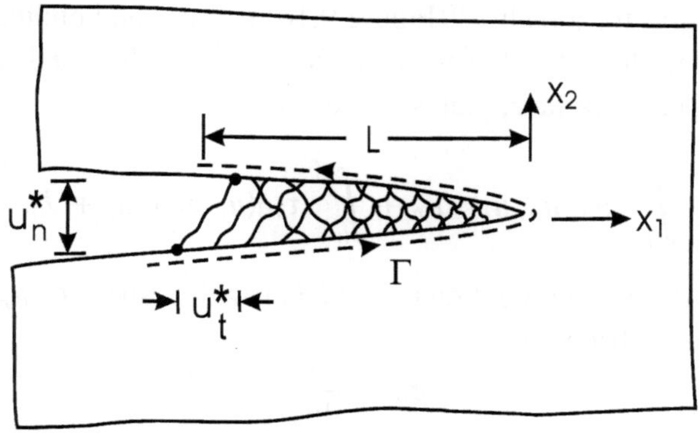

Figure 9.20 Crack with large scale bridging subject to mixed mode loading. After Sorensen and Kirkegaard (2006).

The cohesive law, $\sigma_n = \sigma_n(u_n)$, may be determined from the measured resistance curve. Differentiation of Equation (9.35) yields (Li and Ward, 1989)

$$\sigma_n(u_n) = \frac{dJ_R}{du_n}. \tag{9.36}$$

Such an approach requires that J_R has been determined as a function of the end opening, u_n^*. Hence, the test must employ special displacement measuring devices that allow recording of u_n^* (Lundsgaard et al., 2008).

So far the description has been limited to mode I fracture. The treatment of mixed mode fracture using a cohesive law approach requires consideration of both normal and shear stresses σ_n and τ_{nt} that act in the cohesive zone, and normal and tangential crack face displacements, u_n and u_t. Sorensen and Kirkegaard (2006) present a consistent approach to establish mixed mode cohesive laws for specimens displaying large-scale bridging. The approach utilizes the J integral applied to the novel test method consisting of a DCB specimen (Section 11.7) where the legs are loaded with edge couples as opposed to shear forces in the traditional, standard DCB test. The normal and shear traction/separation stresses are extracted from the J integral in combination with measured normal and tangential crack face displacements at the end of the cohesive zone. Figure 9.20 illustrates the crack region with a large-scale bridging zone subject to mixed mode loading.

The crack opening and sliding displacements u_n and u_t are defined as the relative displacements of the upper and lower crack surfaces (Figure 9.20). The critical values of u_n and u_t are denoted u_n^c and u_t^c. The normal and shear stresses acting over the cohesive zone are denoted by σ_n and τ_{nt}. Calculation of the J integral for a path, Γ (Figure 9.16) starting and ending outside the bridging zone, along the lower crack face, around the crack tip and back along the upper crack face, yields the result

$$J_R = \int_0^{u_n^*} \sigma_n (u_n, u_t)\, du_n + \int_0^{u_t^*} \tau_{nt} (u_n, u_t) du_t + J_0. \tag{9.37}$$

By partially differentiating Equation (9.37) with respect to u_n and u_t, the cohesive stresses become

$$\sigma_n = \frac{\partial J_R}{\partial u_n}, \tag{9.38a}$$

$$\tau_{nt} = \frac{\partial J_R}{\partial u_\tau}. \tag{9.38b}$$

Hence, in this manner the cohesive stresses, i.e. traction/separation (T/S) law, can be determined directly from the measured resistance curve and measured values of the opening and sliding displacements. At steady-state crack propagation, the bridging zone is fully developed and should translate similarly with the growing crack along the interface.

9.4 An Expression for the Energy Release Rate of Face/Core Debonds in Sandwich Beams

A loading which can be especially detrimental for a sandwich structure is compression because these debonds are susceptible to buckling and subsequent rapid growth during the post-buckling phase. Typically, post-buckling solutions are derived in terms of forces and moments at the debond section (Kardomateas and Huang, 2003). Therefore, expressions for the energy release rate in terms of these quantities are particularly useful. Such expressions were first derived by Yin and Wang (1984) for delaminated monolithic composites, and extended by Suo and Hutchinson (1990) to a delaminated bi-material (thin film on a substrate). The sandwich configuration is, however, a "tri-material", i.e. two face sheets, which need not be the same, and a core. This is the configuration treated in this section. Specifically, we use the J integral to obtain a closed form algebraic expression for the energy release rate, G, for a debonded sandwich beam. The most general case of an "asymmetric" sandwich is considered, i.e. the bottom face sheet is not necessarily of the same material and thickness as the top face sheet. The expression derived is in terms of forces and moments (which are typical outputs of post-buckling solutions) acting on the debonded section.

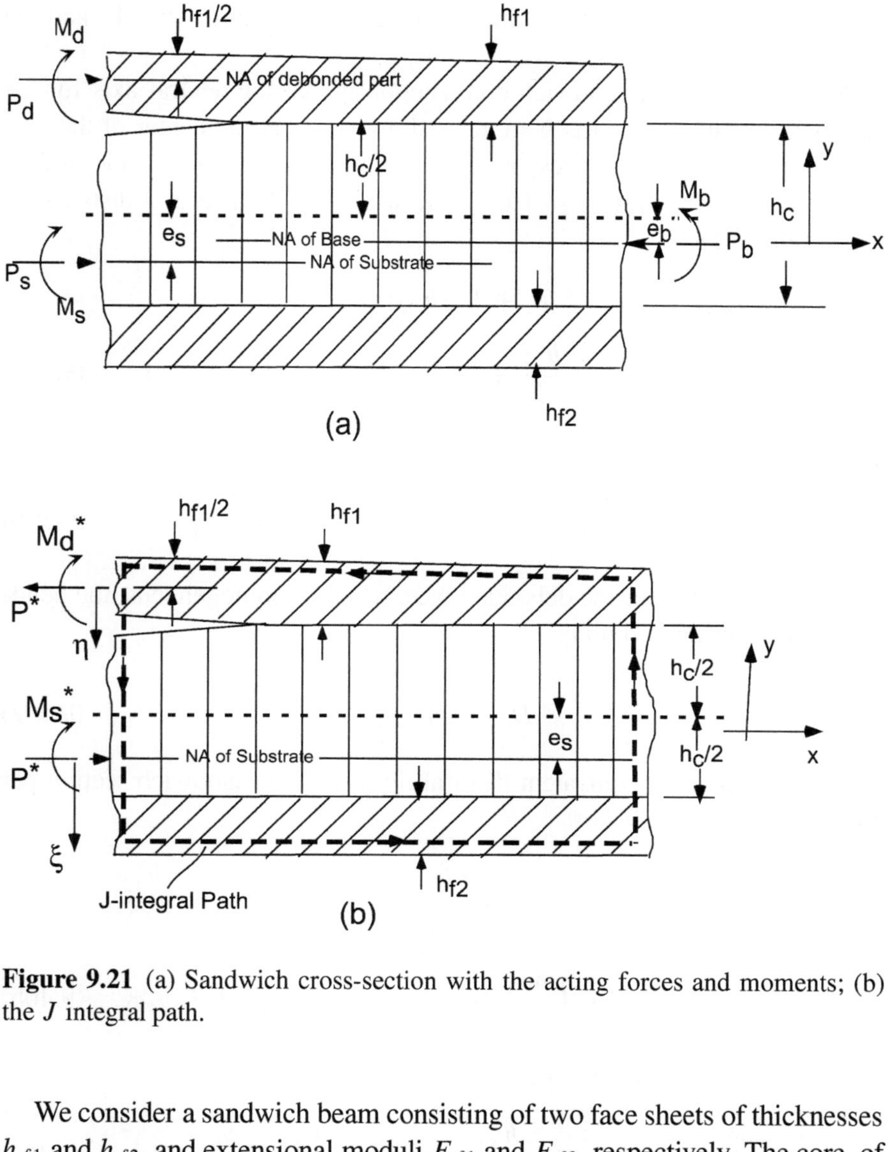

Figure 9.21 (a) Sandwich cross-section with the acting forces and moments; (b) the J integral path.

We consider a sandwich beam consisting of two face sheets of thicknesses h_{f1} and h_{f2}, and extensional moduli E_{f1} and E_{f2}, respectively. The core, of thickness h_c, has an extensional modulus, E_c (Figure 9.21a).

In the region of the debond, the sandwich beam consists of two parts: the debonded upper face sheet (referred to as the "debonded part", of thickness h_{f1}) and the part below the debond ("substrate part", of thickness $h_c + h_{f2}$, which includes the core and the lower face sheet). A unit width is assumed. The region outside the debond is referred to as the "base part" and consists of the entire section of the sandwich beam, i.e., of thickness $h_{f1} + h_c + h_{f2}$.

We shall denote the base part with "b", the debonded part with "d", and the substrate part with "s".

A characteristic of sandwich construction is that the neutral axis for the base and the substrate parts is in general no longer at the center of the corresponding sections. With respect to a reference axis x through the middle of the core, the neutral axis of the base section is defined at a distance e_b (Figure 9.21a), as:

$$e_b \left(E_{f1}h_{f1} + E_c h_c + E_{f2}h_{f2}\right) = E_{f2}h_{f2}\left(\frac{h_{f2}}{2} + \frac{h_c}{2}\right) - E_{f1}h_{f1}\left(\frac{h_{f1}}{2} + \frac{h_c}{2}\right), \quad (9.39a)$$

and that of the substrate part is at a distance e_s, given by

$$e_s \left[E_c h_c + E_{f2}h_{f2}\right] = E_{f2}h_{f2}\left(\frac{h_{f2}}{2} + \frac{h_c}{2}\right). \quad (9.39b)$$

Moreover, while for the debonded face, which is homogeneous, the bending rigidity per unit width is

$$D_d = E_{f1}\frac{h_{f1}^3}{12}. \quad (9.40a)$$

For the base part the equivalent flexural rigidity of the sandwich section per unit width is (Figure 9.21a)

$$D_b = E_{f1}\frac{h_{f1}^3}{12} + E_{f1}h_{f1}\left(\frac{h_{f1}}{2} + \frac{h_c}{2} + e_b\right)^2 + E_{f2}\frac{h_{f2}^3}{12} + E_{f2}h_{f2}\left(\frac{h_{f2}}{2} + \frac{h_c}{2} - e_b\right)^2 + E_c\frac{h_c^3}{12} + E_c h_c e_b^2, \quad (9.40b)$$

and for the substrate (again, per unit width)

$$D_s = E_c\frac{h_c^3}{12} + E_c h_c e_s^2 + E_{f2}\frac{h_{f2}^3}{12} + E_{f2}h_{f2}\left(\frac{h_{f2}}{2} + \frac{h_c}{2} - e_s\right)^2. \quad (9.40c)$$

Figure 9.21a shows a segment of the plate containing the debond front (crack tip). A section of the plate ahead of the crack tip carries the compressive axial force P_b and bending moment M_b, per unit width of the plate (base part loads).

Behind the crack tip, the cross-section above the debond (debonded part) carries the loads P_d and M_d and the cross-section below the debond (substrate part) carries the loads P_s and M_s. It is assumed that these forces and moments have already been determined from the post-buckling solution of the plate. It should be noted that in this derivation we consider only the effects of the axial forces and bending moments; shear forces are neglected. In addition, force and moment (about the neutral axis of the base part) equilibrium conditions hold as follows:

$$P_b = P_d + P_s, \tag{9.40d}$$

$$M_b - M_d - M_s - P_d\left(\frac{h_{f1}}{2} + \frac{h_c}{2} + e_b\right) + P_s(e_s - e_b) = 0. \tag{9.40e}$$

If we set

$$(EA)_b = E_{f1}h_{f1} + E_c h_c + E_{f2}h_{f2}, \tag{9.41a}$$

then the axial stress in the base part is

$$\sigma_{xx}^{(b)} = \begin{cases} -\dfrac{P_b}{(EA)_b}E_{f1} - \dfrac{M_b E_{f1}}{D_b}y & \text{if } e_b + \dfrac{h_c}{2} \le y \le e_b + \dfrac{h_c}{2} + h_{f1} \\ & \text{[debonded face]}, \\ -\dfrac{P_b}{(EA)_b}E_c - \dfrac{M_b E_c}{D_b}y & \text{if } e_b - \dfrac{h_c}{2} \le y \le e_b + \dfrac{h_c}{2} \\ & \text{[core]}, \\ -\dfrac{P_b}{(EA)_b}E_{f2} - \dfrac{M_b E_{f2}}{D_b}y & \text{if } e_b - \dfrac{h_c}{2} - h_{f2} \le y \le e_b - \dfrac{h_c}{2} \\ & \text{[lower face]}. \end{cases} \tag{9.41b}$$

Superimposing $-\sigma_{xx}^{(b)}$ on the stresses behind and ahead of the debond front in the system in Figure 9.21a, would result in the system in Figure 9.21b, whose energy release rate and stress intensity factors would be the same as in the original system (Figure 9.21a) since the system of base part stresses acting alone would produce a non-singular stress field. In this way, we can express the energy release rate in just two parameters.

The forces acting on the system in Figure 9.21b are

$$P^* = \int_{e_b+\frac{h_c}{2}}^{e_b+\frac{h_c}{2}+h_{f1}} (-\sigma_{xx}^{(b)})dy - P_d. \tag{9.42a}$$

Substituting $\sigma_{xx}^{(b)}$ from (9.41b) and performing the integration results in P^* in the form

$$P^* = C_1 P_b + C_2 M_b - P_d, \tag{9.42b}$$

where

$$C_1 = \frac{E_{f1}h_{f1}}{(EA)_b}; \quad C_2 = \frac{E_{f1}h_{f1}}{2D_b}(2e_b + h_{f1} + h_c). \tag{9.42c}$$

Likewise,

$$M_d^* = M_d - \int_{e_b+\frac{h_c}{2}}^{e_b+\frac{h_c}{2}+h_{f1}} (-\sigma_{xx}^{(b)}) \left[y - \left(e_b + \frac{h_c}{2} + \frac{h_{f1}}{2} \right) \right] dy. \quad (9.43a)$$

Again substituting $\sigma_{xx}^{(b)}$ from (9.41b) and performing the integration results in M_d^* in the form

$$M_d^* = M_d - C_3 M_b, \quad (9.43b)$$

where

$$C_3 = \frac{E_{f1}}{D_b} \frac{h_{f1}^3}{12}. \quad (9.43c)$$

The corresponding moment M_s^* in the substrate part is obtained from equilibrium, Figure 9.21b,

$$M_s^* = P^* \left(e_s + \frac{h_c}{2} + \frac{h_{f1}}{2} \right) - M_d^*. \quad (9.43d)$$

Referring now to Figure 9.21b, the stresses in the cross-section of the debonded face produced by the tensile load P^* and the bending moment M_d^* are

$$\sigma_{xx} = \frac{P^*}{h_{f1}} + E_{f1} \frac{M_d^*}{D_d} \eta; \quad -\frac{h_{f1}}{2} \le \eta \le \frac{h_{f1}}{2}, \quad (9.44a)$$

$$\sigma_{yy} = \tau_{xy} \simeq 0, \quad (9.44b)$$

where η is the normal coordinate measured from the mid-plane of the debonded face. In Figure 9.21b, the cross-section of the sandwich ahead of the debond front is subjected to vanishing stress and strain. In the region behind the debond front we assume $\varepsilon_{zz} = 0$ (plane strain). It follows that

$$\sigma_{zz} = \nu_{xz}^{(f1)} \sigma_{xx}; \quad \varepsilon_{xx} = \frac{\sigma_{xx} - \nu_{zx}^{(f1)} \sigma_{zz}}{E_{f1}} = \frac{1 - \nu_{zx}^{(f1)} \nu_{xz}^{(f1)}}{E_{f1}} \sigma_{xx}, \quad (9.44c)$$

where $\nu_{xz}^{(f1)}$ and $\nu_{zx}^{(f1)}$ are the Poisson ratios of the debonded (orthotropic) face sheet.

With the J integral path shown in Figure 9.21b, the following expression holds along the vertical path across the debonded face:

$$dJ = W dy - \mathbf{T} \frac{\partial \mathbf{u}}{\partial x} ds; \quad dy = -ds, \quad (9.44d)$$

where W is the strain energy density, $\mathbf{T}$ is the traction vector and $\mathbf{u}$ the displacement vector (Budiansky and Rice, 1973).

Since

$$W = \frac{1}{2}\left(\sigma_{xx}\varepsilon_{xx} + \sigma_{zz}\varepsilon_{zz}\right) = \frac{1}{2}\sigma_{xx}\varepsilon_{xx}; \quad \mathbf{T}\frac{\partial \mathbf{u}}{\partial x} = -\sigma_{xx}\varepsilon_{xx}, \tag{9.44e}$$

and since $ds = d\eta$, and by use of (9.44c):

$$dJ = -\frac{1}{2}\sigma_{xx}\varepsilon_{xx}ds + \sigma_{xx}\varepsilon_{xx}ds = \frac{1}{2}\sigma_{xx}\varepsilon_{xx}ds = \frac{1 - \nu_{zx}^{(f1)}\nu_{xz}^{(f1)}}{E_{f1}}\sigma_{xx}^2 d\eta, \tag{9.44f}$$

Substituting σ_{xx} from (9.44a), we obtain the contribution of the debonded face to the J integral:

$$J_1 = \int_{-\frac{h_{f1}}{2}}^{\frac{h_{f1}}{2}} \frac{(1 - \nu_{zx}^{(f1)}\nu_{xz}^{(f1)})}{2E_{f1}} \left(\frac{P^*}{h_{f1}} + E_{f1}\frac{M_d^*}{D_d}\eta\right)^2 d\eta$$

$$= \frac{(1 - \nu_{zx}^{(f1)}\nu_{xz}^{(f1)})}{2E_{f1}} \left(\frac{P^{*2}}{h_{f1}} + E_{f1}^2\frac{M_d^{*2}}{D_d^2}\frac{h_{f1}^3}{12}\right). \tag{9.45}$$

Similarly, if we set

$$(EA)_s = E_c h_c + E_{f2} h_{f2}, \tag{9.46a}$$

then the stresses in the cross-section below the debond in Figure 9.21b, which are produced by the compressive force P^* and the bending moment M_s^* are

$$\sigma_{xx} = \begin{cases} -\dfrac{P^*}{(EA)_s}E_c + \dfrac{M_s^* E_c}{D_s}\xi & \text{if } -e_s - \dfrac{h_c}{2} \le \xi \le -e_s + \dfrac{h_c}{2} \\ & \text{[core]} \\ -\dfrac{P^*}{(EA)_s}E_{f2} + \dfrac{M_s^* E_{f2}}{D_s}\xi & \text{if } -e_s + \dfrac{h_c}{2} \le \xi \le -e_s + \dfrac{h_c}{2} + h_{f2} \\ & \text{[lower face]} \end{cases} \tag{9.46b}$$

and

$$\sigma_{yy} = \tau_{xy} \simeq 0, \tag{9.46c}$$

where ξ is the normal coordinate measured from the neutral axis of this cross-section. Again, following the same arguments as before, i.e. that the cross-section of the sandwich ahead of the debond front is subjected to vanishing stress and strain and that in the region behind the debond front we have $\varepsilon_{zz} = 0$ (plane strain), we obtain the the following expression along a vertical path below the debond:

$$dJ = \frac{1}{2}\sigma_{xx}\varepsilon_{xx}ds - \sigma_{xx}\varepsilon_{xx}ds = \frac{1 - \nu_{zx}^{(i)}\nu_{xz}^{(i)}}{E_i}\sigma_{xx}^{(i)2}d\xi, \tag{9.46d}$$

where i refers to the core (c) or the lower face sheet (f_2).

Hence the contribution of the vertical path below the debond to the J integral is

$$J_2 = \int_{-e_s-\frac{h_c}{2}}^{-e_s+\frac{h_c}{2}} \frac{(1-\nu_{zx}^{(c)}\nu_{xz}^{(c)})}{2E_c}\sigma_{xx}^{(c)2}d\xi + \int_{-e_s+\frac{h_c}{2}}^{-e_s+\frac{h_c}{2}+h_{f2}} \frac{(1-\nu_{zx}^{(f2)}\nu_{xz}^{(f2)})}{2E_{f2}}\sigma_{xx}^{(f2)2}d\xi. \tag{9.46e}$$

Substituting the corresponding expressions from (9.46b) and performing the integration leads to

$$J_2 = \frac{P^{*2}}{(EA)_s^2}H_1 + \frac{P^*M_s^*}{(EA)_s D_s}H_2 + \frac{M_s^{*2}}{D_s^2}H_3, \tag{9.47a}$$

where

$$H_1 = \frac{1-\nu_{zx}^{(c)}\nu_{xz}^{(c)}}{2}E_c h_c + \frac{1-\nu_{zx}^{(f2)}\nu_{xz}^{(f2)}}{2}E_{f2}h_{f2}, \tag{9.47b}$$

$$H_2 = \frac{1-\nu_{zx}^{(c)}\nu_{xz}^{(c)}}{2}E_c h_c 2e_s + \frac{1-\nu_{zx}^{(f2)}\nu_{xz}^{(f2)}}{2}E_{f2}h_{f2}\left(2e_s - h_c - h_{f2}\right), \tag{9.47c}$$

$$H_3 = \frac{1-\nu_{zx}^{(c)}\nu_{xz}^{(c)}}{2}E_c h_c\left(\frac{h_c^2}{12}+e_s^2\right) + \frac{1-\nu_{zx}^{(f2)}\nu_{xz}^{(f2)}}{2}E_{f2}h_{f2}\left[\frac{h_{f2}^2}{3}+\left(\frac{h_c}{2}-e_s\right)\left(\frac{h_c}{2}+h_{f2}-e_s\right)\right], \tag{9.47d}$$

The sum of the two integrals (9.45) and (9.47a) represents the energy release rate since the remaining portions of the path make no contribution to the J integral, i.e.

$$G = J_1 + J_2 \tag{9.48}$$

or

$$G = \frac{(1-\nu_{zx}^{(f1)}\nu_{xz}^{(f1)})}{2E_{f1}}\left(\frac{P^{*2}}{h_{f1}} + E_{f1}^2\frac{M_d^{*2}}{D_d^2}\frac{h_{f1}^3}{12}\right) + \left(\frac{P^{*2}}{(EA)_s^2}H_1 + \frac{P^*M_s^*}{(EA)_s D_s}H_2 + \frac{M_s^{*2}}{D_s^2}H_3\right), \tag{9.49}$$

where H_1, H_2 and H_3 are given in (9.47b–d).

Chapter 10
Analysis of Debond Fracture Specimens

Several test methods for determining the fracture toughness of the face/core interface in sandwich specimens have been proposed. All debond specimens are beam specimens where a debond typically is implanted in the form of a thin Teflon sheet between face and core during manufacture of the sandwich panel, or in some cases the debond is machined or cut with a thin blade or knife. This and several other experimental issues will be discussed in Chapter 11. In this chapter, we will introduce some of the most popular sandwich debond tests and outline analysis of compliance and energy release rate.

10.1 Introduction

The analysis of the fracture test specimens typically focuses on the global specimen compliance, C. Once this quantity is determined as a function of crack length, the energy release rate G is readily obtained by differentiation of C with respect to crack length a (see Chapter 9), i.e.,

$$G = \frac{P^2}{2b}\frac{dC}{da}, \tag{10.1}$$

where P is the load applied and b is the width of the specimen. Several of the fracture specimens in use are quite simple in terms of geometry, loading, and support conditions, allowing for analytical solution of the compliance as a function of crack length, i.e., $C = C(a)$. Differentiation according to Equation (10.1) yields G. Some fracture specimens, however, are more complicated and do not readily allow an analytical solution. For such specimens, it may be possible to experimentally determine the compliance at several crack lengths. The data could be graphed vs. crack length and an empirical

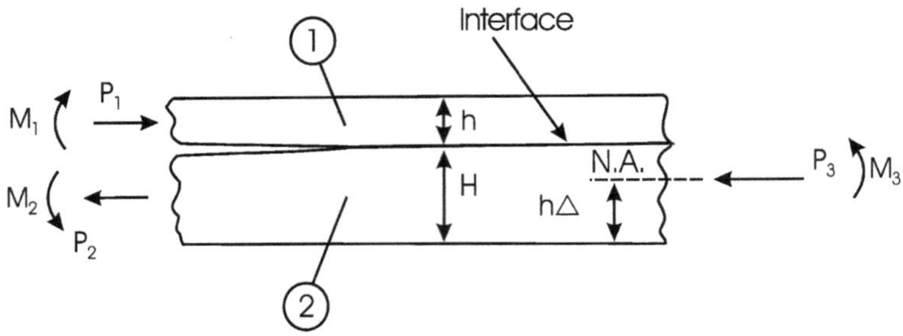

Figure 10.1 Section of cracked bimaterial specimen with an interface crack. Axial forces and moments are applied along the three edges (after Suo and Hutchinson, 1990). The product $h\Delta$ is the distance between the neutral axis and the bottom surface. h is the thickness of the upper beam.

equation, $C = C(a)$ may be obtained by curve-fitting. Ideally, the form of the compliance expression, $C = C(a)$, is known from experience or analysis. Differentiation of $C(a)$ with respect to crack length according to Equation (10.1) yields G. This approach, however, may not work well for tests where the compliance changes very little with crack length, or where the form of $C = C(a)$ is not guided by analysis.

As an alternative to a complete solution for the compliance of a fracture specimen Yin and Wang (1984) proposed consideration of a cut-out section from a cracked laminate where axial loads and moments are applied on the edges of the cut-out sections. They developed an analytical procedure to calculate G for a cracked monolithic composite based on the J integral This method was modified and later extended to bimaterial specimens with an interface crack by Suo and Hutchinson (1990) (see Figure 10.1).

However, a sandwich specimen is not homogeneous or bimaterial. A sandwich specimen could be considered as a trimaterial with two faces that can be different and a core. For this case the analysis presented in Section 9.4 should apply (see also Kardomateas et al., 2010).

Most sandwich test specimens are loaded by transverse shear forces. Until recently such loads have not been considered in crack element formulations. Li et al. (2004) and Thouless (2009) developed a finite element approach where transverse shear forces acting on the cracked element are included (see Figure 10.2). Li et al. (2004) found that shear loading causes "root rotation" of the crack tip. Through extensive finite element computations, they found that such rotations affect not only the energy release rate but also the mode mixity, as quantified by the phase angle ψ

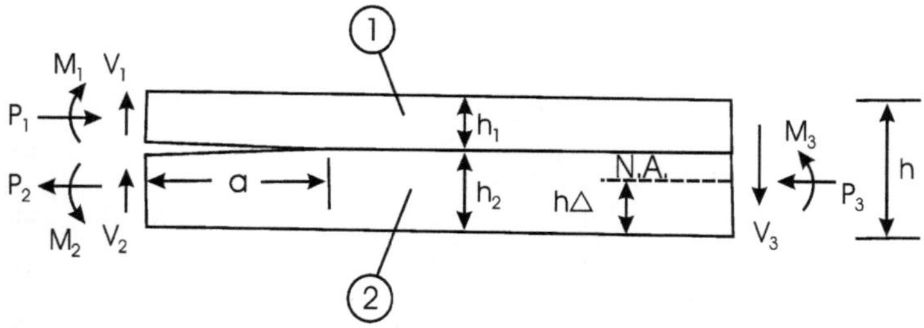

Figure 10.2 Bending moment, axial loads and transverse shear forces acting on segments of a cracked beam geometry. (After Thouless, 2009)

$$\psi = \tan^{-1}\left(\frac{K_{\mathrm{II}}}{K_{\mathrm{I}}}\right), \tag{10.2}$$

where K_{I} and K_{II} are the mode I and mode II stress intensity factors (assuming $\beta = 0$) defined in Chapter 9. The shear effect is greatest for specimens with short crack lengths (compared to the thickness). For long crack lengths, the shear contribution becomes negligible and the previous methods based on axial forces and edge moments only should be valid.

The cracked element approach has many merits. Provided the basic loading parameters, axial load, moment, and transverse shear force are identified, general solutions to difficult problems can be obtained, see a recent paper by Thouless (2009).

10.2 Debond Test Specimens

Several debond sandwich test specimens have emerged. A successful debond test should promote the desired face/core debond propagation failure before any competing failure mode, such as core shear, core crush, indentation failure, face wrinkling, or crack kinking, occurs. Commonly, the test specimens have to be properly designed in order to promote debond growth, and guidelines will be provided in this chapter.

Figure 10.3 shows some of the more widely used debond test specimens for sandwich constructions, viz., the double cantilever beams (DCB), tilted sandwich debond (TSD), cracked sandwich beam (CSB), three-point sandwich beam (TPSB), mixed mode bending (MMB), and DCB-uneven bending moment (DCB-UBM) specimens, each of uniform width (b) and loaded in bending. The specimens shown in Figures 10.3a–d are so-called “fixed mode ratio specimens”, since the mode ratio is fixed by the material combi-

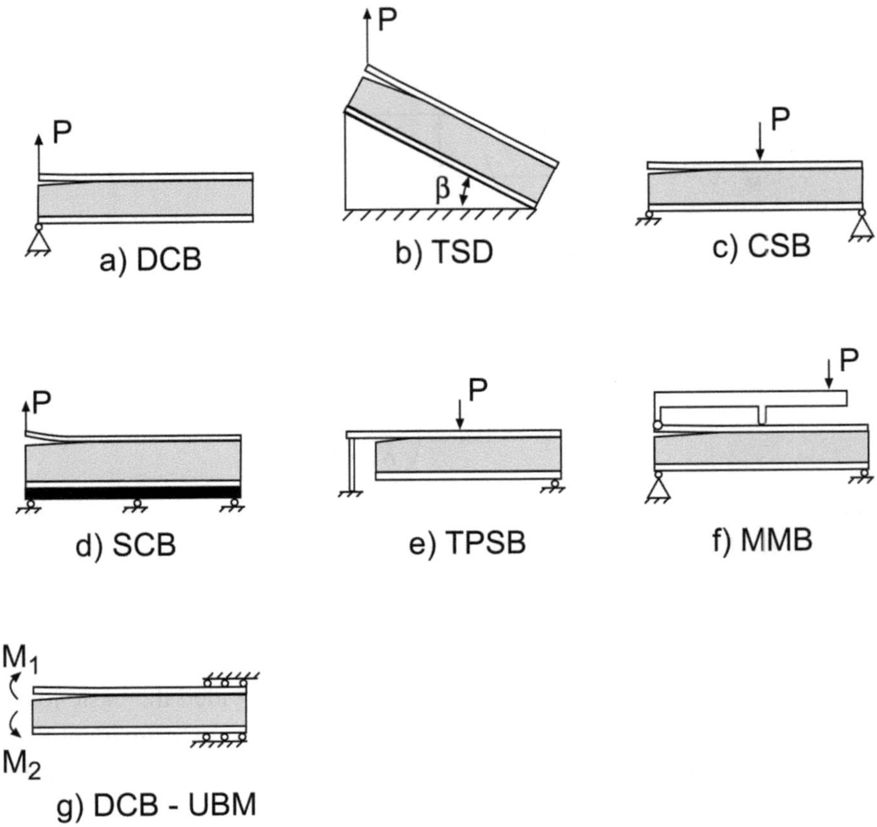

Figure 10.3 Debond sandwich test specimens. (a) DCB, (b) TSD, (c) CSB, (d) SCB, (e) TPSB, (f) MMB, and (g) DCB-UBM.

nation, loading configuration, and specimen geometry, while the specimens in Figures 10.3e and f allow adjustment of the mode mixing. Each of these specimens will be described and available expressions for the compliance and energy release rate will be provided.

10.3 Double Cantilever Beam (DCB) Specimen

The double cantilever beam (DCB) specimen is a very popular test for determining the mode I delamination resistance of laminated composites, and is standardized by ASTM (ASTM D5528, 2001). In DCB testing of composites, the initial delamination is placed symmetrically at the mid-plane. For

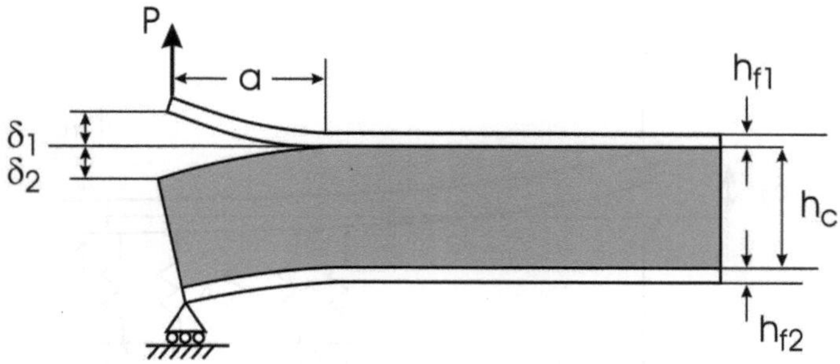

Figure 10.4 Geometry and loading of the sandwich DCB specimen.

the sandwich DCB test, the initial precrack is placed between the upper face sheet and the core to promote face/core debonding (see Figure 10.4). The sandwich DCB specimen was first adopted by Prasad and Carlsson (1994) who conducted testing and finite element analysis of the specimen. They showed that the sandwich DCB specimen is not a pure mode I test as a result of the off-centered crack at a bimaterial interface between widely dissimilar materials. In many cases, experimental testing revealed that crack propagation did not occur at the face/core interface. The crack kinked into the core, as will be discussed later in this section. Shivakumar et al. (2004), however, successfully achieved face/core debond fracture in experimental studies using the sandwich DCB specimen and this will be further discussed in Chapter 11.

In this section, elastic foundation analysis of the compliance and energy release rate of the sandwich DCB specimen will be reviewed. The upper leg of the DCB specimen, i.e. the debonded face sheet (Figure 10.4) is considered as a cantilever beam of effective flexural modulus E_{f1} and thickness, h_{f_1}. The lower leg consists of the lower face, of modulus E_{f2}, and thickness h_{f2}, bonded to a core of modulus, E_c, and thickness h_c. Under load, the loading point (1) displaces an amount $\delta_1 + \delta_2$, where the individual displacements δ_1 and δ_2 are defined in Figure 10.4.

Figure 10.5 illustrates the elastic foundation model (EFM) of the DCB specimen developed by Aviles and Carlsson (2007a). The bonded part of the upper face sheet is supported by the core represented by an elastic foundation. The total specimen length is L and a is the crack length. The elastic foundation is characterized by the foundation modulus k.

The analysis is based on the Winkler foundation model, first applied to isotropic and symmetric DCB specimens by Kaninnen (1973). The Winkler model assumes that the reaction forces in the elastic foundation are propor-

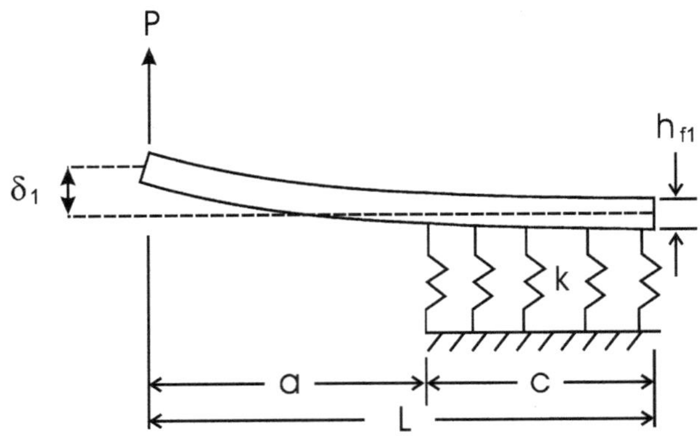

Figure 10.5 Schematic of elastic foundation model (EFM) of a sandwich DCB specimen.

tional to the beam deflection at any point. The foundation modulus, k, is defined as the force required to displace a unit area of the face through a unit distance in the thickness direction. k may be related to the extensional out-of-plane stiffness of the core (Allen, 1969; Kanninen, 1973; see also Chapter 8),

$$k = \frac{2E_c b}{h_c}, \tag{10.3}$$

where b is the width of the specimen. Quispitupa et al. (2009), however, argued that this equation in effect assumes that one half of the core is active as a foundation which is not realistic for thick cores. Quispitupa et al. (2009) proposed the following modified elastic foundation modulus for a sandwich DCB specimen

$$k = \frac{2E_c b}{h_{f1}}. \tag{10.4}$$

Expressions for compliance and energy release rate of a symmetric DCB sandwich specimen were derived by Aviles and Carlsson (2007a),

$$C = \frac{a}{b}\left[\frac{1}{h_c G_{xz}} + \frac{a^2}{3\left(D - \frac{B^2}{A}\right)}\right] + \frac{4}{E_{f1} h_{f1}^3 b}\left[a^3 + 3a^2\eta^{1/4} + 3a\eta^{1/2} + \frac{3}{2}\eta^{3/4}\right], \tag{10.5}$$

$$G = \frac{P^2}{2b^2}\left[\frac{1}{h_c G_{xz}} + \frac{a^2}{\left(D - \frac{B^2}{A}\right)} + \frac{12}{E_{f1}h_{f1}^3}\left[a^2 + 2a\eta^{1/4} + \eta^{1/2}\right]\right], \tag{10.6}$$

$$\eta = \frac{h_{f1}^3 b E_{f1}}{3k}, \tag{10.7}$$

where P is the applied load and the A, B and D terms are the extensional, coupling and bending stiffnesses defined for a general laminated beam in Equations (9.27). For the lower leg of the DCB specimen (lower face bonded to the core)

$$A = E_{f2}h_{f2} + E_c h_c, \tag{10.8a}$$

$$B = h_{f2}h_c \frac{E_c - E_{f2}}{2}, \tag{10.8b}$$

$$D = \frac{1}{12}\left[E_{f2}\left(h_{f2}^3 + 3h_{f2}h_c^2\right) + E_c\left(h_c^3 + 3h_{f2}^2 h_c\right)\right]. \tag{10.8c}$$

To illustrate the foundation effect on the DCB specimen compliance, we will consider a specific (baseline) sandwich DCB specimen. The specimen was obtained from a symmetric sandwich consisting of 2.41 mm thick glass/vinylester face sheets over a 37.9 mm thick H100 PVC foam core. The core is assumed to be isotropic with mechanical properties $E_c = 105$ MPa, $\nu_c = 0.32$, and $G_{xz} = 39.8$ MPa. The mechanical properties of the face sheets are $E_f = 27.6$ GPa and $\nu_f = 0.32$. The face modulus and Poisson ratio refer to loading along the beam axis. In addition, DCB specimens with a range of core moduli $E_c = 10$–800 MPa and a range of total lengths $L = 5$–50 cm were considered. The core shear modulus, G_{xz}, was calculated from E_c using the isotropic relation between the Young's and shear moduli assuming a constant Poisson ratio $\nu_c = 0.32$. The crack length was fixed ($a = 5$ cm).

Figure 10.6a shows the specimen compliance as a function of core modulus. The compliance depends quite strongly on the core modulus and reaches high values for compliant cores ($E_c < 20$ MPa). Figure 10.6b shows the compliance vs. specimen length. The compliance becomes independent of specimen length above a certain length. When the specimen length decreases and becomes comparable to the crack length, however, the compliance increases sharply due to lack of support of the loaded upper face sheets.

This analysis may be used to determine an upper limit on the crack extension for a given test specimen. Calculations by Aviles and Carlsson (2007a) reveal that end-effects are negligible if the crack length is below

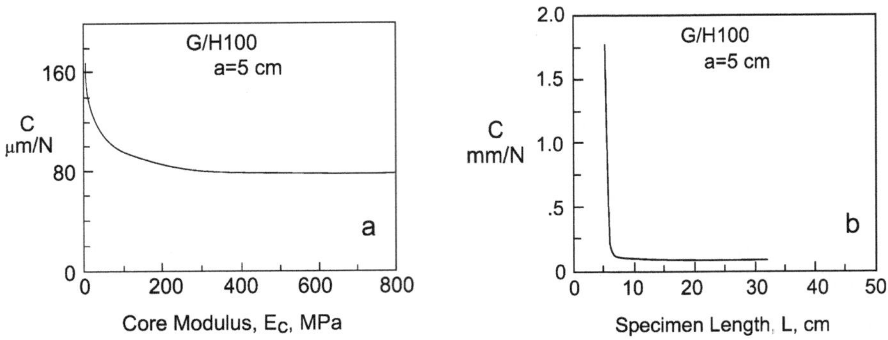

Figure 10.6 DCB specimen compliance. (a) Influence of core modulus, (b) influence of specimen length ($a = 5$ cm).

$$a \leq L - \frac{4.5}{\lambda}, \tag{10.9}$$

where λ is given by

$$\lambda = \sqrt[4]{\frac{3k}{E_{f1}bh_{f1}^3}}. \tag{10.10}$$

For the specific DCB specimen considered here, Equation (10.8) yields $a/L < 0.63$. Hence, to avoid influence of end-effects on the compliance, fracture testing should stop once the crack length, a, reaches 0.63 L.

The accuracy of the analytical model for the DCB compliance was evaluated by Quispitupa et al. (2009), using detailed two-dimensional finite element analysis (FEA). DCB specimens of total length $2L \approx 150$ mm, $b = 35$ mm, $h_{f1} = h_{f2} = 2$ mm and core thicknesses h_c of 10, 20, and 30 mm were analyzed over a range of crack lengths from 5 to 65 mm. The face and core material were E-glass/polyester and H100 PVC foam. The face and core moduli were $E_{f1} = E_{f2} = E_f = 16.4$ GPa, and $E_c = 135$ MPa. The compliance C and energy release rate G were determined from the elastic foundation model, Equations (10.5) and (10.6), and FEA.

Compliance and energy release rate results are shown vs. crack length in Figure 10.7. The energy release rate was calculated using a unit load, $P = 1$ N/mm. Predictions of C and G using the foundation model are in excellent agreement with FEA for the range of face and core materials and geometries considered.

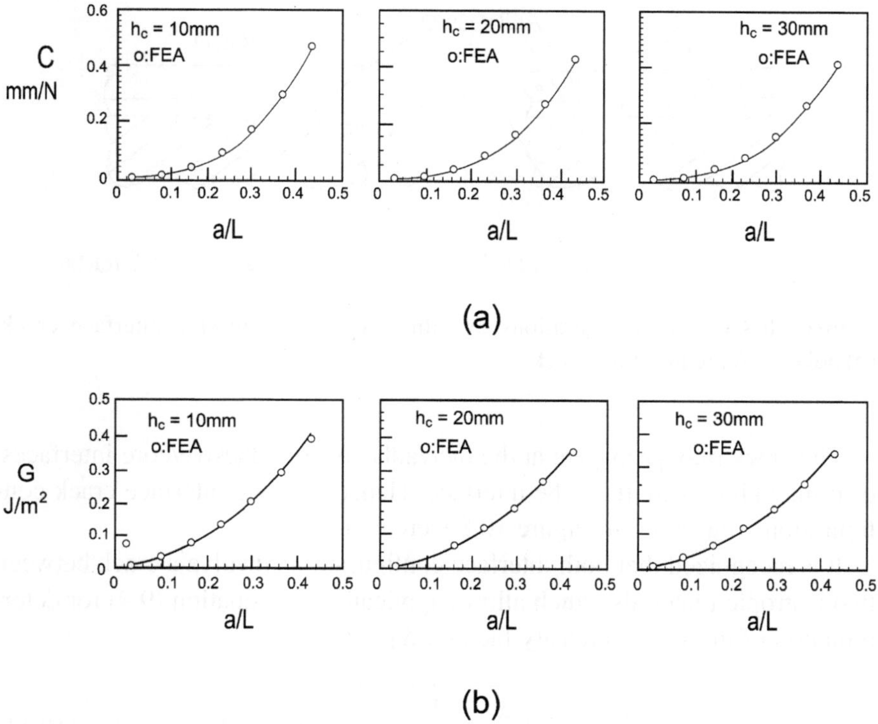

Figure 10.7 Compliance and energy release rate of sandwich DCB specimens. Open circles represent FEA, and the continuous line the elastic foundation model (after Quispitupa et al., 2009). For the calculation of G, a load $P = 1$ N/mm was used.

10.3.1 Crack Kinking Analysis

Crack kinking analysis of a DCB sandwich specimen will be discussed for some specific specimens. Details and assumptions of the crack kinking analysis are outlined in Section 9.2. Two-dimensional, plane strain finite element models of foam cored DCB specimens with aluminum face sheets were constructed to calculate the crack tip stress intensity factors, K_{I} and K_{II} (Prasad and Carlsson, 1994a). The face sheets were 2.2 mm thick and had the following mechanical properties: $E_f = 70$ GPa and $\nu_f = 0.3$. The adhesive layer between face and core was assumed to be 0.1 mm thick, with $E_a = 3.5$ GPa and $\nu_a = 0.35$. The core was 20 mm thick and was considered to have a range of properties from "stiff" to "soft"; $E_c = 9.7 - 0.28$ GPa, and $\nu_c = 0.35$. The crack length was 25.4 mm, and the total specimen length was 152 mm.

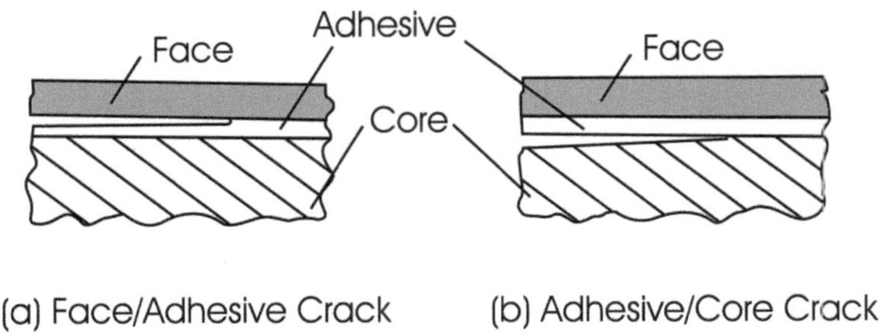

Figure 10.8 Crack configurations examined. (a) Face/adhesive interface crack, (b) adhesive/core interface crack.

The crack may propagate at the face/adhesive or adhesive/core interfaces, or it may kink away from the interface. Hence, the two interface crack configurations illustrated in Figure 10.8 were considered.

It is recognized that both interface configurations involve a crack between two isotropic materials which allows application of Equation (9.9) for determination of the stress intensity factors, K_{I} and K_{II},

$$\delta_{\mathrm{I}} + i\delta_{\mathrm{II}} = 4\sqrt{\frac{x}{2\pi}\left(\frac{1}{\overline{E}_1} + \frac{1}{\overline{E}_2}\right)}(K_{\mathrm{I}} + iK_{\mathrm{II}}). \tag{10.11}$$

Subscripts 1 and 2 denote the material number above and below the crack plane. For the face/adhesive crack configuration, material #1 is aluminum while material #2 is epoxy, and for the adhesive/core configuration, material #1 is epoxy and material #2 is the core. δ_I and δ_{II} are the opening and sliding crack face displacements illustrated in Figure 9.4, and $\overline{E} = E$ for plane stress, and $E/(1-\nu^2)$ for plane strain. Each specimen was loaded by a unit load ($P = 1$ N/mm).

A complete analysis to determine whether or not the crack tip would continue to propagate as an interface crack, or if it would kink, would involve Equation (9.26),

$$\frac{G_{\max}^K}{G} > \frac{G_{IC}}{G_c}, \tag{10.12}$$

where $G_{\max}^k$ is the maximum energy release rate for the kinked crack, G is the energy release rate for the interface crack, G_{IC} the mode I fracture toughness of the core, and G_c the interface fracture toughness.

Hence, characterization of crack kinking requires elaborate analysis and key material toughnesses. It should be pointed out that the determination of the interface toughness, G_c, may not be possible if the interface is tough and if kinking occurs. In such a case, it is still possible to assess kinking, but in a more qualitative manner by examining the angle, Ω, the interface crack would deflect if it were to kink. A negative kink angle, $\Omega \leq 0$, would indicate interface growth (no kinking) or kinking up into face sheet, which is not physically impossible if the face sheets are tough. On the other hand, a positive angle $\Omega > 0$, indicates the tendency for kinking into the core which is possible for brittle polymer foams.

The kink angle, Ω, of the foam core DCB specimens considered is first calculated from Equation (9.24), derived for a mixed mode crack in a homogeneous, isotropic brittle material. The kink angle was also determined from the rigorous analysis of He and Hutchinson (1989), with kink angles depicted in graphical form in their paper. The He and Hutchinson kinking analysis requires specification of Dundurs' (1969) elastic bimaterial mismatch parameter, α,

$$\alpha = \frac{\overline{E}_1 - \overline{E}_2}{\overline{E}_1 + \overline{E}_2}, \tag{10.13}$$

where $\overline{E}$ is defined under Equation (10.11). The parameter α ranges from -1 to 1, where the limits are approached when one material is much stiffer than the other. If the materials 1 and 2 above and below the crack plane are the same, $\alpha = 0$. For the foam core sandwich DCB specimens considered here, α is close to the upper limit ($\alpha \approx 1$). The analysis of He and Hutchinson (1989) provides the kink angle Ω for material combinations with α within $-0.75 < \alpha < 0.75$. For some of the extreme cases considered here, α falls outside this range. For such cases the results for $\alpha = 0.75$ are used.

Figure 10.9 displays kink angle results for the foam core DCB specimens with face/adhesive and adhesive/core interface cracks over the range of core moduli investigated. For DCB specimens with a stiff core ($E_f/E_c \leq 20$) the kink angle is negative and such specimens are not expected to display kinking down into the core ($\Omega < 0$). Kinking up into the tough aluminum face sheets is highly unlikely. For a DCB specimen with a low modulus core, however, the positive value of Ω indicates that an interface crack may leave the interface and enter into the core at an angle which somewhat depends on the actual crack configuration (Figure 10.8). The kink angle is larger for the face/adhesive interface crack than for the adhesive/core crack. It is furthermore observed in Figure 10.9 that the crack kinking analysis of He and Hutchinson (1989), labeled "bimaterial", consistently predicts a larger kink angle than Equation (9.24), labeled "homogeneous", but the difference is less than 5° for all cases investigated.

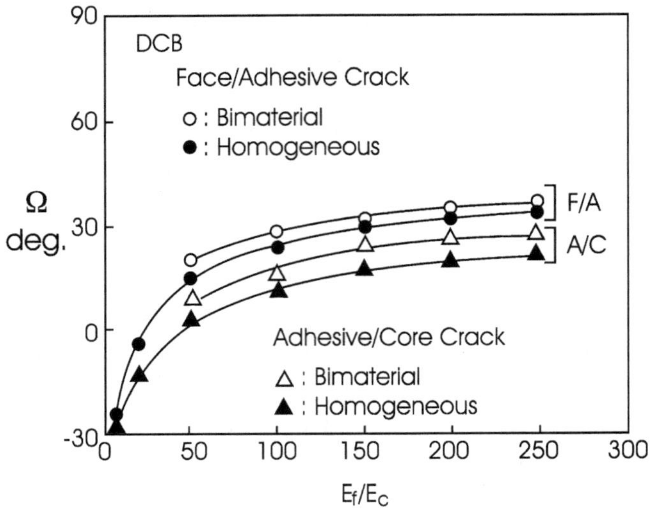

Figure 10.9 Kink angle vs. face/core modulus ratio for DCB sandwich specimens with aluminum face sheets ($a = 25.4$ mm).

10.4 Tilted Sandwich Debond (TSD) Specimen

The tilted sandwich debond (TSD) specimen, shown in Figure 10.10 was introduced as a debond test for foam cored sandwich specimens by Li and Carlsson (1999). The specimen is tilted at an angle, θ, and loaded by a vertical force, P. This force may be resolved into axial and normal components, P_A, and P_N,

$$P_A = P \sin\theta, \tag{10.14a}$$

$$P_N = P \cos\theta. \tag{10.14b}$$

It was initially thought that the TSD specimen would allow mixed mode debond testing. By changing the tilt angle, θ, the mode ratio, e.g. K_{II}/K_I, would also change. As will be discussed, however, this idea is not supported by detailed analysis. Still, testing of foam cored sandwich specimens reveals that this specimen configuration is less prone to crack kinking than the DCB specimen discussed in Section 10.4.

Analysis of the TSD specimen based on elastic foundation modeling has been presented by Li and Carlsson (2000). Figure 10.11 defines several of the geometry symbols, such as the crack length a and the bonded length l. The loaded face sheet (Figure 10.11) is considered a beam on an elastic foundation. The applied load P may be resolved into axial and normal components, Equations (10.14). The analysis is based on superposition of solutions for the

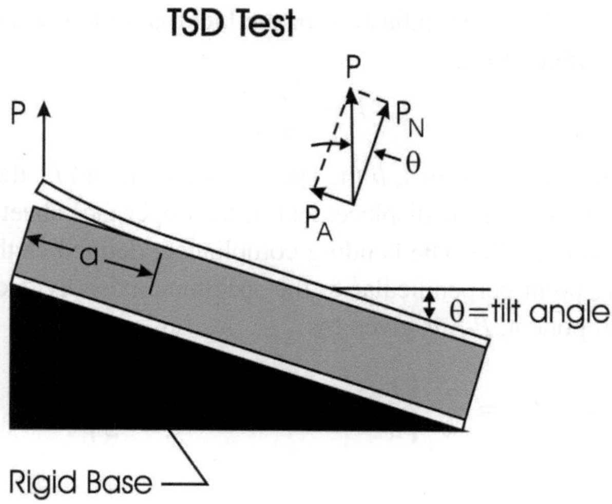

Figure 10.10 TSD specimen. θ is the tilt angle.

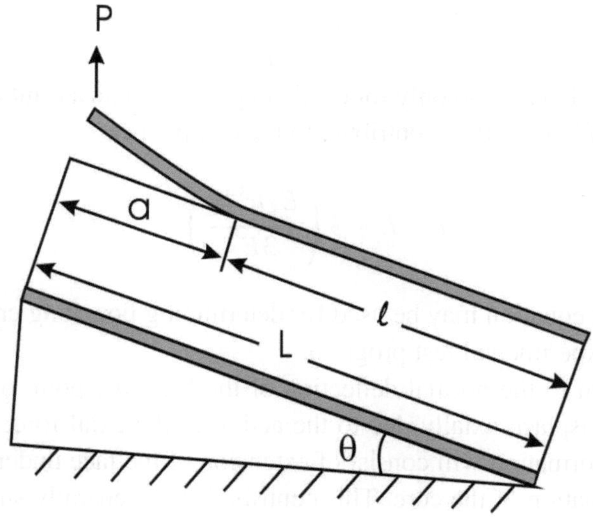

Figure 10.11 TSD specimen loading and geometry.

face sheet being subjected to an edge force and edge moment at the left end of the bonded region (at the crack tip). In addition, the face will deflect due to the normal force component, P_N, and the core will deform in shear due to the axial component, P_A.

Li and Carlsson (2000) used a one-parameter foundation model for the bonded region, with a foundation modulus, k, given by the classical expression, Equation (10.3), here adjusted for the fact that only one face is loaded in the TSD configuration,

$$k = \frac{E_c b}{h_c}, \tag{10.15}$$

where E_c is the core modulus, b the specimen width, and h_c the core thickness. An expression for the displacement of the upper face sheet was derived (Li and Carlsson, 2000). The bending compliance, defined as the deflection of the loading point perpendicular to the specimen axis, divided by the normal force component, P_N, is given by

$$C_1 = \frac{4\beta}{k}\left\{\frac{1}{3}\beta^3 a^3 + \beta^2 a^2 + \beta a + \frac{1}{2}\right\}, \tag{10.16}$$

where

$$\beta = \left(\frac{k}{4E_f I_f}\right)^{1/4}, \tag{10.17a}$$

$$I_f = \frac{h_f^3}{12} \tag{10.17b}$$

Equation (10.16) is valid only for crack lengths less than a limit crack length, where end-effects start to contribute to the compliance

$$a \leq L - 3\left(\frac{E_f h_f^3 h_c}{3E_c}\right)^{1/4}. \tag{10.18}$$

This explicit equation may be used for determining how long cracks may be used in an experimental test program.

In addition to the normal deflection of the face, the point of load application will displace axially due to the action of the axial force component, P_{II}. This deformation will consist of extension of the face under tension and shear deformation of the core. This contribution is generally small and may be neglected. For such a case, the load point compliance, $C = \delta/P$, where δ is the vertical displacement component, becomes

$$C = \frac{4\beta}{k}\left\{\frac{1}{3}\beta^3 a^3 + \beta^2 a^2 + \beta a + \frac{1}{2}\right\}\cos^2\theta. \tag{10.19}$$

Differentiation of Equation (10.19) with respect to crack length, yields the energy release rate

$$G = \frac{4\beta P^2}{2bk}(\beta a + 1)^2 \cos^2\theta. \tag{10.20}$$

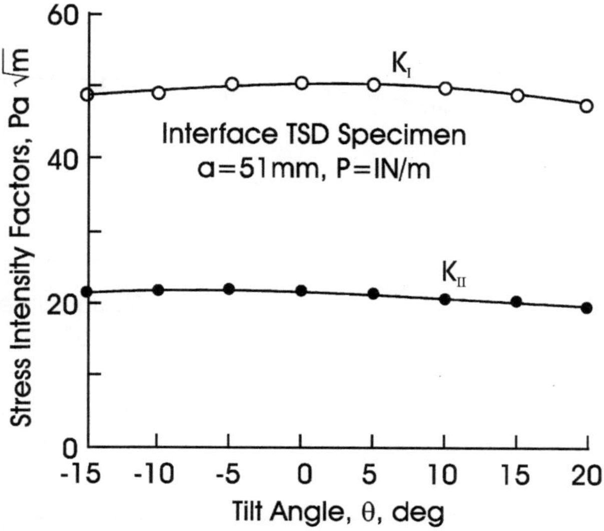

Figure 10.12 Interface stress intensity factors vs. tilt angle for the TSD specimen.

10.4.1 Mode Mixity Analysis

Several TSD configurations were analyzed by Li and Carlsson (2001). We will here reproduce results for the "interface configuration" where a 51 mm long crack is supposed to lie on the interface between a 3.6 mm thick face sheet and a 50 mm thick core. The glass/epoxy face and H100 core moduli were: $E = 21.2$ GPa and $E_c = 99$ MPa. To examine the influence of tilt angle, θ on the interface stress intensity factors, K_{I} and K_{II}, finite element calculations were conducted over a range of tilt angles (–15° to 20°).

Figure 10.12 displays K_{I} and K_{II} vs. tilt angle θ. It is observed that K_{I} and K_{II} remain essentially independent of the tilt angle. Because $K_{\mathrm{II}} > 0$, for the tilt angles considered, the face/core crack would have a tendency to kink down into the core ($\Omega > 0$) and would do so unless the interface is weak and the core is tough.

Further analysis of the stress intensity factors was conducted for a TSD specimen at zero tilt angle over a range of core stiffnesses. In this analysis, the face sheet thickness and modulus were kept as above while the core modulus was varied. The kink angle, Ω, was calculated from Equation (9.24) based on the stress intensity factors K_{I} and K_{II}. Figure 10.13 shows the kink angle plotted vs. the face-to-core modulus ratio, E_f/E_c. For modulus ratios greater than about 20, the kink angle changes sign from negative to positive. Above this modulus ratio, crack kinking into the core is a possibility.

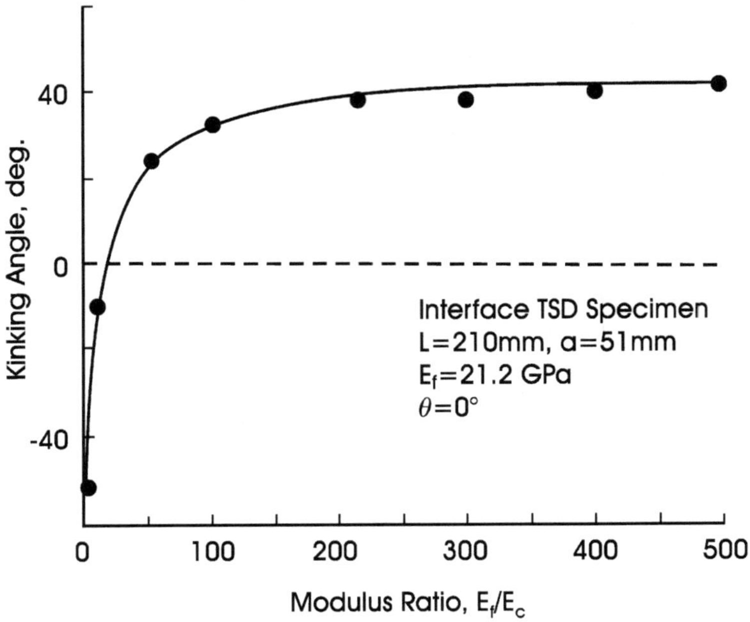

Figure 10.13 Kink angle for interface TSD specimen vs. face-to-core modulus ratio.

It should be pointed out that, in practice, most foam core sandwich panels have large face-to-core modulus ratios $E_f/E_c > 20$, and such panels may be prone to kinking behavior for such loading.

10.5 Cracked Sandwich Beam (CSB) Specimen

One of the earliest proposed debond tests is the cracked sandwich beam (CSB). This test was introduced by Carlsson et al. (1991) in an effort to determine the mode II fracture toughness, $G_{\mathrm{II}c}$, of the face/core interface. The test is an extension of the mode II end-notched-flexure (ENF) test introduced by Barrett and Foshi (1977) for testing wooden beams, and later applied to composite laminates by Russell and Street (1982). Figure 10.14 illustrates the test principle and the state of stress in an element near the crack tip.

The sign of the interlaminar shear stress, τ_{xz}, is negative ($K_{\mathrm{II}} < 0$) and, hence, crack kinking into the core is not an issue. The core element shown in Figure 10.14, however, is loaded in compression and may fail in a crushing mode. Such failure may be avoided by a specific design of the test specimen.

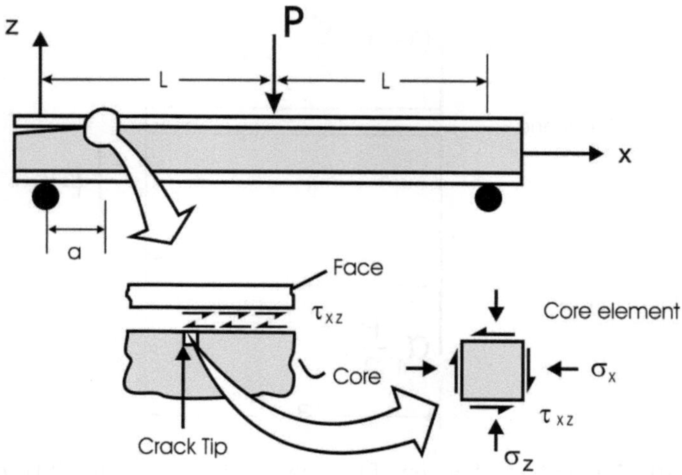

Figure 10.14 The cracked sandwich beam (CSB) test, and state of stress near the crack tip.

The CSB specimen was analyzed by Carlsson et al. (1991) using first-order shear deformation theory, Chapters 3 and 4. The analysis is limited to a symmetric sandwich, but could be extended to sandwich specimens with different face sheets on top and bottom. The presence of a debond will make the specimen more compliant which provides the crack driving force, G. To determine the energy release rate, G, an expression for the CSB compliance as a function of crack length, a, was derived:

$$C = \frac{L^3}{6bD_i} + \frac{L}{2h_c bG_{xz}} + \frac{a^3}{12b}\left[\frac{1}{D_d} - \frac{1}{D_i}\right], \tag{10.21}$$

where L is the half-span length, D_i and D_d are the flexural stiffnesses per unit width of the intact and debonded regions of the specimen, b is the width of the beam, h_c is the core thickness, and G_{xz} is the core shear modulus. The flexural stiffness of the intact region is (Chapter 4)

$$D_i = \frac{E_f h_f}{2}\left(h_c + h_f\right)^2 + \frac{E_f h_f^3}{6} + \frac{E_c h_c^3}{12}, \tag{10.22}$$

where E_f and E_c are the core face and core Young's moduli and h_f is the face thickness. To determine the effective flexural stiffness of the debonded region, D_d, consider the free-body diagram (Figure 10.15). The two regions, 1 and 2, represent the upper and lower faces bonded to the core. The shear forces carried by the upper and lower faces are determined from force equilibrium and compatibility of deformation at the left end.

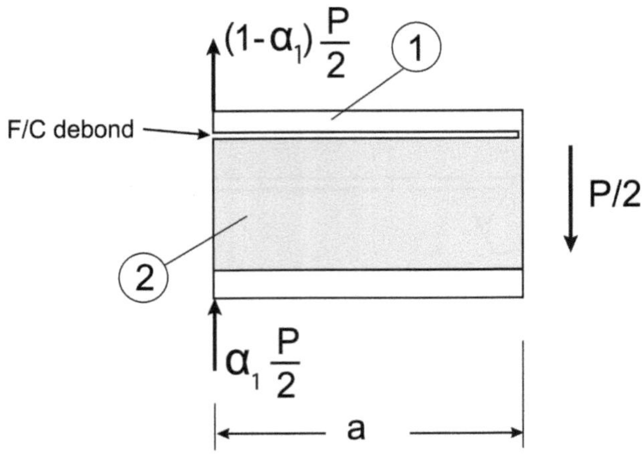

Figure 10.15 Free-body diagram of the debonded region of the CSB specimen.

Analysis of the load partitioning in this region leads to the following expression for the effective flexural stiffness of the debonded region:

$$D_d = (1 - \alpha_1) D_2, \tag{10.23}$$

where α_1 is the load partitioning parameter given by

$$\alpha_1 = \frac{1 + \frac{3D_2}{a^2 h_c G_{xz}}}{1 + \frac{3D_2}{a^2 h_c G_{xz}} + \frac{D_2}{D_1}}, \tag{10.24}$$

in which D represents flexural stiffness per unit width and the subscripts 1 and 2 refer to the upper and lower sub-beams of the debonded region (Figure 10.15)

$$D_1 = \frac{E_f h_f^3}{12}, \tag{10.25a}$$

$$D_2 = D - \frac{B^2}{A}. \tag{10.25b}$$

The A, B, and D terms are the extensional, coupling, and bonding stiffnesses defined in Chapter 9. The energy release rate of the CSB specimen is obtained by differentiation of Equation (10.21) with respect to crack length

$$G = \frac{P^2 a^2}{8b^2} \left[\frac{1}{D_d} - \frac{1}{D_i} \right]. \tag{10.26}$$

The accuracy of the compliance and energy release rate predictions for the CSB specimen have been examined by Quispitupa et al. (2009). The same

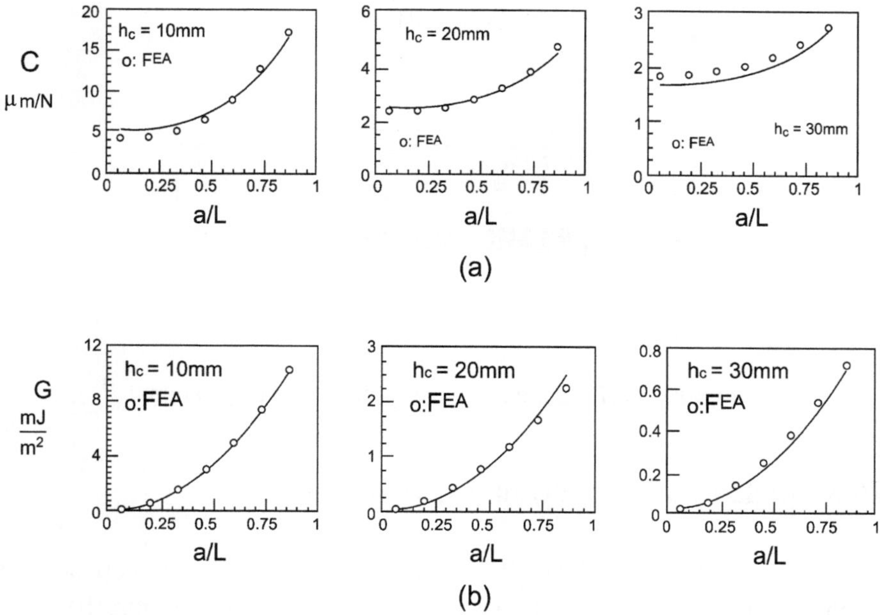

Figure 10.16 Influence of core thickness on the compliance (a) and energy release rate (b) of the CSB specimen. Open circles represent FEA and continuous lines the beam model.

materials and geometries as in the analysis of the DCB specimen (Section 10.3) were used in a detailed finite element analysis of the CSB specimen. Analytical and finite element results for the compliance and energy release rate ($P = 1$ N/mm) are displayed vs. crack length in Figure 10.16. Compliance values determined analytically and numerically are in close agreement, except for short crack lengths (Figure 10.16a). Differences could be attributed to the contact pressure developed between the upper and lower sub-beams in the debonded region. The analytical formulation models load transfer between the lower part of the beam to the upper face sheet through a concentrated force, while the finite element model includes frictionless contact surfaces between the upper and lower sub-beams in order to achieve load transfer for the debonded region. The two contact definitions are not identical and small variations between these two models might be expected.

For the energy release rate, Figure 10.16b shows that the beam theory and finite element results agree closely, lending confidence to the beam theory modeling.

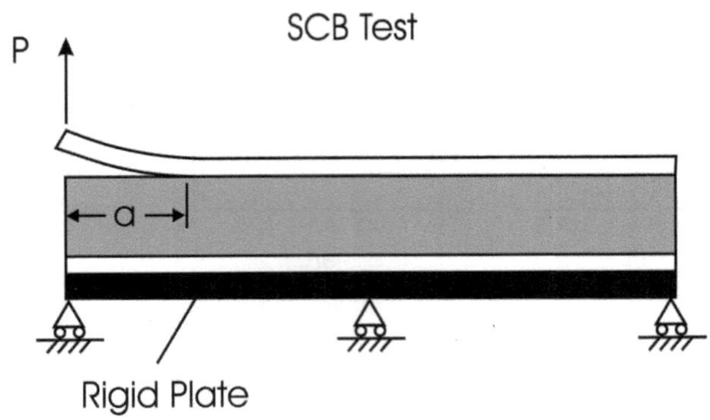

Figure 10.17 Single cantilever beam test configuration.

10.6 Single Cantilever Beam (SCB) Specimen

Cantwell and Davies (1994, 1996), introduced a sandwich debond test method called the "single cantilever beam" (SCB) shown in Figure 10.17.

The bottom surface of the SCB specimen is bonded to a rigid steel plate mounted on a carriage supported by linear roller bearings. Load is applied using a long hinged vertical bar. Since the specimen is attached to the roller-supported base, horizontal forces will not be introduced in the specimen as the upper loaded face deflects. No fracture mechanics of this test has been presented, although the analysis presented for the TSD specimen in Section 10.4 should also be valid for the SCB specimen by setting the tilt angle, $\theta = 0$ in the TSD analysis. Cantwell and Davies (1994, 1996) used this specimen in experimental studies and determined the face/core debond fracture toughness, G_c, from the experimental compliance calibration method. Discussion of the experimental aspects of the test, data reduction, and test results will be presented in Chapter 11.

10.7 Three-Point Sandwich Beam (TPSB) Specimen

Cantwell et al. (1999) proposed a debond test called "three-point bend sandwich beam (TPSB) test", with a support placed under the upper face sheet as shown in Figure 10.18. This arrangement was made possible by removing a section of the core and lower face at the left end of the specimen. This loading arrangement creates a mixed mode I and mode II loading at the crack tip.

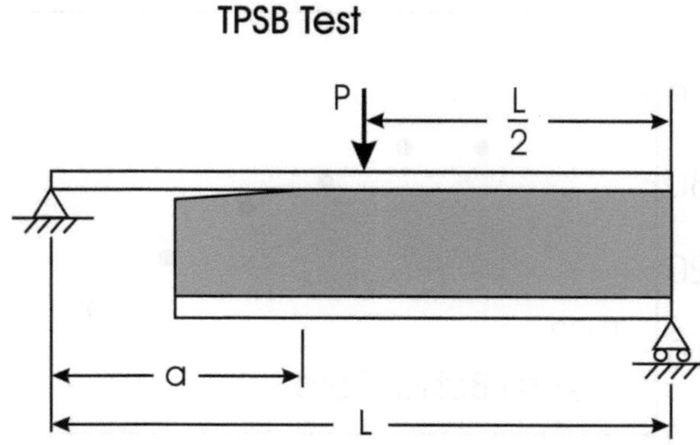

Figure 10.18 Three-point bend sandwich beam (TPSB) test.

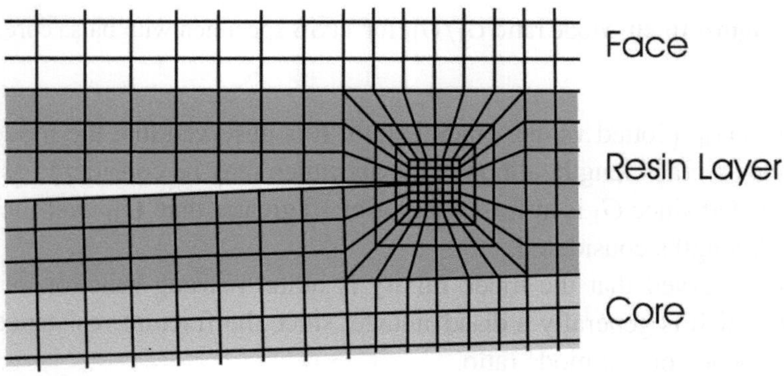

Figure 10.19 Finite element mesh of crack tip region for TPSB specimen.

To determine the strain energy release rate components, G_{I} and G_{II}, and the mode mixity, expressed as $G_{\mathrm{I}}/G_{\mathrm{II}}$, Cantwell et al. (1999) conducted plane strain finite element analysis of the TPSB specimen, with a refined mesh near the crack tip, see Figure 10.19. The crack tip is embedded in a thin resin layer between face and core so as to circumvent complications due to the bimaterial crack tip. A range of crack lengths, a, from 30 to 55 mm was considered. The sandwich beams consisted of 2.5 mm thick glass/polyester faces over a 15 mm thick end-grain balsa core of 175 kg/m^3 density. Cantwell et al. (1990) did not provide the material properties of the face sheets and core, but they should be close to those listed for similar materials in Chapter 1.

The opening and sliding components, G_{I} and G_{II} of the energy release rate, G, were determined by the virtual crack closure technique described in Section 9.1.1. Figure 10.20 shows the mode mixity results presented as the

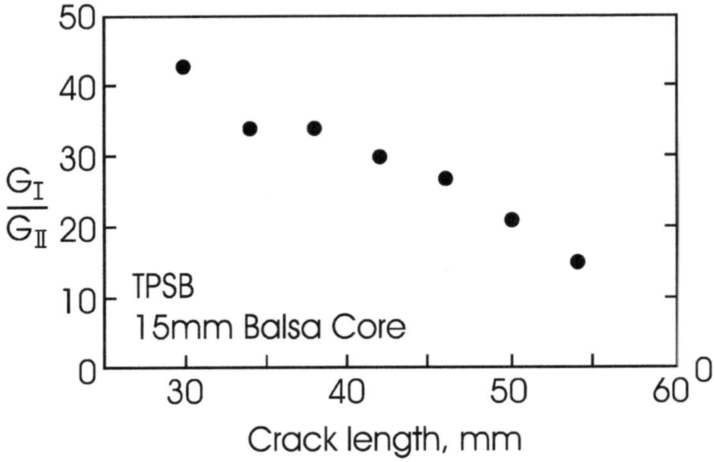

Figure 10.20 Mode ratio $G_{\mathrm{I}}/G_{\mathrm{II}}$ for TPSB specimen with balsa core.

ratio $G_{\mathrm{I}}/G_{\mathrm{II}}$, plotted vs. the crack length. It is observed that the mode ratio depends on crack length, although the specimen may be considered as mode I-dominated since G_{I} is at least a factor of 10 greater than G_{II} over the range of crack lengths considered.

It is observed that the mode mixity is quite a strong function of crack length which is generally a disadvantage, since the fracture resistance typically depends on the mode ratio.

10.8 Mixed Mode Bending (MMB) Specimen

The mixed mode bending (MMB) test was originally developed by Reeder and Crews (1990) for mixed mode delamination fracture characterization of unidirectional composites. This test was recently modified to accommodate sandwich specimens by Quispitupa et al. (2009), see Figure 10.21. The MMB sandwich specimen incorporates a through-width face/core crack at the left edge of the specimen. A vertical, downward load, P, applied to the lever arm, provides an upward directed load at the left end of the debonded face sheet and a downward directed load at the center.

The MMB specimen can be considered as a superposition of the previously discussed CSB and DCB specimens, see Figure 10.22. Analytic expressions for the MMB compliance and energy release rate for symmetric sandwich specimens ($h_{f1} = h_{f2} = h_f$ and $E_{f1} = E_{f2} = E_f$) were derived

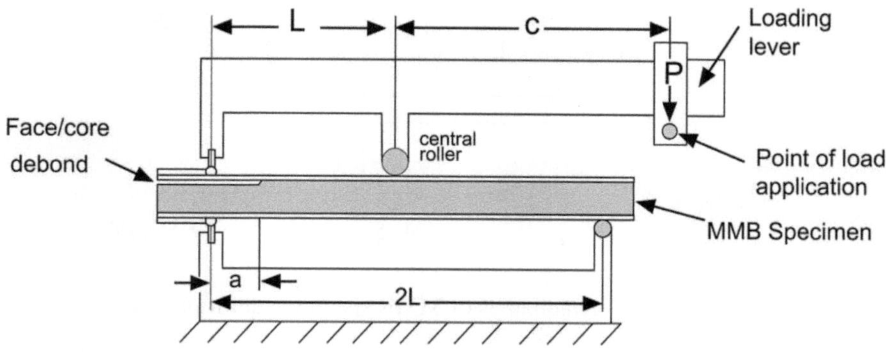

Figure 10.21 MMB test principle and sandwich specimen.

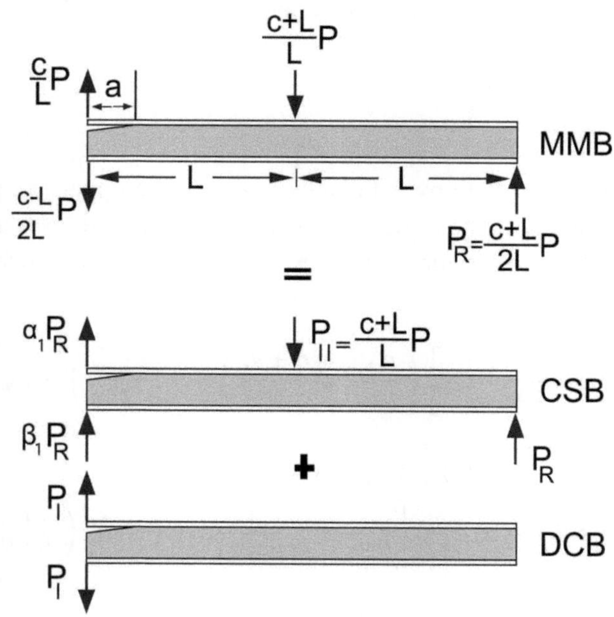

Figure 10.22 Mixed mode loading decomposed into CSB and DCB loadings. α_1 and β_1 are parameters quantifying the load share between the upper and lower parts of the debonded region ($\alpha_1 + \beta_1 = 1$).

based on the load partitioning shown in Figure 10.22 and previous solutions for the CSB and DCB specimens.

The MMB specimen compliance, C, is defined according to

$$C = \frac{\delta_{\text{MMB}}}{P}, \tag{10.27}$$

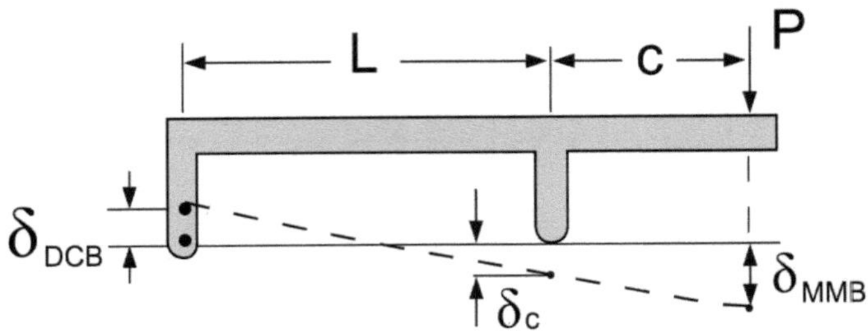

Figure 10.23 Kinematics of MMB specimen deformation.

where δ_{MMB} is the displacement of the load application point calculated from the kinematics of deformation as illustrated in Figure 10.23.

By similar triangles

$$\delta_{\text{MMB}} = \delta_c + \frac{c}{L}\left(\delta_c + \delta_{\text{DCB}}\right). \tag{10.28}$$

where δ_{DCB} refers to the upward displacement of the upper left point of the loading lever and δ_c is the downward displacement of the lever at the center of the beam. The expressions for the MMB compliance and energy release rate are

$$C = \left[\frac{c}{L}C_1 + \frac{c-L}{2L}C_2\right]\left(\frac{c}{L} - \alpha_1\frac{c+L}{2L}\right) + \left(\frac{c+L}{L}\right)^2 C_{\text{CSB}}, \tag{10.29}$$

$$G = \frac{P^2}{2b^2}\left(\begin{array}{l}\frac{c}{L}\left(\frac{c}{L} - \alpha_1\frac{c+L}{2L}\right)\frac{12}{E_f h_f^3}\left[a^2 + 2a\eta^{1/4} + \eta^{1/2}\right] + \\ \frac{c-L}{2L}\left(\frac{c}{L} - \alpha_1\frac{c+L}{2L}\right)\left[\frac{1}{h_c G_{xz}} + \frac{a^2}{\left(D - \frac{B^2}{A}\right)}\right] + \left(\frac{c+L}{L}\right)^2\left(\frac{a^2}{8}\left[\frac{1}{D_d} - \frac{1}{D_i}\right]\right)\end{array}\right), \tag{10.30}$$

where $2L$ is the span length, c is the lever arm distance, a is the crack length, and A, B and D are the extensional, coupling, and bending stiffnesses defined for the DCB specimen in Equations (10.8). η is the elastic foundation modulus parameter defined in Equation (10.7) and D_d and D_i are the flexural stiffness of the debonded and intact region of the beam, defined in Section 10.5. The parameter α_1 is given by Equation (10.24). C_1 and C_2 represent the compliances of the upper and lower legs (1 and 2) of the DCB specimen, Figure 10.4, i.e.

$$C_1 = \frac{\delta_1}{P_1}, \tag{10.31a}$$

$$C_2 = \frac{\delta_2}{P_1} \tag{10.31b}$$

given by

$$C_1 = \frac{4}{E_f h_f^3 b}\left[a^3 + 3a^2\eta^{1/4} + 3a\eta^{1/2} + \frac{3}{2}\eta^{3/4}\right], \tag{10.32a}$$

$$C_2 = \frac{a}{b}\left[\frac{1}{h_c G_{xz}} + \frac{a^2}{3\left(D - \frac{B^2}{A}\right)}\right], \tag{10.32b}$$

where the symbols are defined earlier in this section.

The loads acting on the DCB and CSB specimens (Figure 10.22) are

$$P_I = \frac{c}{L}P - \alpha_1 P_R, \tag{10.33a}$$

$$P_{\text{II}} = \left(1 + \frac{c}{L}\right)P, \tag{10.33b}$$

$$P_R = \frac{c+L}{2L}P. \tag{10.33c}$$

The mode I and II components of the total energy release rate, G (Equation (10.30)) are obtained by substitution of the loads P_{I} and P_{II} into the expressions for G for the DCB and CSB specimens, i.e., $P = P_{\text{I}}$ in Equation (10.6) for the DCB specimen and $P = P_{\text{II}}$ in Equation (10.26) for the CSB specimen. The mode ratio $G_{\text{II}}/G_{\text{I}}$ is

$$\frac{G_{\text{II}}}{G_I} = \left(\frac{P_{\text{II}}a}{2P_I}\right)^2 \frac{\left[\frac{1}{D_d} - \frac{1}{D_i}\right]}{\frac{1}{h_c G_{xz}} + \frac{a^2}{\left(D - \frac{B^2}{A}\right)} + \frac{12}{E_f h_f^3}[a^2 + 2a\eta^{1/4} + \eta^{1/2}]}. \tag{10.34}$$

Notice here that the mode ratio is a global mode ratio. Due to the asymmetric bimaterial character of the sandwich specimen, the local mode mixities expressed in terms of stress intensity factors, will differ (as shown later).

The methodology presented above is not valid when contact between crack faces is present. Contact arises at a lever arm distance, c, when the mode I load (Equation (10.33a)) vanishes. The minimum lever arm distance c, which is required to avoid contact is given by

$$c > \frac{\alpha_1 L}{2 - \alpha_1}. \tag{10.35}$$

For a symmetric specimen, $\alpha_1 = 1/2$, and Equation (10.35) gives $c > L/3$, which is generally used as a limit for MMB testing of monolithic composites. However, for the MMB sandwich specimen, α_1 is very small and therefore the minimum c distance is also very small, which is convenient in order to expand the range crack lengths in the test program.

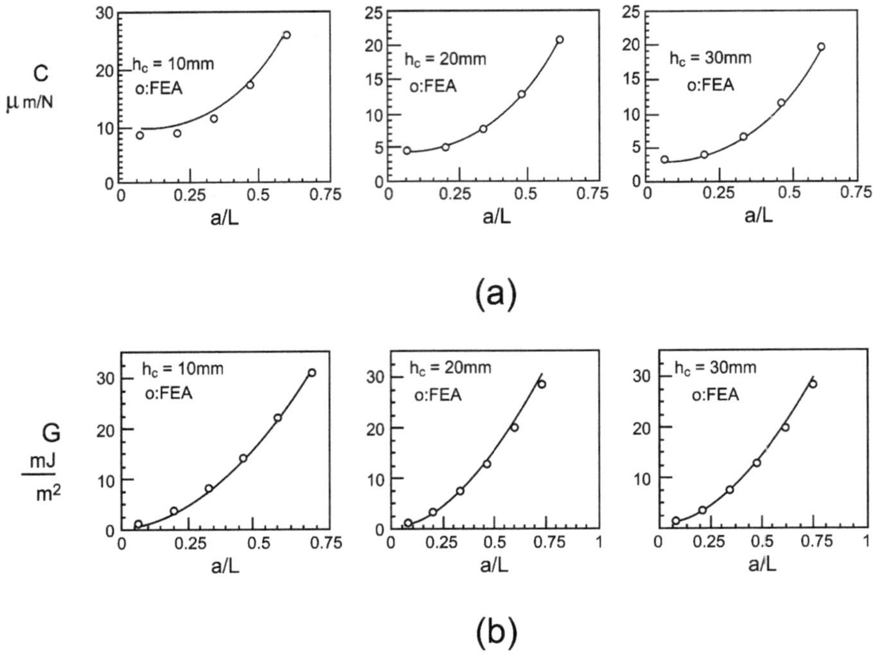

Figure 10.24 Compliance (a) and energy release rate (b) vs. crack length for MMB specimens with core thicknesses of 10, 20 and 30 mm ($c = 25$ mm).

10.8.1 Analytical and Finite Element Results

Quispitupa et al. (2009) conducted extensive parametric finite element analysis of the influence of various geometry and material parameters on the MMB compliance and energy release rate. The energy release rate, G, was extracted from the FEA results using the opening and sliding relative crack flank displacements (Figure 9.4). The same material properties, specimen geometries, and FE mesh, used for the DCB and CSB specimens examined in Sections 10.3 and 10.5, were used for the MMB specimen. The reaction loads from the loading lever were applied at the left cracked end and the center of the MMB specimens, as shown in Figure 10.24.

Figure 10.24 shows MMB compliance and energy release rate for three core thicknesses ($h_c = 10$, 20 and 30 mm) calculated over a range of crack lengths using FEA and the beam analysis (Equations (10.29) and (10.30)) at a fixed lever distance ($c = 25$ mm).

The finite element and beam analysis results are in good agreement. The discrepancies in the compliance for the thinnest specimen ($h_c = 10$ mm)

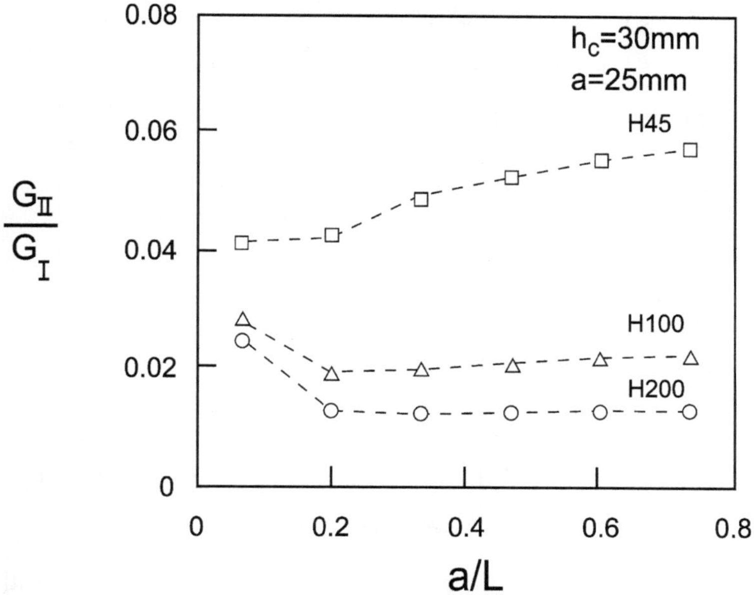

Figure 10.25 Global mode ratio vs. crack length for a set of core materials ($2L$ = 150 mm, $b = 35$ mm).

may be due to its small core-to-face thickness ratio ($h_c/h_f = 5$). This thickness ratio is slightly below the "thin face criterion", Equation (1.7). The energy release rate calculated from beam theory and FEA are in very good agreement.

When testing for debond toughness, it is desirable to maintain a constant mode ratio when the crack advances. Hence, the mode ratio $G_{\mathrm{II}}/G_{\mathrm{I}}$ should, ideally, be independent of the crack length. Figure 10.25 shows the global mode mixity ratio, $G_{\mathrm{II}}/G_{\mathrm{I}}$, defined in Equation (10.34) for a sandwich specimen with 2 mm thick glass/polyester face sheets and 30 mm thick H45, H100 and H200 cores at a fixed loading lever distance (c = 40 mm) vs. crack length.

The results show that the global mode mixity for the specimens with H100 and H200 cores is approximately constant for $a/L > 0.32$. For the specimens with a H45 core, however, the mode ratio increases with crack length.

As discussed in Chapter 9, a more realistic assessment of the crack loading is obtained by considering the relative crack flank displacements (Figure 9.4). This provides the "local mode mixity" expressed by the phase angle ψ (Equation (10.2)). Sandwich specimens with 2 mm thick glass/polyester face sheets (E_f = 16.4 GPa) and a H100 PVC foam core (E_c = 135 MPa)

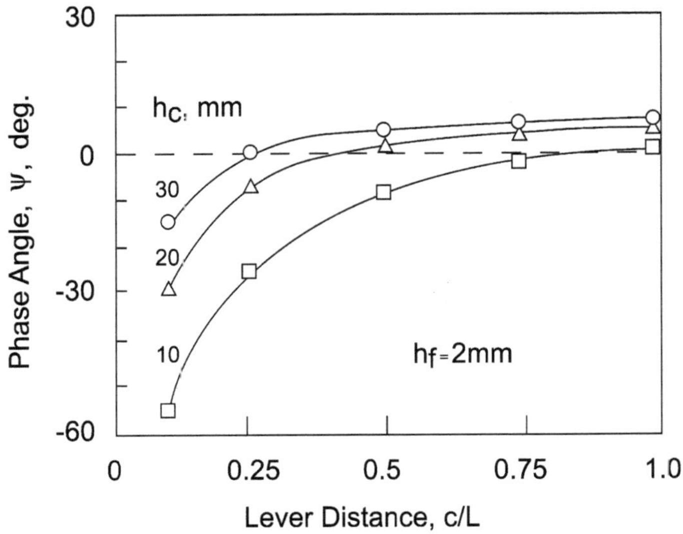

Figure 10.26 Phase angle ψ vs. lever arm distance c/L for a sandwich with $h_f =$ 2 mm, $E_f = 16.4$ GPa, and H100 core ($E_c = 135$ MPa) with $h_c = 10$, 20 and 30 mm. $2L = 150$ mm, $b = 35$ mm, and $a = 25$ mm ($a/L = 1/3$).

of 10, 20, and 30 mm were analyzed over a range of lever arm distances ($c/L = 0.1$ to 1). The results are shown in Figure 10.26.

The results show, as expected, that the phase angle increases with increasing lever arm distance (c/L). Furthermore, the results show that increasing core thicknesses (h_c/h_f) leads to increased phase angle. For large lever arm distances, the thickest cores start to display mode I dominance, and crack kinking (Section 9.2) may occur.

10.9 Double Cantilever Beam-Uneven Bending Moments (DCB-UBM) Specimen

The DCB-UBM test principle, shown schematically in Figure 10.27, was introduced by Sorensen et al. (2006) in an effort to measure the debond fracture toughness of composite and sandwich specimens over a large range of mode mixities. The moments M_1, and M_2 are introduced using two arms, adhesively bonded to the cracked end of the DCB specimen. The arms are loaded through a wire/roller arrangement connected to the arms. The assembly is

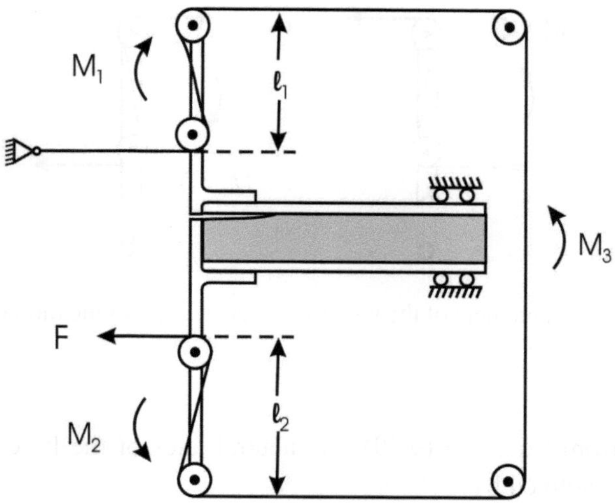

Figure 10.27 Schematic of DCB-UBM test principle for a debonded sandwich specimen.

mounted with the specimen axis oriented vertically in a tall test frame where a moving cross-head provides the pulling force of the wire.

If the wire is flexible and the rollers turn with negligible resistance, the force F in the wire is uniform along the wire and the moments become

$$M_1 = Fl_1, \tag{10.36a}$$

$$M_2 = Fl_2, \tag{10.36b}$$

where the lengths l_1 and l_2 are defined in Figure 10.27 and the signs of M_1 and M_2 are defined positive if both M_1 and M_2 tend to open the crack as shown. A crack loading dominated by mode I is achieved for $M_1/M_2 \approx 1$, i.e. equal opening moments. Mode II-dominated loading is achieved for negative moment ratios M_1/M_2. The magnitudes of the moments may be changed by changing the distances l_1 and l_2. By rearranging the wire as shown in Figure 10.28, the sign of the moment will change.

The uncracked end of the specimen is supported by a roller system (Figure 10.27) that provides rotational constraint by a moment $M_3 = M_1 - M_2$. Further specific details on the test set-up are provided by Sorensen et al. (2007).

The implied loading on the specimen thus consists of the pure moments M_1 and M_2. Following Section 9.4, for this situation we have

$$M_d = M_1; \quad M_s = -M_2; \quad M_b = M_1 - M_2; \quad P_d = P_s = P_b = 0. \tag{10.37}$$

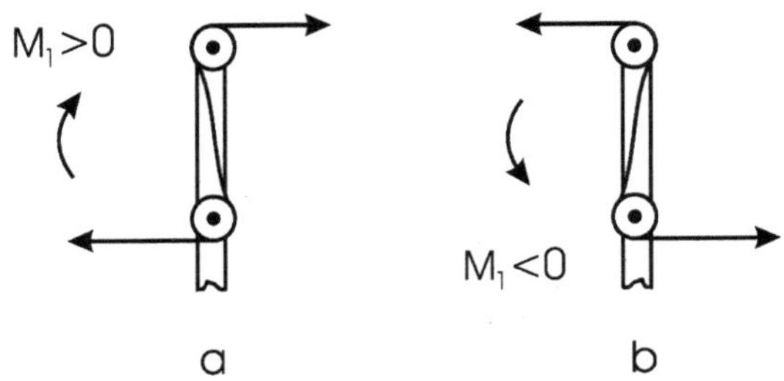

Figure 10.28 Arrangement of the wire to change the sign of the moment a) $M_1 > 0$, b) $M_1 < 0$.

Moreover, from Equation (9.39), the neutral axes of the base part and the substrate are defined by

$$e_b = 0; \quad e_s = \frac{E_f h_f (h_f + h_c)}{2(E_f h_f + E_c h_c)}. \tag{10.38}$$

The load factors on the reduced system are from (9.42b, c) and (9.43b, c)

$$P^* = \frac{E_f h_f}{2D_b}(h_f + h_c)(M_1 - M_2), \quad M_d^* = M_1 - \frac{E_f}{D_b}\frac{h_f^3}{12}(M_1 - M_2), \tag{10.39}$$

and

$$M_s^* = P^* \left(e_s + \frac{h_f + h_c}{2} \right) - M_1, \tag{10.40}$$

where D_b is the rigidity of the base part, given by (9.40b)

$$D_b = \frac{E_f h_f^3}{6} + E_f h_f \frac{(h_f + h_c)^2}{2} + \frac{E_c h_c^3}{12}. \tag{10.41}$$

Then the energy release rate of the debond is obtained from (9.49):

$$G = \frac{1 - \nu_f^2}{2E_f} \left(\frac{P^{*2}}{h_f} + 12 \frac{M_d^{*2}}{h_f^3} \right) + \frac{P^{*2}}{(E_f h_f + E_c h_c)^2} H_1$$
$$+ \frac{P^* M_s^*}{(E_f h_f + E_c h_c) D_s} H_2 + \frac{M_s^{*2}}{D_s^2} H_3, \tag{10.42}$$

where

$$H_1 = \frac{1 - \nu_f^2}{2} E_f h_f + \frac{1 - \nu_c^2}{2} E_c h_c, \tag{10.43}$$

$$H_2 = (1 - \nu_c^2) E_c h_c e_s + (1 - \nu_f^2) E_f h_f \left(e_s - \frac{h_c}{2} - \frac{h_f}{2} \right), \tag{10.44}$$

$$H_3 = \frac{1 - \nu_c^2}{2} E_c h_c \left(\frac{h_c^2}{12} + e_s^2 \right) + \frac{1 - \nu_f^2}{2} E_f h_f \left[\frac{h_f^2}{3} + \left(\frac{h_c}{2} - e_s \right) \left(\frac{h_c}{2} + h_f - e_s \right) \right], \tag{10.45}$$

and D_s is the rigidity of the substrate part, given by (9.40c)

$$D_s = \frac{E_f h_f^3}{12} + E_f h_f \left(\frac{h_f + h_c}{2} - e_s \right)^2 + \frac{E_c h_c^3}{12} + E_c h_c e_s^2. \tag{10.46}$$

Chapter 11
Debond Fracture Testing

Results from experimental studies using the debond test specimens introduced in Chapter 10, i.e. the double cantilever beam (DCB), tilted sandwich debond (TSD), cracked sandwich beam (CSB), single cantilever beam (SCB), three-point sandwich beam (TPSB), mixed mode bending (MMB), and double cantilever beam-uneven bending moments (DCB-UMB) specimens are discussed. Results of particular interest are the compliance and energy release rate, and the manner in which the crack propagates, i.e., interface propagation or crack kinking and the determination of debond fracture toughness.

11.1 Double Cantilever Beam (DCB) Specimen Testing

Prasad and Carlsson (1994b) conducted experiments on DCB specimens (Figure 11.1) with 2.2 mm thick aluminum face sheets over 15 and 20 mm thick PVC and polymethacrylimid (PMI) polymer foam cores. The specimens were 25.4 mm wide and 152 mm long. Steel hinge tabs were adhesively bonded to the precracked end of the specimen, see Figure 11.1.

In order to define a precrack at the upper face/core interface (Figure 11.1), a 25 μm thick Teflon film was placed between face and core prior to bonding. Care was used in order not to introduce the bonding adhesive in this region during panel manufacture. The Teflon film may stick to the face and core. Furthermore, at the end of the insert film, there is typically a resin pocket that may provide excessive local fracture resistance. To achieve a natural crack before the actual fracture tests, load was slowly applied to the specimen until the crack opened up over the region covered by the insert film and slightly extended beyond the end of the Teflon film. After precracking, the location

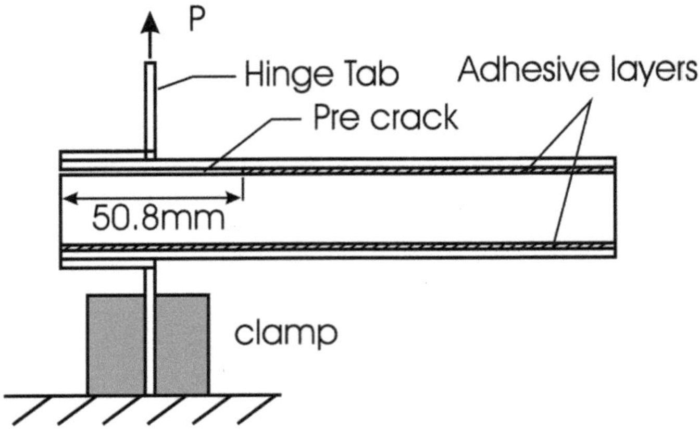

Figure 11.1 DCB specimen test configuration.

of the precrack tip was marked on the specimen edge with a thin marker pen. Keeping the marked point as a reference, subsequent points were marked at 5 mm increments. Each crack length was measured at the edge with a high resolution ruler.

The DCB test specimens were tested under displacement control at a low cross-head speed of 2 mm/min to allow monitoring of crack propagation with a traveling microscope. The load vs. displacement ($P-\delta$) response was recorded using an x–y recorder. Each specimen was loaded until the crack propagated beyond the insert film. The majority of the specimens failed by kinking into the core. If the crack remained interfacial, the cross-head was stopped after 5 mm of crack extension. After the crack growth was completely arrested, the new crack length was measured and recorded on the x–y chart. The DCB specimen was then partially unloaded and loaded again. This sequence was repeated until the crack extended about 50 mm. This test method provided a set of critical loads for interfacial crack growth for each test specimen.

Shivakumar and Smith (2004) conducted testing of DCB specimens with PVC foam and balsa cores with glass/vinylester and carbon/epoxy face sheets. The PVC foams were H80, H100, H130, and H200, where the nominal density (in kg/m^3) is indicated by the number following the "H". The carbon/epoxy face sheets were used only for the H100 PVC foam core. Baltek D-grade end-grain balsa wood cores of nominal densities of 100 and 150 kg/m^3 were used in combination with the glass/vinylester face sheets. The test specimens were 25.4 cm long and 3.81 cm wide. Each specimen was loaded at a cross-head rate of 1.3 mm/min while monitoring the crack

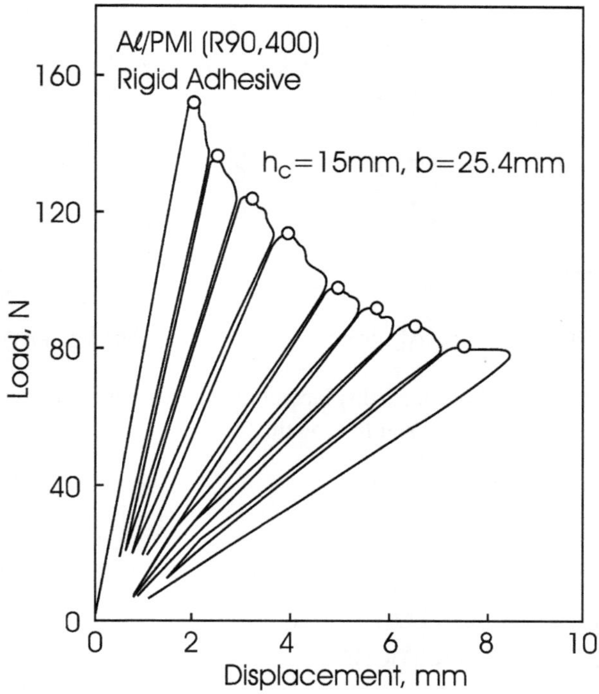

Figure 11.2 Load-displacement diagram for sandwich DCB specimen with a 15 mm thick R90.400 core.

tip. The crack was allowed to grow slowly until it extended about 2.5 cm, unless unstable growth occurred, and the cross-head was reversed to obtain the load-displacement record during unloading. This procedure was repeated until the total crack extension was about 10 cm. Figure 11.3 shows typical load-displacement graphs for specimens where the crack propagated unstably and stably, respectively.

The specimens with H130 and H200 foam cores and balsa cores displayed stable growth, while the lower density cores (H80 and H100) displayed unstable growth (denoted as "stick slip behavior"). The crack in the foam-cored specimens grew in the core, at a distance *d* from the face/core interface that increased with face-to-core modulus ratio, see Figure 11.4.

For the balsa-cored DCB specimens, the crack grew at the face/core interface in regions with weaker bonding, while the crack grew in the core at some distance from the actual face/core interface in regions of strong face/core adhesion. The crack sometimes kinked 90° between two blocks of balsa and grew through the entire core.

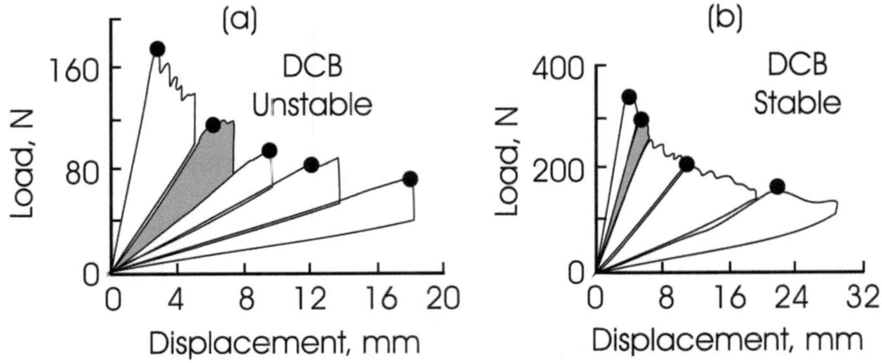

Figure 11.3 Representative load-displacement records for sandwich DCB specimens: (a) unstable growth, (b) stable growth.

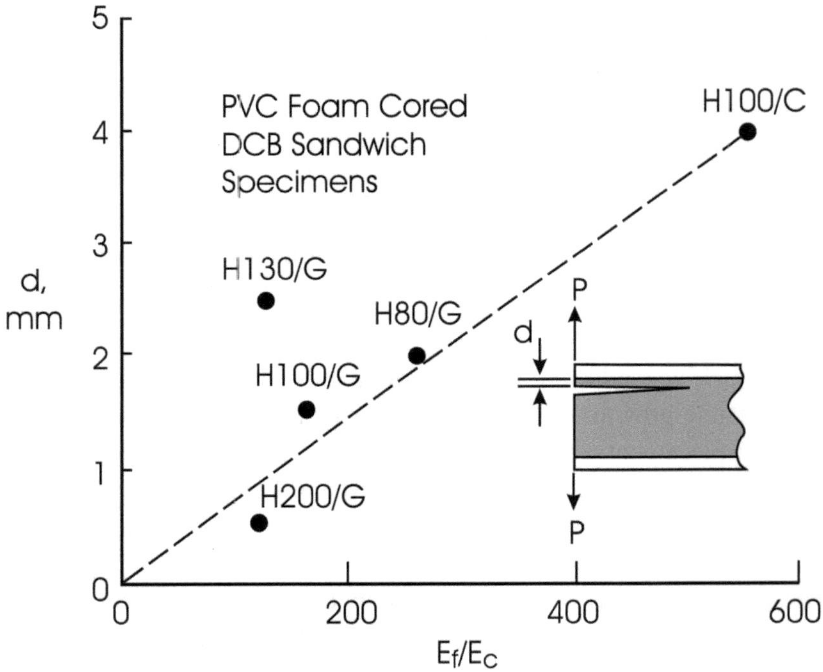

Figure 11.4 Depth of sub-interface crack in foam-cored sandwich DCB specimens.

The debond toughness, G_c, of DCB specimens may be determined from the experimentally determined load-displacement records such as those shown in Figures 11.2 and 11.3. Shivakumar et al. (2005) examined a variety of methods to determine G_c for foam core DCB sandwich specimens. Several of these methods are currently used to determine the delamination tough-

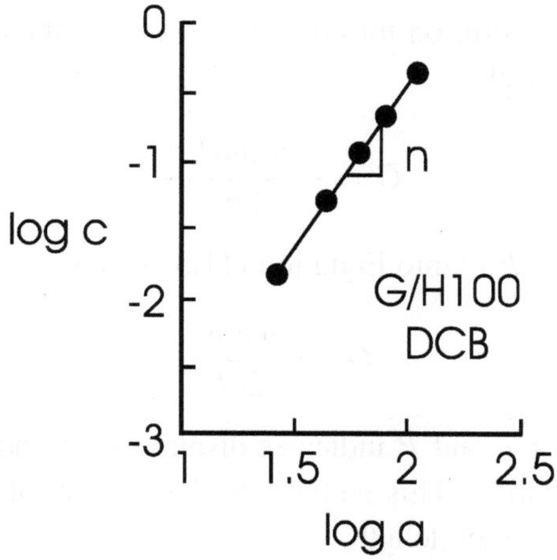

Figure 11.5 Compliance-crack length plot used to determine the exponent n in the Berry expression.

ness of unidirectional composite DCB specimens (ASTM Standard D5528, 2001). The methods are compliance calibration (CC), modified beam theory (MBT), and the area method. The compliance calibration (CC) method utilizes the following expression for the compliance (Berry, 1963):

$$C = c_0 a^n, \tag{11.1}$$

where c_0 and n are empirical constants. Notice that the exponent in Equation (11.1), $n = 3$, if classical beam theory is valid. As recognized by Berry (1963) and several others, the legs in a DCB specimen are not rigidly built into the uncracked part of the specimen in front of the crack tip. If the ends of the beams are elastically built-in, they tend to rotate at the crack tip which causes deviation from the classical beam theory manifested in values of the exponent, $n < 3$.

For experimental determination of the parameters c_0 and n in Equation (11.1) the compliance of the DCB specimen is measured at several crack lengths (Figures 11.12 and 11.13). Compliance and crack-length data are graphed in a logarithmic form as shown schematically in Figure 11.5.

Taking the logarithm of both sides of Equation (11.5) yields

$$\log C = \log c_0 + n \log a. \tag{11.2}$$

Hence plotting compliance vs. crack length in a double logarithmic graph should provide a linear relationship. The exponent n is the slope of the line as shown in Figure 11.5. Differentiation of Equation (11.1) with respect to crack length and substitution into the expression for the energy release rate, G, Equation (10.1) yields

$$G = \frac{P^2 c_0 n a^{n-1}}{2b}. \tag{11.3}$$

Substitution of $\delta_c = P_c C$ into Equation (11.3) yields

$$G_c = \frac{n P_c \delta_c}{2ab}, \tag{11.4}$$

where subscript c on δ and P indicates displacement and load values at onset of crack propagation. This method produces a set of fracture toughness values, one for each crack length.

Another common method to evaluate G_c from a set of load-displacement curves is the modified beam theory (MBT) method. For a rigidly built-in beam with negligible shear deformation, $n = 3$ in Equation (11.1). Equation (11.4) with $n = 3$ becomes

$$G_c = \frac{3 P_c \delta_c}{2ab}. \tag{11.5}$$

This equation overestimates G_c since n is less than 3. The MBT corrects this by adding a length, Δ, to the actual crack length, a, and Equation (11.5) becomes

$$G_e = \frac{3 P_c \delta_c}{2(a + \Delta)b}. \tag{11.6}$$

The length Δ is determined by plotting the third root of compliance vs. crack length as illustrated in Figure 11.6. A linear fit is constructed, and the line is extrapolated to $C^{1/3} = 0$ to yield the parameter Δ. Once Δ is determined, G_c is calculated at each crack length using Equation (11.6).

Shivakumar and Smith (2004) also determined the fracture toughness, G_c, from the load-displacement records using the "area method"

$$G_c = \frac{\Delta U}{b \Delta a}, \tag{11.7}$$

where ΔU is the area enclosed within a loading-unloading cycle, see Figure 11.3, and Δa is the amount of crack extension during such a cycle. This method was originally proposed for evaluation of G_c in delamination testing of composite laminates by Whitney et al. (1982). This procedure, however, is generally not accepted because it provides an integrated value of G_c.

Shivakumar et al. (2005) determined G_c from DCB test results using the various methods described above and presented the results in the form of R

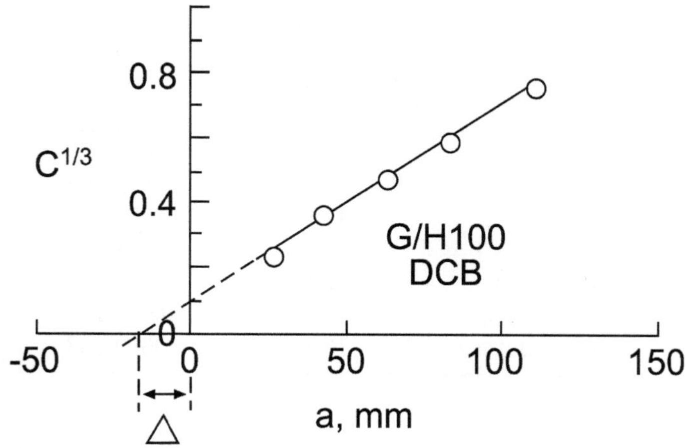

Figure 11.6 Compliance plot for experimental determination of the parameter Δ in the MBT approach.

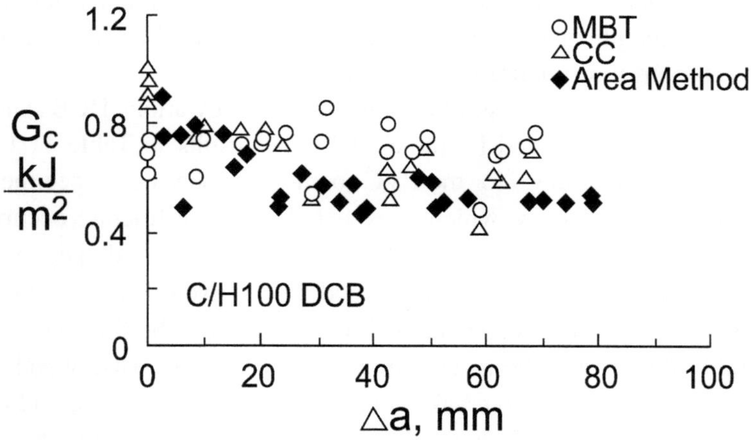

Figure 11.7 R curves for DCB sandwich specimens with carbon/vinylester face sheets and H100 core determined using various methods (after Shivakumar et al., 2005).

curves (G_c vs. Δa where Δa is the extension of the crack from the initial artificial precrack), see Figure 11.7.

The MBT method yields quite consistent toughness values, while the CC results in large scatter. The area method yileds G_c values below those determined using the other methods. The area method provides an average value of G_c for each crack increment (Δa), and such an average does not have a very clear interpretation when the crack propagates in a dynamic mode be-

Table 11.1 Debond toughness of sandwich DCB specimens. The ± range refers to the standard deviation (Shivakumar and Smith, 2004).

Face	Core	Core Density (kg/m^3)	G_c (J/m^2)
G/VE	PVC	80	367±59
G/VE	PVC	100	558±106
C/VE	PVC	100	560±110
G/VE	PVC	130	878±211
G/VE	PVC	200	1350±270
G/VE	Balsa	100	693±243
G/VE	Balsa	150	1008±262

tween onset and arrest. The area method was eliminated from the ASTM DCB test standard D5528 (2001), for this and other reasons. Shivakumar et al. (2005) recommend the MBT method because of its accuracy and simplicity. They further recommend that testing is aborted once the crack length-to-specimen length ratio (a/L) exceeds 0.53. Beyond this range specimen end-effects start to introduce nonlinearities.

Average debond toughness values and standard deviations DCB sandwich specimens are summarized in Table 11.1. The results in Table 11.1 show that the debond toughness for the PVC foam-cored sandwich specimens increases with foam density, which is expected since the fracture occurred in the core. Similarly, the debond toughness of the balsa-cored specimens increases with the density of the balsa wood. For a given density of the core, the balsa core provides higher debond toughness than the PVC cores. This is probably due to the fibrous structure of the end-grain balsa wood which provides more resistance to crack propagation than the isotropic and homogeneous cross-linked PVC foams. Table 11.1 further reveals that the specimens with different face sheets (glass/vinylester and carbon/epoxy) and same core (H100) have almost identical debond toughnesses. This shows that the debond toughness does not depend on face stiffness and the type of resin used, provided that the actual face/core interface is tough enough to promote fracture of the core. Figure 11.4 shows that debond growth in the specimens with glass/vinylester face sheets occurred much closer to the interface than for those with carbon/epoxy faces, but this difference does not seem to influence the toughness.

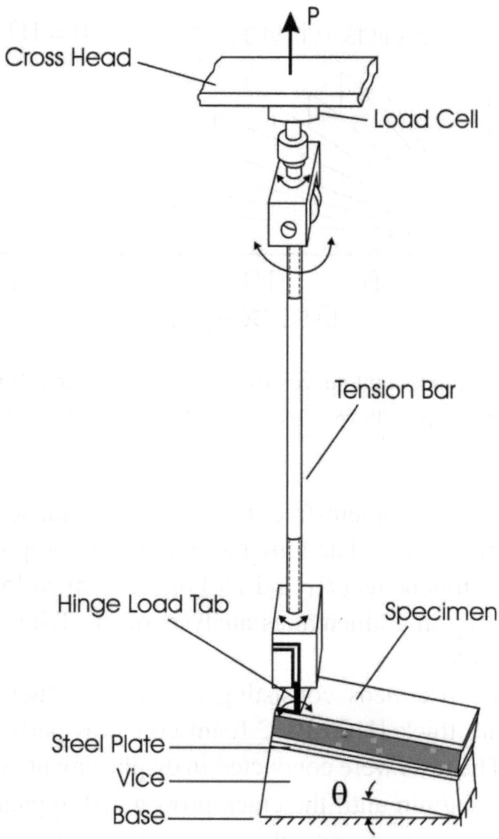

Figure 11.8 Schematic of TSD specimen test configuration.

11.2 Tilted Sandwich Debond (TSD) Specimen Testing

The TSD specimen (Figure 11.8), discussed in Section 10.4, was used in experimental studies of face/core debond toughness by Li and Carlsson (1999), and Viana and Carlsson (2003). This specimen was employed also by Majumdar et al. (2003) who examined the effects of resin penetration on the face/core debond toughness.

The bottom face of the sandwich specimen is adhesively bonded to a 2.5 mm thick steel plate bolted to a rotatable vice resting on the horizontal base of the test frame. The vice arrangement allows testing over a range of tilt angles, $-60° \leq \theta \leq 60°$. Vertical loading without the introduction of moments and horizontal force is accomplished by bonding a hinge load tab to the end of the specimen, on the partially separated face sheet, connected

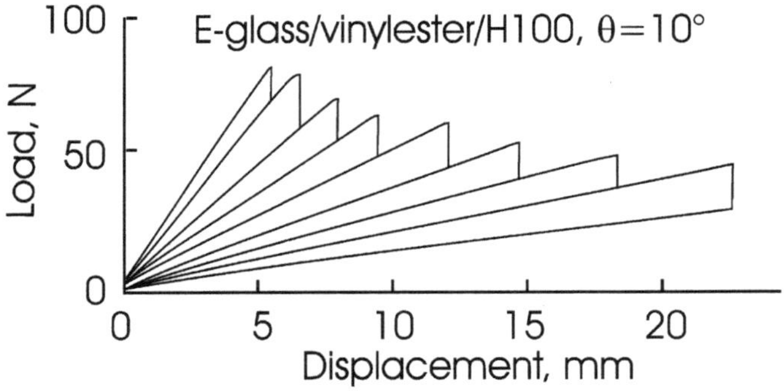

Figure 11.9 Load-displacement curves for a glass/polyester/H100 TSD specimen. Tilt angle $\theta = 10°$. The specimens were 25.4 cm long and 3.8 cm wide.

to a long rod mounted, moment-free, to a load cell attached to the moving cross-head of the test frame. The long bar guaranteed proportionality of the axial and normal components (P_A and P_N) of the vertical load as defined in Figure 10.11, verified in a kinematics analysis of the TSD configuration by Li and Carlsson (2000).

Testing of TSD specimens consisting of 3.6 mm thick glass/polyester faces over a 50 mm thick H100 PVC foam core was performed by Li and Carlsson (1999). The tests were conducted in displacement control at a cross-head rate of 1.27 mm/min until the crack propagated, typically followed by rapid (unstable) propagation and subsequent arrest. After arrest, the specimen was partially or fully unloaded. This procedure was repeated until the crack front approached the specimen end to achieve multiple compliance and critical load values. Six to eight measurements of compliance and critical load were recorded for the 23–25 cm long specimens. Figure 11.9 shows a typical load-displacement record for a glass/polyester/H100 specimen with a 6.4 cm long precrack, where in this case the crack length refers to the edge of the specimen.

The response is almost linear prior to crack propagation which occurs in a stick-slip manner at each increment of loading. The amount of unstable growth increased with increasing crack length and typically ranged from 5–30 mm. The crack front was observed from one of the specimen edges. It was observed that the initial face/core crack (cut by a sharp knife) deflected slightly into the core and reached a depth of about 1 to 5 mm and then returned to a location near the upper face just below the 0.3–0.9 mm thick face/core interphase region, see the schematic illustration in Figure 11.10.

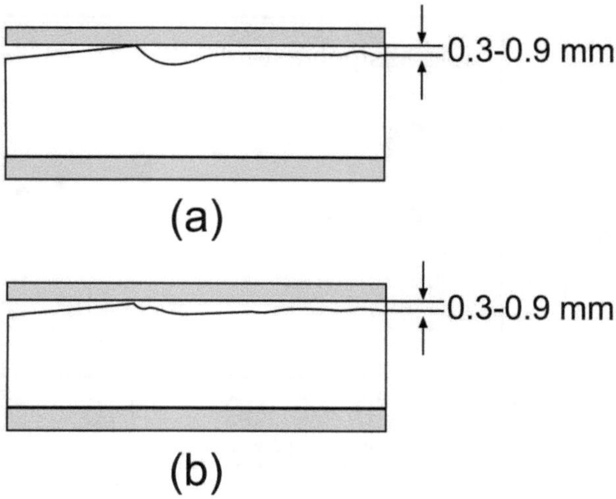

Figure 11.10 Schematic of crack propagation paths in foam-cored TSD specimens. (a) Deep sub-interface crack, (b) shallow sub-interface crack.

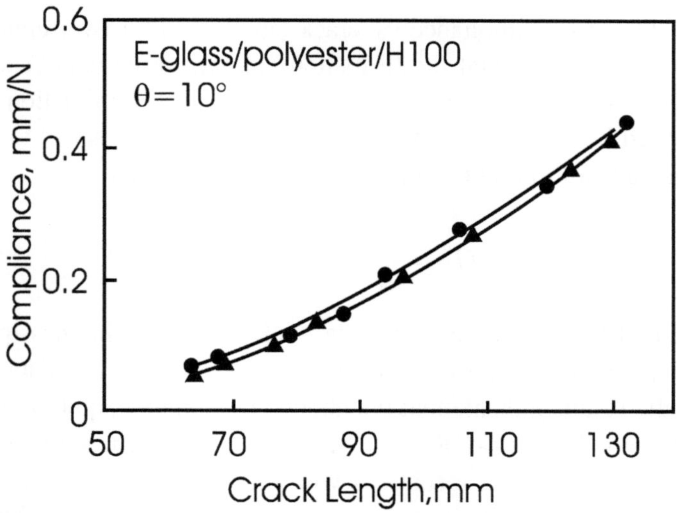

Figure 11.11 Compliance vs. crack length for two glass/polyester/H100 TSD specimens ($\theta = 10°$). The solid triangles and circles represent experimental data while the curves represent fits to the data.

Debond fracture toughness G_c was determined using the experimental compliance calibration method by fitting a polynomial to the compliance-crack length data,

$$C = m_0 + m_1 a + m_2 a^2. \tag{11.8}$$

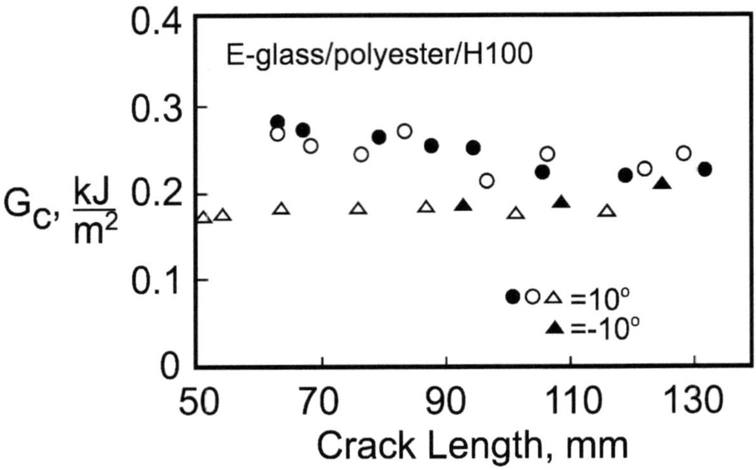

Figure 11.12 Fracture resistance curves for face/core separation in TSD specimens with H100 core.

Figure 11.11 shows compliance vs. crack length data for two replicate TSD specimens, and polynomial fits (Equation (11.8)), to the data. The two fitted curves reveal consistency between specimens. Determination of G_c is accomplished by differentiation of Equation (11.8) with respect to the crack length and substitution into Equation (10.1), which yields

$$G_c = \frac{P_c^2}{2b}(m_1 + 2m_2 a). \tag{11.9}$$

Figure 11.12 shows some representative R curves for the glass/polyester/H100 specimens tested at tilt angles, $\theta = -10°$ and $10°$. The data scatter is large but the results indicate that the debond toughness is quite independent of crack length. The lack of R curve behavior is consistent with results from DCB testing of sandwich specimens with H100 core, see Figure 11.7. The toughness, however, is significantly less than measured with the DCB specimen. This difference is not expected to be due to the different test methods. Both are mode I-dominated and the fracture process similar. The difference is most likely due to batch-to-batch variations between core panels. Figure 11.12 shows that the specimens with $\theta = 10°$ and $\theta = -10°$ produce similar R curves (considering the relatively large specimen-to-specimen scatter). The insensitivity of the toughness to such large difference in tilt angle is consistent with the insensitivity of the mode mixity to changes of the tilt angle discussed earlier (see, e.g., Figure 10.13).

Table 11.2 Debond toughness measured with the TSD specimen. The ± range denotes the standard deviation.

Specimen	G_c (J/m^2)
G/PE/H30	196±48
G/VE/R75	325±12
Al/H100	368±57
Al/H200	1270±57
G/VE/R400	1300±97

Viana and Carlsson (2003) determined the debond toughness of sandwich TSD specimen with a range of PVC foam cores. Both composite and aluminum face sheets were examined. The TSD specimens were tested at a tilt angle of 10° and a cross-head rate of 1 mm/min. Fracture of the sandwich specimens with H30, R75, H100, and H200 cores occurred in the core at a depth of about 1–2 mm below the actual face/core interface. For the specimens with the highest density core (R400), the crack "meandered", i.e., propagated in the core for some distance and then returned very close to the actual face/core interface.

Debond toughness G_c was evaluated from the experimental load-displacement curves using the experimental compliance calibration method and a polynomial fit (Equation (11.8)) as described earlier. Overall, the resistance curves were flat, within a relatively large scatter band. Table 11.2 summarizes average toughness values.

Comparison with the DCB test data in Table 11.1 shows that the DCB and TSD tests produce similar results. As discussed above, face/core separation in TSD testing of sandwich specimens with a PVC foam core tends to occur in the core, parallel to the actual face/core interface. Shivakumar and Smith (2004) found this also to be the case for DCB sandwich specimens.

If the crack propagates in the core it is expected that the face/core debond toughness G_c measured with the mode I-dominated TSD specimen, should be similar to the fracture toughness of the foam (G_{Ic}). Figure 11.13 shows fracture toughness for face/core separation measured using the DCB specimen (Shivakumar and Smith, 2004), the TSD specimen (Viana and Carlsson, 2003), and fracture toughness of the PVC foam measured using the single-edge notch bend (SENB) specimen (Viana and Carlsson, 2002a), plotted vs. foam density.

Notice that the DCB and TSD tests produce similar toughness values which would be expected from the similarity in fracture mechanisms and

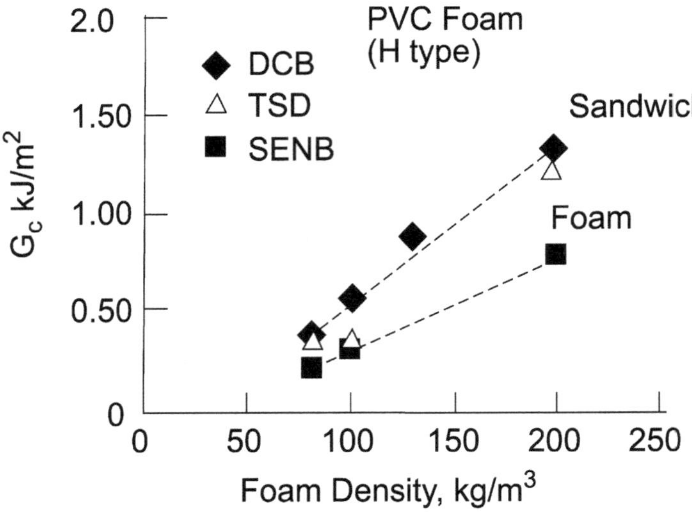

Figure 11.13 Fracture toughness for face/core separation G_c and foam fracture toughness G_{IC} vs. foam density.

their mode I dominance. The core toughness, however, falls significantly below the debond toughness at any given foam density. The difference may be due to several reasons, as discussed by Viana and Carlsson (2002b), such as a gradient of foam density in the core with a higher density (and toughness) near the surfaces of the foam sheet, i.e., near the face/core interface where the crack propagates. It is also possible that the core is not completely isotropic. In the DCB and TSD tests, the crack propagates in-plane, while in the SENB test, the crack propagates through the core panel thickness, perpendicular to the plane of the core panel.

11.3 Cracked Sandwich Beam (CSB) Specimen Testing

Experimental studies have been conducted on the CSB specimen shown in Figure 11.14 by Carlsson et al. (1991) and Shipsha et al. (1999). Carlsson et al. (1991) examined sandwich specimens consisting of glass/polyester face sheets over a balsa core. Sandwich panels were wet laid-up at Viking Yachts in St. Petersburg, Florida, representative of hull constructions presently in use. A precrack was achieved by placing a folded thin (0.025 mm) vacuum bagging film between the core and one of the face sheets. A major difference between the actual boat construction and the present panels is in the

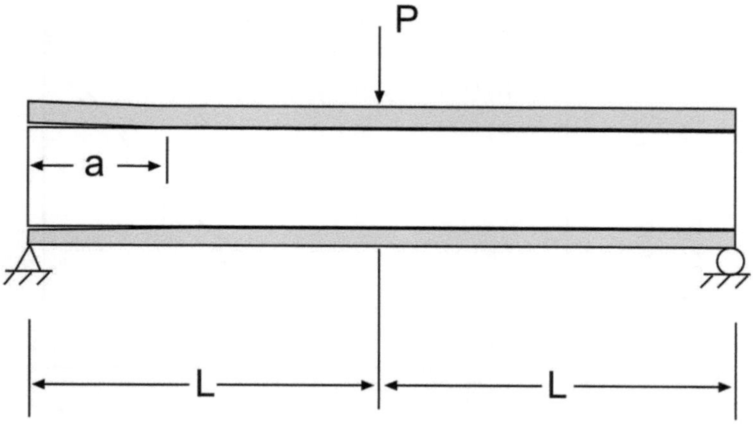

Figure 11.14 Schematic of CSB specimen.

omission of chopped strand matting (CSM) from the face sheet-core interfaces. The CSM was omitted because the CSM layers are difficult to model precisely *in situ* and to avoid the possible complexities associated with a relative thick composite interlayer between face and core. The face laminates employed zero degree plies next to the core in order to promote co-planar crack growth. The lay up was a $[0/\pm 45/\pm 45/0/\overline{\text{core}}]_s$ where the overbar denotes that the plane of symmetry of the sandwich is at the core center. Mechanical properties of the face sheets were determined by standard test methods and properties of end-grain balsa wood core material were obtained from the manufacturer's data sheet. It should be pointed out here that the end-grain balsa wood core material is composed of wooden blocks with the longitudinal (L) axis along the z axis of the CSB specimen, with the radial (R) and tangential (T) directions randomly oriented in the plane of the sandwich panel. The mechanical properties in the radial and tangential planes may differ significantly and there are large block-to-block density variations which lead to scatter in the properties of the core. The beam width was 12.7 cm, the total beam length was 55 cm, and the debond length was 20 cm.

The CSB compliance was determined in a three-point flexure fixture mounted in a general-purpose test frame. The span length, $2L$, in Figure 10.15, was 30.5 cm. The cylindrical steel rods used for load introduction and supports were 25.4 mm in diameter. Displacements at the central loading point were monitored with a linear voltage differential transformer (LVDT).

The long beam allowed positioning of the beam on the three-point flexure fixture so that any desired crack length, a in Figure 11.14, between 0

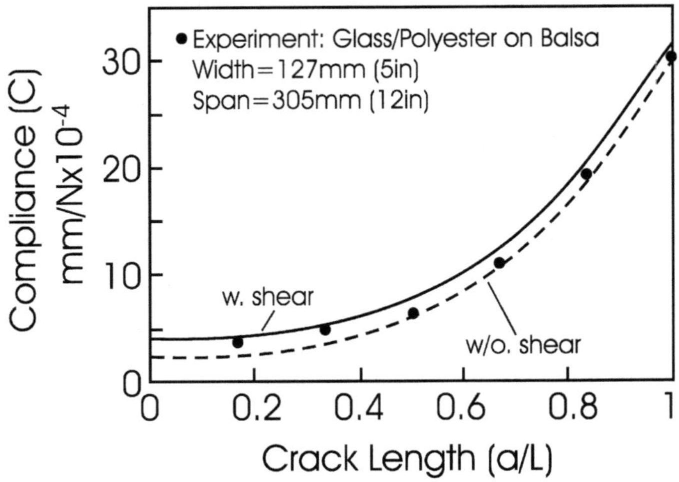

Figure 11.15 Compliance vs. crack length (a/L) for a CSB specimen with glass/polyester faces on balsa core. Solid line: CSB analysis with shear. Dashed line: CSB analysis without shear. Filled circles: experiment.

and 15.2 cm could be achieved. Hence, any given specimen could be tested "without crack" and at any crack length. It was possible to load the same specimen several times by keeping the load small, within the linear elastic region of response. Seven crack lengths per specimen and five replicate specimens were tested. The specimens were loaded at a constant displacement rate of 1.27 mm/min, and unloaded after a linear response curve was recorded.

Figure 11.15 shows compliance plotted vs. crack length, normalized by the half span length L. The scatter (c.o.v) in any compliance value was less than 12%. Such scatter is largely due to the nature of the core material. Compliance predictions using Equation (10.21) with and without inclusion of transverse shear deformation are also shown.

It is observed that the experiments are bounded by the analytical predictions and that shear deformation must be included to realistically predict the compliance, especially at short crack lengths where neglecting the shear deformation would lead to substantial percentage error.

Compliance calibration was performed by fitting a third-order polynomial to experimental compliance vs. crack length data,

$$C = C_0 + ma^3, \tag{11.10}$$

where C_0 is the compliance at zero crack length and m is a curve fit parameter. Equations (11.10) and (10.1) yield the following expression for the

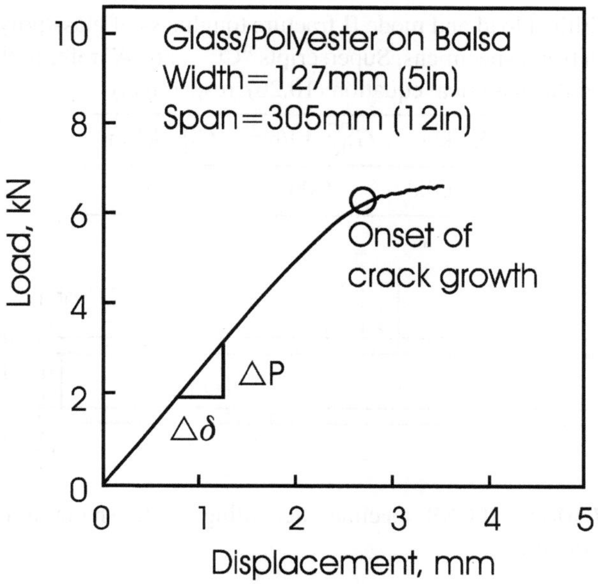

Figure 11.16 Load-displacement curve for a CSB sandwich specimen with glass/polyester face sheets and with a 5.1 cm precrack.

fracture toughness:

$$G_c = \frac{3ma^2P_c^2}{2b}. \tag{11.11}$$

After the completion of the compliance measurements each specimen was loaded until the crack propagated. A 5.1 cm crack length was found to be optimum. Shorter crack lengths would not allow crack growth to occur prior to core shear failure and longer crack lengths resulted in core crushing near the crack tip. The onset of crack growth was determined by inspection of the crack front by a microscope. The point where direct observation of the crack tip revealed that the crack started to propagate was marked on the load vs. displacement graph, see Figure 11.16. After initiation, the crack propagated in a stable manner as indicated. The average measured critical load P_c for the test specimens was used to calculate the fracture toughness, G_c, based on experimental compliance calibration, Equation (11.11) and direct application of the beam analysis (Equation (10.26)). Critical load and the fracture toughness for face sheet-core debonding are listed in Table 11.3.

The compliance calibration and analytical method to determine G_{IIc} are in close agreement. This is to be expected for this case because of the consistent analytical and experimental compliance data (Figure 11.15).

Table 11.3 Critical load and mode II fracture toughness of glass/polyester on balsa core sandwich beam specimens. Superscripts "CC" and "A" refer to the compliance calibration and the analytical equation (10.26), respectively.

P_c, kN	$G_{\text{IIc}}^{\text{CC}}$, kJ/m^2	$G_{\text{IIc}}^{\text{A}}$, kJ/m^2
6.7±0.8	1.06	1.07

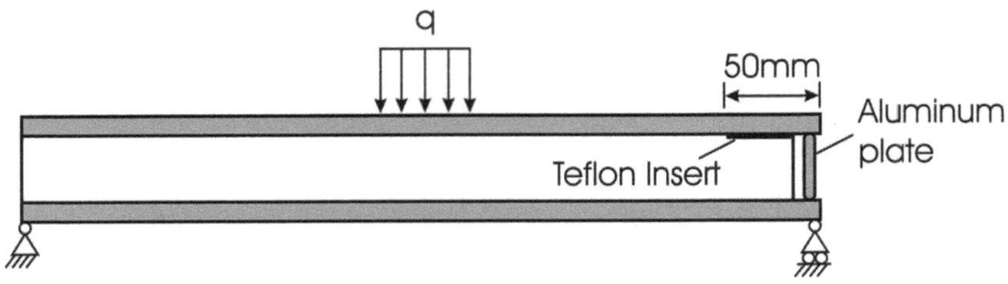

Figure 11.17 Modified CSB specimen according to Shipsha et al. (1999). All dimensions are in mm.

Shipsha et al. (1999) conducted analysis and testing on CSB specimens consisting of 3.3 mm thick glass fabric reinforced polyester faces and 25 mm thick Rohacell WF51 and H100 cores, respectively. The face/core debond was simulated using a 50 mm long and 125 μm thick Teflon film placed between face and core prior to wet lay-up so as to prevent bonding. Figure 11.17 shows the CSB configuration. By removing a thin vertical slice of the core at the cracked end and inserting an aluminum plate with rounded edges, compressive load transfer was achieved between the lower and upper face sheets, without constraining the sliding displacements associated with mode II crack loading. A 50 mm wide wooden pad was used to distribute the central load on the upper face sheet and 60 mm diameter steel cylinders were used as supports.

Static tests were performed at a displacement rate of 2 mm/min. The crack propagated unstably from the end of the insert to the inner support in one increment. Crack propagation occurred near the face/core interface one or two cells down in the core below the resin-impregnated cells. Finite element analysis was conducted on the two specimen configurations to determine stress intensity factors, K_{I} and K_{II}, over a range of crack lengths from 50 to about 80 mm. Eight-node quadratic membrane elements were used in the two-dimensional linear numerical models. The face and core materials were treated as isotropic, and linear elastic fracture mechanics (LEFM) was applied. The crack was assumed to be located two cells down in the core material. Contact elements were used on the two crack flanges which were

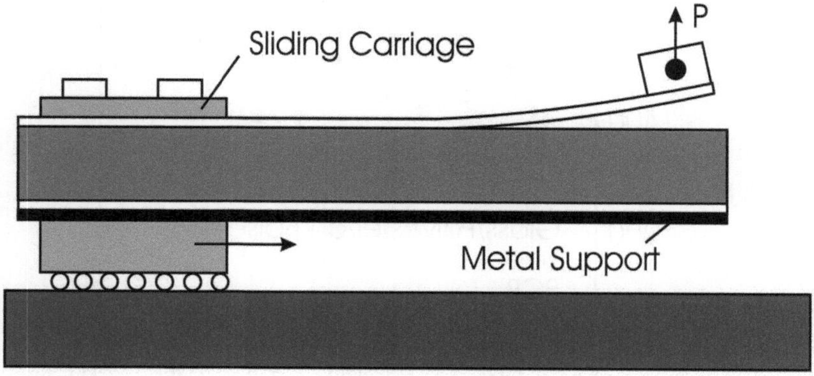

Figure 11.18 SCB test fixture.

assumed to slide without friction. Stress intensity factors were determined by using the displacement extrapolation method (Shipsha et al., 1999).

Crack loading was found to be highly dominated by shear (mode II), although there was a small amount of opening present, apparently a consequence of the strong material mismatch and the use of an aluminum plate at the end. Based on their unit load solutions for the stress intensity factors and the measured critical load for crack propagation, Shipsha et al. (1999) determined the mode II fracture toughness expressed as the critical stress intensity factors and obtained $K_{IIC} = 0.5$ and $0.31\ \text{MPa}\sqrt{\text{m}}$ for the sandwich specimens with H100 and WF51 cores, respectively. With the following data for the H100 and WF51 foams, provided by Burman (1998): $E = 105$ MPa, $\nu = 0.31$ (H100), and $E = 85$ MPa, $\nu = 0.42$ (WF51), conversion to energy release rate, Equation (9.12) yields $G_{IIc} = 2.15\ \text{kJ/m}^2$ (H100) and $0.93\ \text{kJ/m}^2$ (WF51).

11.4 Single Cantilever Beam (SCB) Specimen Testing

The SCB specimen test principle proposed by Cantwell and Davies (1994), illustrated in Figure 11.18, was used to examine face/core adhesion in sandwich specimens consisting of glass/polyester face sheets and balsa and honeycomb cores.

The specimens were 20 cm long and 2 cm wide. A precrack was introduced by advancing a sharp razor blade a couple of mm at the face/core interface, and then pulling the face away from the core with the aid of a steel

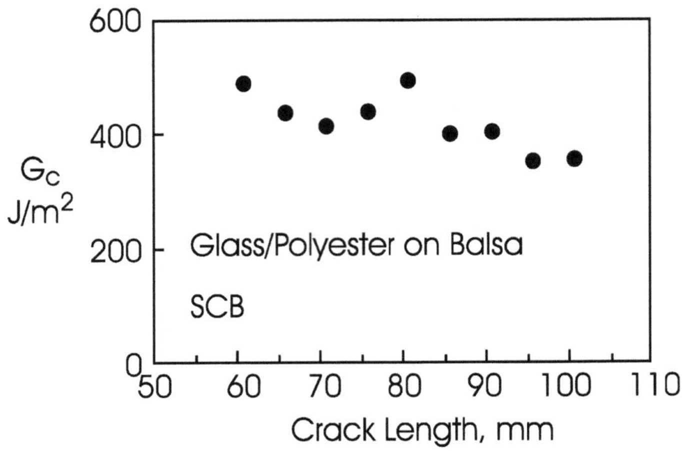

Figure 11.19 Fracture resistance curve for a SCB specimen with glass/polyester face sheets and balsa core.

bar inserted between face and core. The total length of the precrack was about 50 mm. After precracking, the lower face was adhesively bonded to a 10 mm thick steel plate, placed on a movable chariot. An aluminum block or a hinge tab was bonded to the face at the end of the precracked region. A load was applied at the end tab until crack propagation occurred. The specimen was then unloaded and loaded again until a new increment of crack growth occurred. This procedure was repeated until the crack propagated close to the specimen end. The crack length was determined by a mm scale attached to the edge of the specimen. Fracture toughness for separation of the face was determined using the Berry (1963) experimental compliance calibration method, see Section 11.1. For end-grain balsa-cored specimens (density = 100 kg/m^3) it was found that the crack growth was stable, always occurring at the face/core interface.

Figure 11.19 shows a R curve for the balsa-cored specimen. The toughness varies between 360 and 500 J/m^2 indicating some significant scatter. Cantwell and Davies (1994) pointed out that the integrity of the face/core interface of balsa-cored sandwich panels is sensitive to the moisture content in the balsa present during bonding of the faces. The toughness is quite small compared to values determined for co-cured balsa-cored DCB sandwich specimens by Shivakumar and Smith (2004) (see Figure 11.20).

Shivakumar and Smith (2004) consistently observed crack propagation in the core (sub-interface crack) as a result of good face/core adhesion. Cantwell and Davies (1994) also examined a sandwich consisting of glass/polyester face sheets over a carbon/epoxy honeycomb core (unspec-

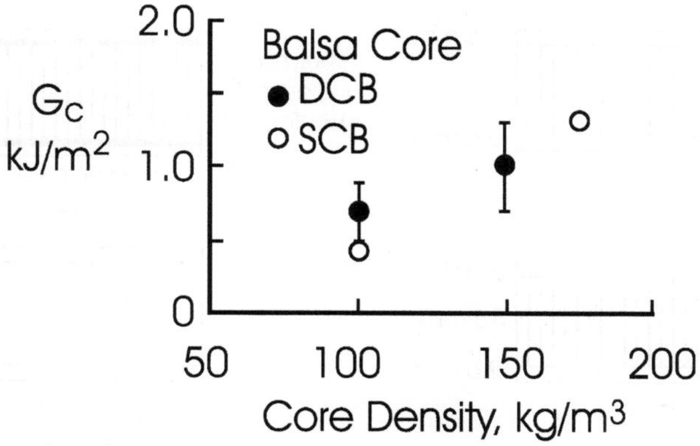

Figure 11.20 Debond toughness vs. density of balsa core.

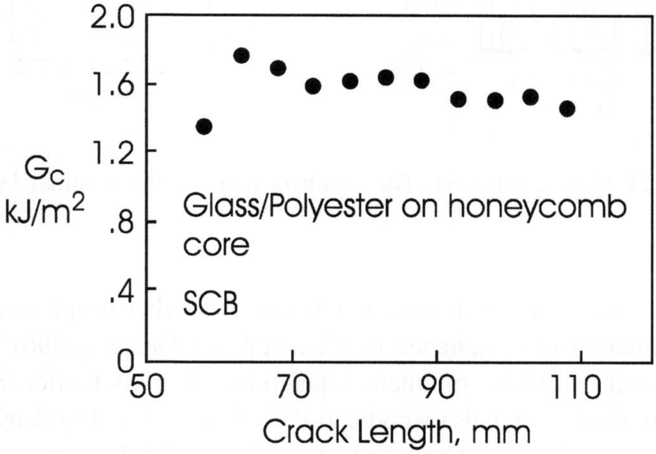

Figure 11.21 Fracture resistance curve determined using the SCB test of a sandwich with glass/polyester face sheets over a honeycomb core.

ified grade). Based on the photographs provided, the cell size was about 2 mm. Debond testing revealed stable crack propagation behavior. Figure 11.21 shows a R curve for this sandwich. Although there is some scatter, the fracture toughness is substantially higher for the honeycomb-cored sandwich than for the balsa-cored sandwich (Figure 11.19). This may not be expected due to the discontinuous structure of the core in a honeycomb-cored sandwich. Inspection of the failure surfaces of debonded honeycomb core revealed that the face separation did not occur at the actual face/core interface, but fractions of a mm away from the interface, inside the core, which indicates a strong interface bond.

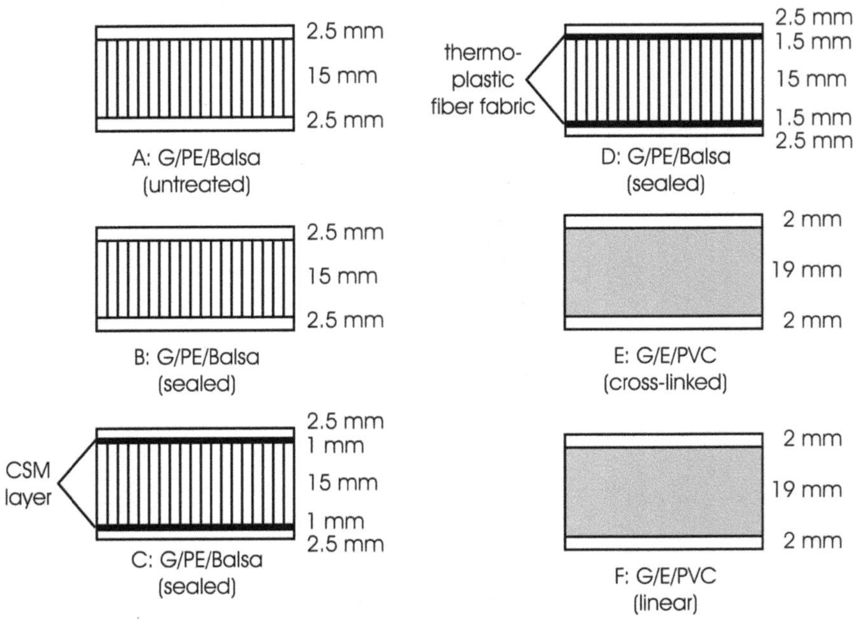

Figure 11.22 SCB test panels. The numbers indicate the nominal layer thicknesses.

Further studies of the debond toughness of sandwich specimens using SCB specimens were conducted by Cantwell and Davies (1996). The face sheets utilized two layers of stitched quadriaxial E-glass fabrics. Sandwich panels were made with balsa wood and PVC foam cores. The density of the balsa core was 175 kg/m^3. Cross-linked and linear PVC foam cores of densities of 80 and 75 kg/m^3, respectively, were used. Figure 11.22 illustrates the various sandwich panels examined.

Sandwich panel A constitutes a reference in which the face sheets were laid-up on an untreated (as-received) balsa core. The sandwich panels B, C and D utilized a core that was "sealed" before wet lay-up by application of a coating of the polyester resin used as a matrix for the face sheets. The resin was allowed to gel before the face sheets were added. Sandwich panel C contained a layer of chopped strand mat (areal weight 450 g/m^2) at the face/core interfaces, while sandwich D utilized a 130 g/m^2 thermoplastic polyester fiber mat at both face/core interfaces. The sandwich panels E and F utilized thermoset and thermoplastic PVC foam cores, respectively.

Face/core debond testing was conducted using the SCB specimen. Crack propagation occurred in a stable manner in all specimens, except for the specimens with a cross-linked PVC core, panel E in Figure 11.22, which showed

Table 11.4 Face separation toughness for sandwich specimens with glass/polyester (G/PE) and glass/epoxy (G/E) face sheets on balsa and PVC foam cores measured using the SCB specimen. "S"represents sealed core, "X" cross-linked, and "L" linear.

Sandwich	Face	Core	Density (kg/m^3)	G_c (kJ/m^2)
A	G/PE	Balsa	175	1.32
B	G/PE	Balsa (S)	175	1.13
C	G/PE	Balsa (S)	175	0.96
D	G/PE	Balsa (S)	175	1.30
E	G/E	PVC (X)	80	0.51
F	G/E	PVC (L)	75	2.77*

*G_c determined using area method.

stick-slip behavior. In sandwich A, fibers from the face sheets debonded and formed a bridging zone behind the propagating crack. In sandwich B, less fiber bridging was observed although the crack remained at the interface. In sandwich C, the crack remained within the CSM layer. The face/core adhesion in sandwich D was so high that the crack propagated as a delamination in the composite face sheet. For the sandwich with a cross-linked PVC core, sandwich E, the crack meandered between the face/core interface and within the core, while for sandwich F the crack propagated in the faom core. Face separation toughness values determined using the compliance calibration method are summarized in Table 11.4.

Of the four balsa wood sandwich systems, the one an with untreated core (A) displayed the highest toughness, although fracture occurred at the face/core interface and not within the core. The high toughness of panel A seems to be associated with the debonding and pull-out contributions from bridged fibers between face and core. The lower debond toughness of panel B was explained by fewer bridged fibers and a weaker face/core bond as the sealed core layer restricts penetration of the face resin into the core. Panel C displayed less toughness than the other balsa-cored panels. This was attributed to fracture governed by the relatively weak CSM layer. The high toughness of panel D, with a layer of thermoplastic polyester fibers in polyester resin between face and core, is attributed to the very tough thermoplastic fiber material and strong adhesion of the layer to the core.

For the cross-linked PVC foamed core sandwich panel (E), the toughness falls below that of the balsa core specimens, but the core density is much less. The very high toughness of the sandwich specimens with a thermoplastic core (F) is attributed to the high ductility of the thermoplastic foam,

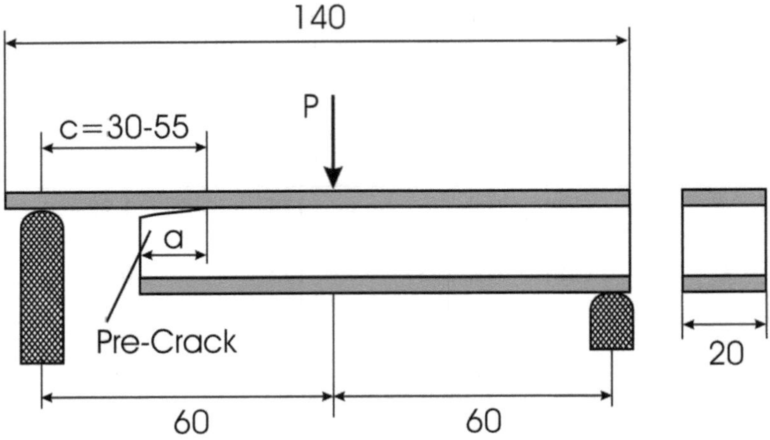

Figure 11.23 TPSB test configuration used by Cantwell et al. (1999).

although the area method may overestimate the actual toughness due to the possible influences of large displacements and plasticity during the fracture testing. The specimens exhibited considerable permanent deformation after unloading.

11.5 Three-Point Sandwich Beam (TPSB) Specimen Testing

The TPSB specimen (Figure 11.23) provides mixed mode crack loading, see Section 10.7. This test method was used by Cantwell et al. (1999) to characterize debond fracture of balsa-cored sandwich panels A–D (Figure 11.22). Beams of 14 cm length and 2 cm width were tested on a three-point flexure fixture at a span length of 12 cm. The specimens contained a precrack approximately 30 mm long. Testing was conducted under displacement control at a cross-head rate of 1 mm/min, except for some tests where the influence of rate on fracture toughness was investigated.

The debond fracture toughness was determined using the compliance calibration method by fitting a third-order polynomial to the compliance data

$$C = C_0 + ka^3, \tag{11.12}$$

where C_0 and k are constants, and a is the crack length. Figure 11.24 shows a typical compliance calibration plot for a sandwich specimen from panel A. The fracture toughness was determined from the critical load, P_c, according to

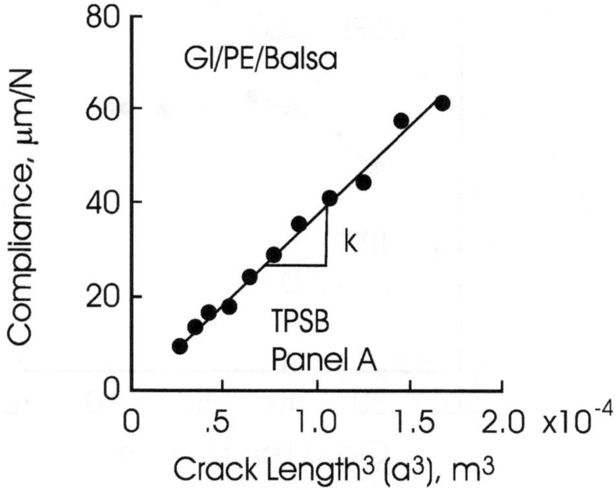

Figure 11.24 Compliance vs. crack length cube graph for a TPSB specimen (panel A).

$$G_c = \frac{3kP_c{}^2a^2}{2b}, \tag{11.13}$$

where k is the slope of the C vs. a^3 graph (Figure 11.24).

Fracture testing revealed that the crack generally propagated in a stable manner, except for panel B. The failure process at the interface was similar to that observed for the SCB specimens from panels A–D discussed in Section 11.4, i.e., debond failure involved a substantial degree of fiber bridging. Figure 11.25 shows a typical R curve for a specimen from panel D. The average toughness is about 1.1 kJ/m^2.

This value may be compared to the value of 1.3 kJ/m^2 determined for panel D using the SCB specimen (Table 11.4). The values are similar. The TPSB specimen (Figure 11.23) is expected to be more mode II-dominated than the SCB specimen (Figure 11.18). Generally, increasing mode II dominance is expected to elevate the toughness of the interface. The fracture process, however, is greatly influenced by fiber bridging which makes local concepts such as mode mixity less meaningful.

The face separation toughness values determined for sandwich panels A–D are summarized in Figure 11.26. The toughness values for panels A, C and D are similar, while the toughness of panel B is only about 500 J/m^2. Cantwell et al. (1999) attributed the low toughness of this panel to a defective panel. They also examined the effect of loading rate

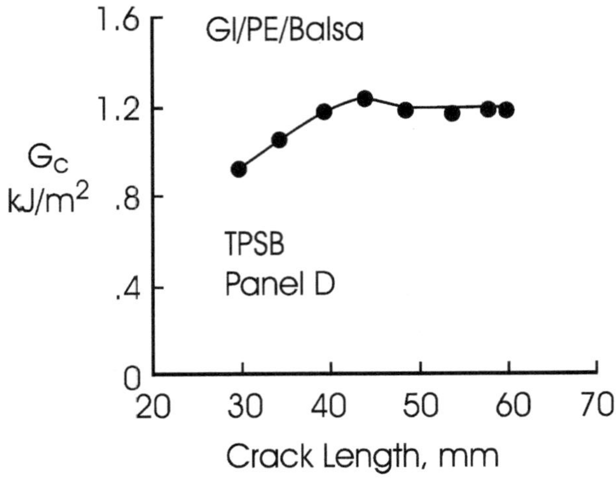

Figure 11.25 Typical R curve for a sandwich specimen from panel D determined using the TPSB test.

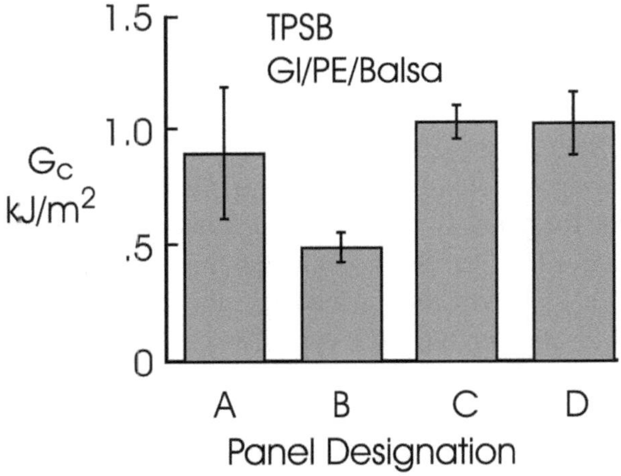

Figure 11.26 Face separation toughness of balsa-cored sandwich specimens determined using the TPSB specimen.

on the debond toughness. Increased loading rate reduced the toughness, see Figure 11.27.

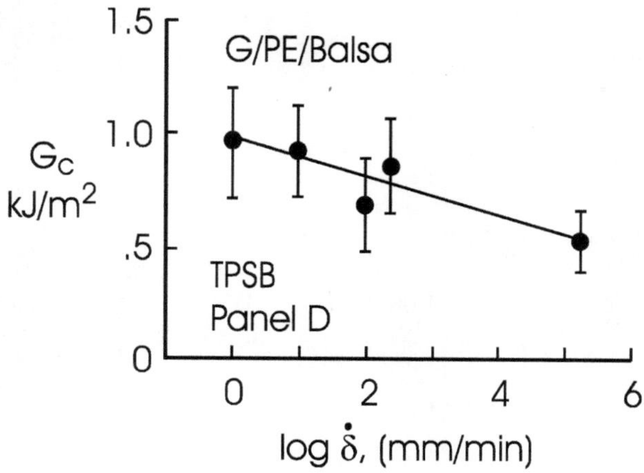

Figure 11.27 Debond fracture toughness of panel D at various cross-head rates.

11.6 Mixed Mode Bending (MMB) Specimen Testing

Analysis of the MMB specimen (Section 10.8), provides expressions for the compliance and energy release rate, Equations (10.29) and (10.30). This test configuration does not allow compliance calibration and testing is conducted straight from the precrack. Figure 11.28 shows a photograph of the test fixture with a sandwich specimen.

MMB sandwich beams were prepared from a sandwich panel consisting of a 30 mm thick H100 PVC foam core and 2 mm E-glass/polyester face sheets. The face and core moduli are: $E_f = 16.4$ GPa and $E_c = 135$ MPa. The specimens were 35 mm wide and the span length was 150 mm. A face/core debond was created by cutting the core cells at the upper face/core interface with a very thin razor blade of 0.35 mm thickness. Steel hinges were bonded to the MMB specimen to allow load introduction. The initial debond length (a), measured from the loading line to the crack tip was 25 mm. The mode mixity was varied by changing the lever arm distance, c ($c = 30$, 40 and 50 mm). The specimens were loaded at a cross-head rate of 1 mm/min. Typical load vs. displacement curves are shown in Figure 11.29. All specimens failed through debond propagation as marked with an open circle "o"). Very good overall agreement between the experimentally measured and predicted (Equation (10.29)) initial stiffness of the MMB specimen is noted.

In all cases, the crack propagated in the core below the resin-rich foam cell region. The face/core fracture toughness G_c was determined by substitution of the debond fracture load into Equation (10.30): $G_c = 580$ J/m^2. This

Figure 11.28 MMB test fixture and sandwich test specimen.

result agrees reasonably with the fracture toughness (700 J/m^2) for a similar sandwich measured by Shivakamar and Chen (2005) using the DCB test. The phase angle, ψ, determined from Equation (10.2) ranges from -18 to $-21°$ for the lever arm distances used ($c = 30$, 40, and 50 mm).

11.7 Double Cantilever Beam-Uneven Bending Moments (DCB-UMB) Specimen Testing

Lundsgaard et al. (2008) utilized the DCB-UBM test described in Section 10.9 to determine cohesive laws for face/core separation in sandwich specimens consisting of quasi-isotropic glass/polyester face sheets and H200 PVC foam core. The face sheets are assumed to be isotropic in the plane of the sandwich with $E_f = 14.1$ GPa and $\nu_c = 0.32$. The corresponding mechanical properties of the isotropic core are $E_c = 250$ MPa and $\nu_c = 0.32$. In addition to the face and core materials, a thin layer of "chopped strand mat" (CSM) was added at each face/core interface to promote toughening through

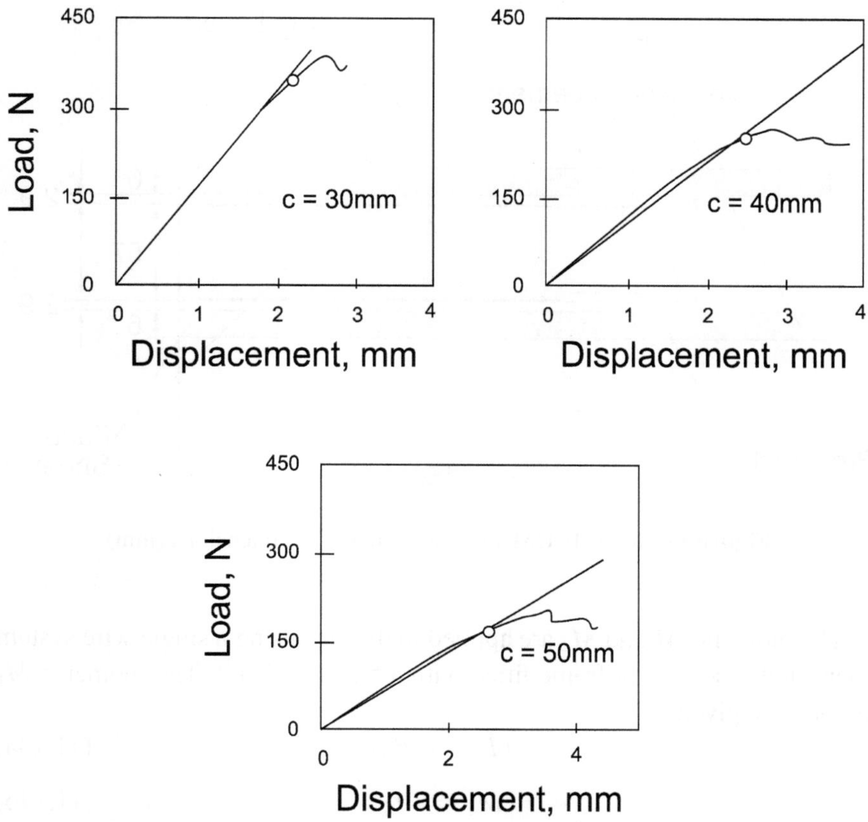

Figure 11.29 Experimental and analytical load vs. displacement for MMB specimens with glass/polyester face sheets bonded to a H100 PVC foam core (“○ marks the onset of crack growth, $h_c = 30$ mm, $h_f = 2$ mm, $2L = 150$ mm and $b = 35$ mm). After Quispitupa et al. (2009).

bridging of the chopped fibers. Figure 11.30 shows the sandwich test specimen and dimensions.

The nominal width of the test specimens was 30 mm. A 70 mm edge precrack was defined at the lower face/core interface by a 0.2 mm thick non-stick Teflon film inserted before processing. To avoid excessive bending deformation and failure of the DCB-UBM specimen legs, 6 mm thick high yield strength steel bars were adhesively bonded to the upper and lower face sheets of the sandwich test specimen, see Figure 11.30. The DCB-UBM test fixture, discussed in Section 10.9, and shown schematically in Figure 11.31, was used in the fracture testing.

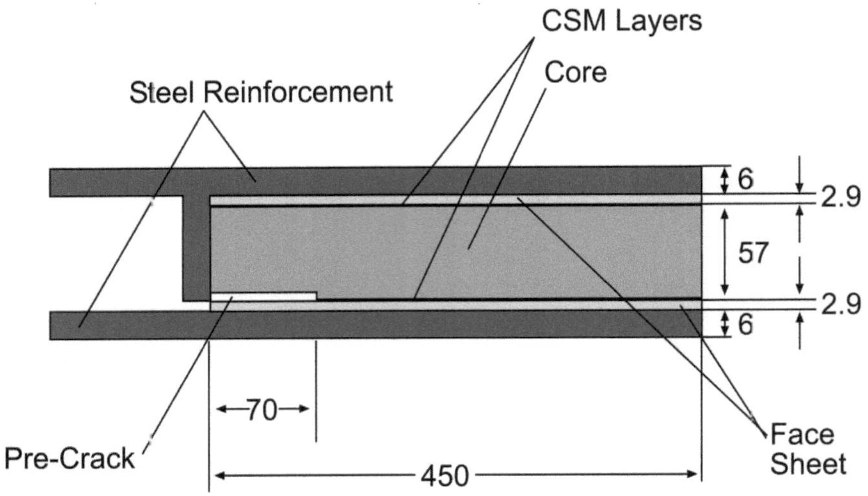

Figure 11.30 DCB-UMB test specimen and dimensions (mm).

The moments M and M_2 are applied to the legs using a single wire system connected to a tall test frame fitted with a 5 kN load cell. The moments M_1 and M_2 are given by

$$M_1 = -P\ell_1, \tag{11.14a}$$

$$M_2 = P\ell_2, \tag{11.14b}$$

where P is the tension in the wire and ℓ_1 and ℓ_2 are the lengths shown in Figure 11.31. The moments are defined positive if they tend to open the crack. As discussed in Section 10.9, mode I-dominated crack loading is achieved by positive moments ($M_1 > 0$, $M_2 > 0$), while mode II-dominated loading corresponds to opposite signs of the moments.

With this loading and test specimen configuration, it is possible to vary the ratio between the opening and the sliding crack flank displacement δ_t/δ_n defined in Figure 9.4. The opening and sliding deformations of the initial crack tip were measured throughout the test. In early experiments on crack propagation in monolithic composites, reference pins were inserted into the material and the displacements were measured using extensometers and LVDTs. For sandwich specimens, displacements were measured with a commercial digital photogrammetry system ARAMIS. A speckled pattern is spray-painted to the specimen edges, which allows the system software to track the displacements. The set-up including the test specimen and 2 Mpixel cameras is shown in Figure 11.32.

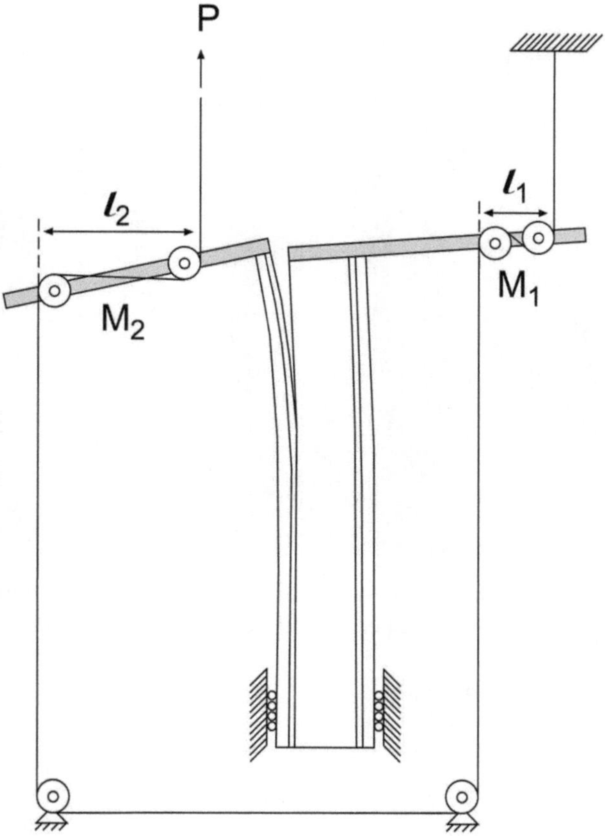

Figure 11.31 DCB-UBM test principle.

11.7.1 Double Cantilever Beam-Uneven Bending Moments (DCB-UBM) Specimen Test Results

The test was conducted by Lundsgaard et al. (2008) at a moment ratio $M_1/M_2 = -0.53$. It was found that this loading corresponds to almost pure mode I loading, i.e. the sliding displacement $\delta_t \approx 0$. The J integral, defined in Equation (9.32) is calculated from the applied moments for the five-layer test specimen shown in Figure 11.30. For calculation of the J integral, the beam is divided into three sub-beams (Figure 11.33). The cracked region is represented by sub-beams #1 and 2, while the intact region is represented by beam #3. Calculation of the J integral requires selection of a path, Γ, around the crack tip. In this case, for simplicity, a path around the external boundaries of the specimen is selected. As shown in Figure 11.33, the path

Figure 11.32 The deformation of the sandwich specimen is tracked by digital cameras with a frequency of 1 frame per second. (Courtesy of C. Lundsgaard (2009).

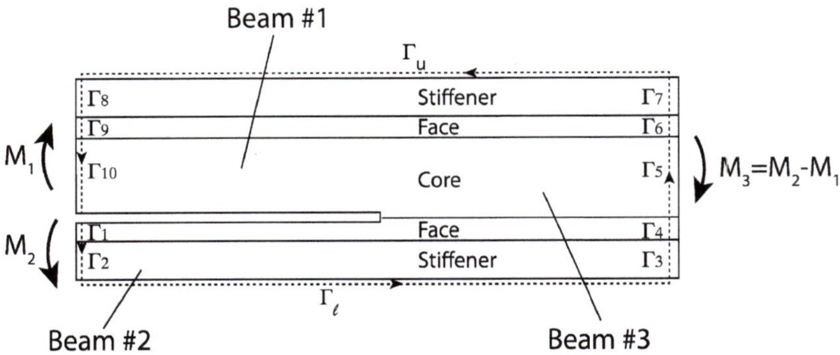

Figure 11.33 J integral path for element of five-layered DCB-UBM specimen. Only the vertical segments of the path, Γ, provide non-zero contributions to J.

consists of vertical and horizontal segments. For the vertical segments on the left edge, dS in Equation (9.32) is $dS = -dy$, and the outward normal vector, $n_j = (-1, 0)$, whereas for the vertical segments on the right edge $dS = dy$ and $n_j = (1, 0)$. For the horizontal segments on top and bottom of the beam element, the outward normals are $n_j = (0, 1)$ and $n_j = (0, -1)$ for the upper (u) and lower (l) segments. Since the top and bottom surfaces are traction free and $dy = 0$, there is no contribution to the J integral from the horizontal segments of the path. The J integral becomes a sum of the

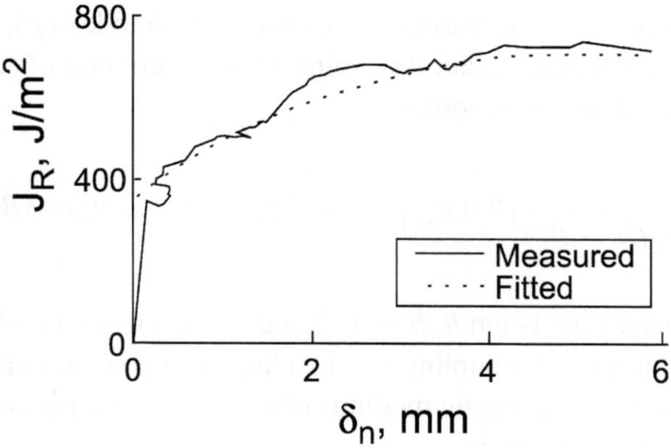

Figure 11.34 J_R vs. crack opening for a sandwich specimen with glass/polyester face sheets and H200 foam core.

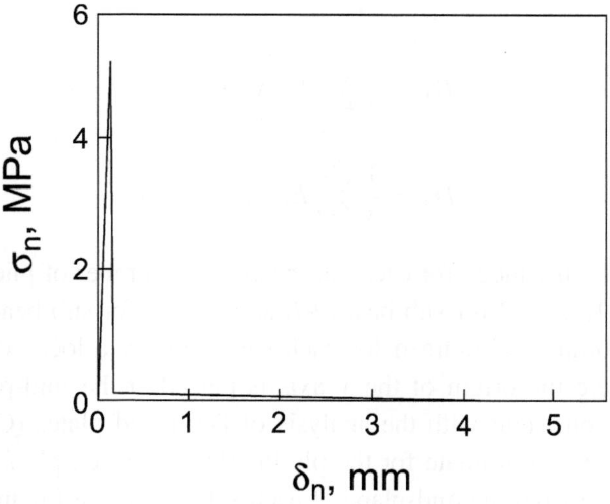

Figure 11.35 Cohesive law for face/core separation under mode I-dominated loading when fiber bridging is present.

10 contributions, two from each layer in the five-layer beam, which results in (Lundsgaard et al., 2008)

$$J = \sum_{p=1}^{10} -\frac{1}{2}\bar{E}_p(\varepsilon_{xx})_p^2 dy, \tag{11.15}$$

where p is the path segment #, Figure 11.33 ($p = 1, 2, \ldots, 10$), $\bar{E}_p = E_p$ for plane stress, and $\bar{E}_p = E_p/(1 - \nu_p^2)$ for plane strain, where E_p is the elastic modulus of the layer associated with the path index (p), and ν_p is the corresponding Poisson's ratio. Assuming a linear variation of strain, ε_{xx}, the expression for J finally becomes

$$J=\sum_{p=1}^{10}\frac{\bar{E}_p M_b^2}{6(A_b D_b - B_b^2)}[A_b^2(y_{p-1}^3-y_p^3)-3A_b B_b(y_{p-1}^2-y_p^2)+3B_b^2(y_{p-1}-y_p)] \tag{11.16}$$

where b indicates the beam #, $b = 1$, 2 and 3, see Figure 11.33. A_b, B_b and D_b are the extensional, coupling and bending stiffnesses for each sub-beam (b) calculated from the elastic modulus of each layer and ply coordinates for each sub-beam according to

$$A_b = \sum_{u=1}^{n} \bar{E}_k(y_k - y_{k-1}) \tag{11.17a}$$

$$B_b = \frac{1}{2}\sum_{u=1}^{n} \bar{E}_k(y_h^2 - y_{k-1}^2) \tag{11.17b}$$

$$D_b = \frac{1}{3}\sum_{u=1}^{n} \bar{E}_k(y_k^3 - y_{k-1}^3) \tag{11.17c}$$

where k is the ply index for each sub-beam. The number of plies, $n = 3$ for sub-beam #1, $n = 2$ for sub-beam #2, and $n = 5$ for sub-beam #3. y_k are the "ply coordinates" defined for each sub-beam by a local xy coordinate system, where the origin of the y axis is placed at the mid-plane of each sub-beam. Consistent with the analysis of laminated plates (Chapter 3) y_k represents the y coordinate for the ply interface between ply $k-1$ and ply k as measured from the mid-plane of each sub-beam, see Figure 3.2. Notice that J is independent of the crack length.

Because of the mode I-dominated loading case, only normal tractions and displacements are considered. J_R vs. the opening displacement δ_n is shown in Figure 11.34. J_R increases rapidly with δ_n until the crack growth initiates at about 400 J/m^2. From this point, the crack propagates in a stable manner where the increased fracture resistance is due to the development of a bridging zone behind the crack tip.

The data points are fitted by third-order least-squares splines and the fitted date are represented by the curve. To determine the normal tractions, σ_n, acting over the cohesive zone (Equation (9.38a)) is applied:

$$\sigma_n = \frac{\partial J_r}{\partial \delta_n}. \tag{11.18}$$

The cohesive law determined from the results in Figure 11.34 is shown in Figure 11.35. It is noted that the fracture process consists of two major parts. Initially, the stress increases rapidly to 5.2 MPa when the material ahead of the crack tip separates. Since fiber bridging is present, the stress decreases gradually to close to zero as the crack opens from 0.15 to 5.7 mm.

Chapter 12
Face/Core Debond Buckling and Growth

12.1 Introduction

The face/core debond is justifiably considered to be a weak link in the use of sandwich structures. This is because such debonds tend to grow and eventually completely delaminate the face sheet. The most common cause of these defects is poor or missing bonding due to careless manufacturing or a mismatch in the geometry. Similar defects may also arise during service due to thermo-mechanical loads, impact events, or structural fatigue. Debonds or delaminations are susceptible to the phenomenon of "delamination buckling" which occurs when local compressive loading is introduced at the debond site. It is not actually necessary for the structure to be compressed, as even a pure bending loading would introduce axial compression on one of the face sheets. Therefore, the mechanics of debonds considers both structural (buckling) and fracture behavior (stress intensity factors, etc.).

This chapter presents solutions for both the debond buckling and the fracture mechanics problems. The problem is first treated for the one-dimensional case of a wide plate (or beam) and both the debond buckling and the debond initial post-buckling behavior are studied through a perturbation procedure that is based on the nonlinear beam equations with transverse shear included (Kardomateas and Huang, 2003). Subsequently, Section 12.4 presents experimental studies and finite element analysis on the buckling and growth of face/core debonds.

12.2 Buckling of a Face/Core Debond in a Wide Plate/Beam

The mechanics of debond buckling differ from the corresponding studies of delamination buckling in composites due to the fact that the substrate in a debonded sandwich structure includes a much different kind of material, namely a flexible core, typically foam or honeycomb. To this extent, the contribution of the shear stresses and shear deformations of the core are noteworthy and therefore should be included in the formulation.

We consider a compression loaded sandwich beam, of length $2L$, and unit width, consisting of face sheets of thicknesses h_{f1} and h_{f2}, extensional moduli E_{f1} and E_{f2}, and shear moduli G_{f1} and G_{f2}, respectively. The core, of thickness h_c, has an extensional modulus, E_c, and shear modulus G_c (Figure 12.1). The debond, of length $2a$, is symmetrically located at the interface of the top face sheet and the core. Over the region of the debond, the sandwich beam consists of two parts: the debonded layer of the upper face sheet (referred to as the "debonded part") of thickness h_{f1} and the part below the debond "substrate part" of thickness $h_c + h_{f2}$, which includes the core and the lower face sheet. The region outside the debond is referred to as the "base part" and consists of the entire section of the sandwich beam, i.e., of thickness $h_{f1} + h_c + h_{f2}$. We shall also denote the base part with "b", the debonded part with "d", and the substrate part with "s". Let us also assume that the beam is clamped-clamped.

A characteristic of sandwich construction is that the neutral axis for the base and the substrate parts is in general no longer at the middle of the corresponding sections. With respect to a reference axis x through the middle of the core, the neutral axis of the base section is defined at a distance e_b (Figure 12.2), as

$$\begin{aligned} & e_b \left(E_{f1}h_{f1} + E_c h_c + E_{f2}h_{f2}\right) \\ & \quad = E_{f2}h_{f2}\left(\frac{h_{f2}}{2} + \frac{h_c}{2}\right) - E_{f1}h_{f1}\left(\frac{h_{f1}}{2} + \frac{h_c}{2}\right), \end{aligned} \tag{12.1a}$$

and that of the substrate part is at a distance e_s given by

$$e_s\left[E_c h_c + E_{f2}h_{f2}\right] = E_{f2}h_{f2}\left(\frac{h_{f2}}{2} + \frac{h_c}{2}\right). \tag{12.1b}$$

Moreover, while for the debonded face layer, assumed homogeneous, the bending rigidity per unit width is

$$D_d = E_{f1}\frac{h_{f1}^3}{12}. \tag{12.2a}$$

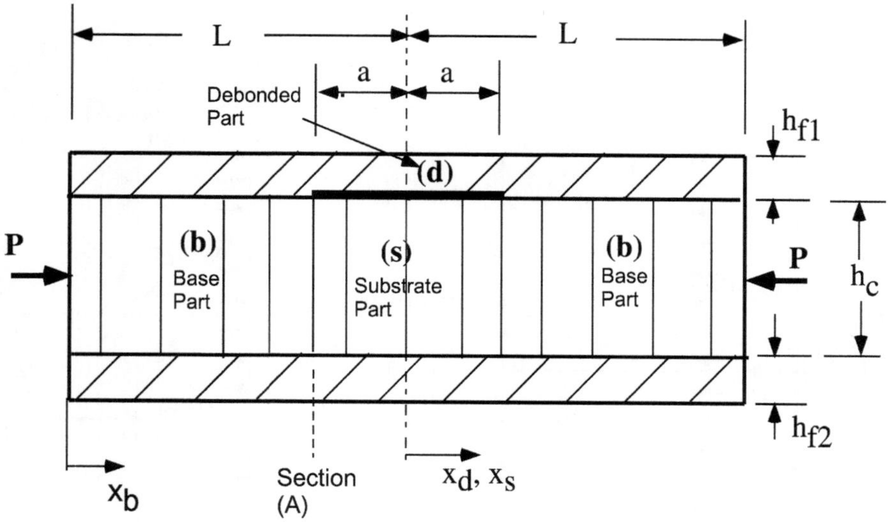

Figure 12.1 A sandwich beam/wide plate with a debond at the face sheet/core interface.

The equivalent flexural rigidity of the base part of the sandwich section per unit width is (Figure 12.2)

$$D_b = E_{f1}\frac{h_{f1}^3}{12} + E_{f1}h_{f1}\left(\frac{h_{f1}}{2} + \frac{h_c}{2} + e_b\right)^2 + E_{f2}\frac{h_{f2}^3}{12} + E_{f2}h_{f2}\left(\frac{h_{f2}}{2} + \frac{h_c}{2} - e_b\right)^2 + E_c\frac{h_c^3}{12} + E_c h_c e_b^2, \quad (12.2b)$$

and for the substrate (per unit width)

$$D_s = E_c\frac{h_c^3}{12} + E_c h_c e_s^2 + E_{f2}\frac{h_{f2}^3}{12} + E_{f2}h_{f2}\left(\frac{h_{f2}}{2} + \frac{h_c}{2} - e_s\right)^2. \quad (12.2c)$$

The nonlinear differential equations including transverse shear for the three parts of the sandwich beam (Figure 12.1), namely the base part (b), debonded part (d), and substrate part (s), are (Huang and Kardomateas, 2002)

$$D_i\frac{d^2\theta_i}{ds^2} + P_i\left(\frac{\beta_i P_i}{2A_i\bar{G}_i}\sin 2\theta_i + \sin\theta_i\right) = 0, \quad i = b, d, s, \quad (12.3a)$$

where θ_i is the rotation angle of the cross-section due to bending and s is the distance along the deflected beam. After Taylor series expansion of $\sin\theta$, (12.3a) becomes

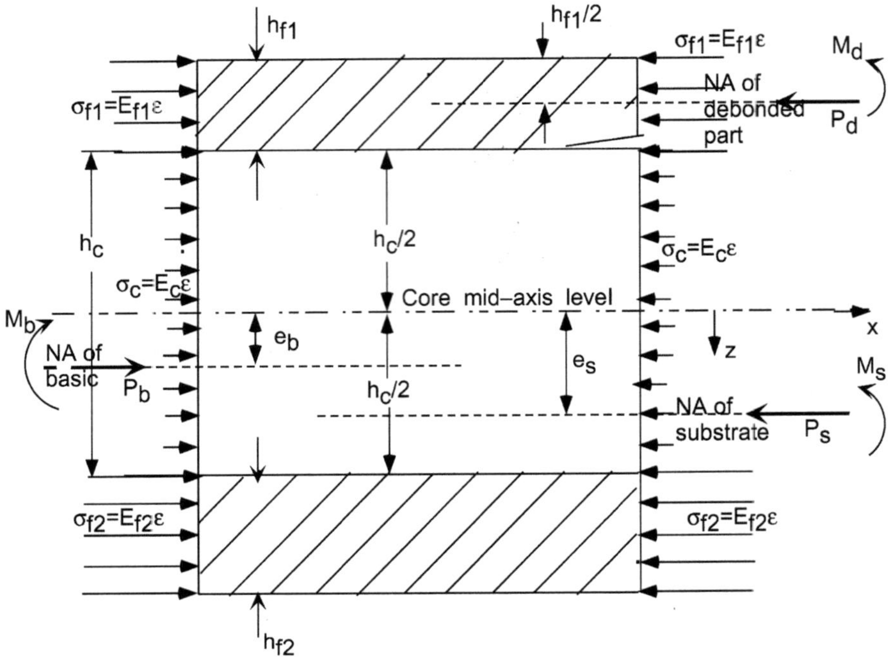

Figure 12.2 Sandwich cross-section with the acting forces and moments.

$$D_i \frac{d^2\theta_i(x_i)}{dx_i^2} + \left(\frac{\beta_i P_i^2}{A_i \bar{G}_i} + P_i\right)\theta_i(x_i) - \left(\frac{2\beta_i P_i^2}{3A_i \bar{G}_i} + \frac{P_i}{6}\right)\theta_i^3(x_i) = 0, \quad (12.3b)$$

where $\theta_i(x)$ is the rotation of the normal to the cross-section, D_i is the bending rigidity, β_i is the shear correction factor, P_i is the axial load per unit width, A_i is the cross-sectional area, and $\bar{G}_i$ is the "effective" shear modulus of each part. The latter may be calculated as follows:

$$\frac{h_{f1} + h_c + h_{f2}}{\bar{G}_b} = \frac{h_{f1}}{G_{f1}} + \frac{h_c}{G_c} + \frac{h_{f2}}{G_{f2}}, \quad (12.3c)$$

$$\bar{G}_d = G_{f1}, \quad (12.3d)$$

$$\frac{h_c + h_{f2}}{\bar{G}_s} = \frac{h_c}{G_c} + \frac{h_{f2}}{G_{f2}}. \quad (12.3e)$$

Formulas for the shear correction factors in sandwich structures were derived by Huang and Kardomateas (2002) based on shear energy equivalency. For the base part (b):

$$\frac{\beta_b}{\bar{G}_b A_b} = \sum_{i=1,2} \frac{E_{fi}^2}{4D_b^2 G_{fi}} \left[a_i^4 h_{fi} - \frac{2}{3} a_i^2 \left(a_i^3 - b_i^3\right) + \frac{1}{5}\left(a_i^5 - b_i^5\right) \right]$$

$$+ \frac{E_{fi}^2}{D_b^2 G_c} h_{fi}^2 c_i^2 b_i + \frac{2}{15} \frac{E_c^2}{D_b^2 G_c} b_i^5 + \frac{2}{3} \frac{E_{fi} E_c}{D_b^2 G_c} h_{fi} c_i b_i^3, \quad (12.3f)$$

where

$$a_1 = h_{f1} + \frac{h_c}{2} + e_b; \quad b_1 = \frac{h_c}{2} + e_b; \quad c_1 = \frac{h_{f1}}{2} + \frac{h_c}{2} + e_b, \quad (12.3g)$$

$$a_2 = h_{f2} + \frac{h_c}{2} - e_b; \quad b_2 = \frac{h_c}{2} - e_b; \quad c_2 = \frac{h_{f2}}{2} + \frac{h_c}{2} - e_b. \quad (12.3h)$$

Since the debonded part (d) is homogeneous,

$$\frac{\beta_d}{\bar{G}_d A_d} = \frac{6}{5}\left(\frac{1}{G_{f1} h_{f1}}\right). \quad (12.3i)$$

For the substrate part (s), the corresponding formula for the shear correction is in terms of

$$a_3 = h_{f2} + \frac{h_c}{2} - e_s; \quad b_3 = \frac{h_c}{2} - e_s; \quad c_3 = \frac{h_{f2}}{2} + \frac{h_c}{2} - e_s; \quad d_3 = \frac{h_c}{2} + e_s, \quad (12.3j)$$

as follows:

$$\frac{\beta_s}{\bar{G}_s A_s} = \frac{E_{f2}^2}{4D_s^2 G_{f2}} \left[a_3^4 h_{f2} - \frac{2}{3} a_3^2 (a_3^3 - b_3^3) + \frac{1}{5}(a_3^5 - b_3^5) \right]$$

$$+ \frac{2}{15} \frac{E_c^2}{D_s^2 G_c} (d_3^5 + b_3^5) + \frac{E_{f2}^2}{D_s^2 G_c} h_{f2}^2 c_3^2 b_3 + \frac{2}{3} \frac{E_{f2} E_c}{D_s^2 G_c} h_{f2} c_3 b_3^3. \quad (12.3k)$$

These expressions for the $\beta_i/(\bar{G}_i A_i)$ are for unit width; for a width other than unit, these expressions need to be divided by the width of the beam.

The way the geometry is configured gives the following conditions at $x_i = 0$:

$$\theta_i(0) = 0, \quad i = b, d, s. \quad (12.4)$$

The above condition is valid for $i = b$ (base part) because of the clamped-end and for $i = d, s$ (debonded and substrate parts) because of symmetry.

Furthermore, a kinematic condition of common slope, θ_A, for the parts of the section where the debond starts and ends reads

$$\theta_b(L-a) = \theta_d(-a) = \theta_s(-a) = \theta_A. \tag{12.5}$$

The force and moment (about the neutral axis of the base part) equilibrium conditions are (Figure 12.2)

$$P_b = P_d + P_s, \tag{12.6}$$

$$M_b - M_d - M_s - P_d\left(\frac{h_{f1}}{2} + \frac{h_c}{2} + e_b\right) + P_s(e_s - e_b) = 0. \tag{12.7}$$

Finally, the axial displacement continuity condition at the tip A (Figure 12.1) is

$$u_d^A = u_s^A, \tag{12.8}$$

where

$$u_d^A = \frac{1}{2}\int_{-a}^{0} \theta_d^2 dx_d + \frac{P_d a}{E_{f1}h_{f1}} + \theta_A \frac{h_{f1}}{2}, \tag{12.9}$$

$$u_s^A = \frac{1}{2}\int_{-a}^{0} \theta_s^2 dx_s + \frac{P_s a}{E_c h_c + E_{f2}h_{f2}} - \theta_A\left(e_s + \frac{h_c}{2}\right). \tag{12.10}$$

12.2.1 Asymptotic Expansion

The asymptotic expansion is an efficient way of deriving closed form solutions for the critical load and the initial post-buckling behavior. This approach has been used previously by Kardomateas (1993) in a study of delaminations in monolithic composites in conjunction with the elastica theory.

Now, let us expand P_i and θ_i as

$$P_i = P_{i0} + \xi P_{i1} + \xi^2 P_{i2} + \xi^3 P_{i3} + \cdots, \tag{12.11}$$

$$\theta_i(x_i) = \xi\theta_{i1}(x_i) + \xi^2\theta_{i2}(x_i) + \xi^3\theta_{i3}(x_i) + \cdots, \tag{12.12}$$

where the 0 subscript corresponds to the pre-buckling state, 1 to the buckling state and 2, etc., to the post-buckling state. Also, let us set ξ to be the common slope of the section at the debond tip A, i.e.

$$\xi = \theta_A. \tag{12.13}$$

From (12.5) and (12.12), this gives the additional conditions

$$\theta_{b1}(L-a) = 1; \quad \theta_{b2}(L-a) = \theta_{b3}(L-a) = \ldots = 0, \tag{12.14}$$

and

$$\theta_{i1}(-a) = 1; \quad \theta_{i2}(-a) = \theta_{i3}(-a) = \ldots = 0, \quad i = d, s. \tag{12.15}$$

Substituting Equations (12.11) and (12.12) into (12.3) and (12.4)–(12.10) and rearranging the terms based on the order of ξ, we obtain separately the equations and boundary conditions for the pre-buckling, buckling, and initial post-buckling problem.

12.2.2 Pre-Buckling State, $O(\xi^0)$

A major characteristic of the prebuckling state for a sandwich section under uniform compressive strain is that there are non-zero bending moments (as opposed to a monolithic one in which the bending moments are zero) but zero bending deflections.

Under a uniformly applied compressive strain, ε_0, the resultant forces (per unit width) for the base part (b), delaminated part (d) and substrate part (s), are (Figure 12.2)

$$P_{b0} = \varepsilon_0 \left(E_{f1}h_{f1} + E_c h_c + E_{f2}h_{f2}\right), \tag{12.16a}$$

$$P_{d0} = \varepsilon_0 E_{f1}h_{f1}; \quad P_{s0} = \varepsilon_0 \left(E_c h_c + E_{f2}h_{f2}\right). \tag{12.16b}$$

The prebuckling moments (per unit width) are then found as (Figure 12.2):

$$M_{b0} = \varepsilon_0 \left[E_{f1}h_{f1} \left(\frac{h_{f1}}{2} + \frac{h_c}{2} + e_b \right) + E_c h_c e_b - E_{f2}h_{f2} \left(\frac{h_{f2}}{2} + \frac{h_c}{2} - e_b \right) \right], \tag{12.17a}$$

$$M_{d0} = 0; \quad M_{s0} = \varepsilon_0 \left[E_c h_c e_s - E_{f2}h_{f2} \left(\frac{h_{f2}}{2} + \frac{h_c}{2} - e_s \right) \right]. \tag{12.17b}$$

These pre-buckling forces and moments satisfy identically the force and moment equilibrium equation (about the neutral axis of the base part), Equations (12.6) and (12.7). Furthermore, since a state of pure axial compressive strain exists without bending deflections, the compatibility of shortening, Equation (12.8) is also satisfied.

12.2.3 Buckling (First-Order) Equations, $O(\xi^1)$

From (12.3), (12.11) and (12.12), the first-order differential equations for the three parts are

$$D_i \frac{d^2\theta_{i1}(x_i)}{dx_i^2} + \left(\frac{\beta_i P_{i0}^2}{A_i \bar{G}_i} + P_{i0}\right)\theta_{i1}(x_i) = 0, \quad i = b, d, s. \tag{12.18a}$$

The corresponding boundary conditions from (12.4) are

$$\theta_{i1}(0) = 0, \quad i = b, d, s \tag{12.18b}$$

and from (12.5)

$$\theta_{b1}(L-a) = \theta_{d1}(-a) = \theta_{s1}(-a) = \theta_{A1} = 1. \tag{12.18c}$$

The first-order moment equilibrium from (12.7) is

$$D_1 \frac{d\theta_{b1}}{dx_b}\bigg|_{x_b=L-a} - D_2 \frac{d\theta_{d1}}{dx_d}\bigg|_{x_d=-a} - D_3 \frac{d\theta_{s1}}{dx_s}\bigg|_{x_s=-a}$$
$$- P_{d1}\left(\frac{h_{f1}}{2} + \frac{h_c}{2} + e_b\right) + P_{s1}(e_s - e_b) = 0, \tag{12.18d}$$

and the first-order force equilibrium

$$P_{d1} + P_{s1} = P_{b1}. \tag{12.18e}$$

The first-order compatibility equation from (12.8) with $\theta_{A1} = 1$ becomes

$$\frac{P_{s1}a}{E_c h_c + E_{f2} h_{f2}} - \left(e_s + \frac{h_c}{2}\right) = \frac{P_{d1}a}{E_{f1} h_{f1}} + \frac{h_{f1}}{2}. \tag{12.18f}$$

Let us set

$$\lambda_i = \sqrt{\left(\frac{\beta_i P_{i0}^2}{A_i \bar{G}_i} + P_{i0}\right) \Big/ D_i}, \quad i = b, d, s, \tag{12.19}$$

where P_{b0}, P_{d0} and P_{s0} are given in (12.16) in terms of the uniform compressive strain ε_0. Solutions for Equation (12.18a) that satisfy the boundary conditions (12.18b) are

$$\theta_{i1} = C_{i1}\sin(\lambda_i x_i), \quad i = b, d, s. \tag{12.20}$$

Now the constants C_{b1}, C_{d1}, C_{s1} are determined from the common slope equations (12.18c), as

$$C_{b1} = 1/\sin\lambda_b(L-a); \quad C_{d1} = -1/\sin\lambda_d a; \quad C_{s1} = -1/\sin\lambda_s a. \tag{12.21}$$

The characteristic equation in terms of ε_0 is found by eliminating P_{b1}, P_{d1} and P_{s1} from the previous equations. This is done as follows. The moment equilibrium equation (12.18d), becomes

$$D_b\lambda_b\cot\lambda_b(L-a) + D_d\lambda_d\cot\lambda_d a + D_s\lambda_s\cot\lambda_s a = P_{d1}\left(\frac{h_{f1}}{2}+\frac{h_c}{2}+e_b\right) - P_{s1}(e_s-e_b). \tag{12.22a}$$

By using the neutral axis definitions (12.1a) and (12.1b), we obtain

$$e_s - e_b = \left(\frac{h_{f1}}{2}+\frac{h_c}{2}+e_b\right)\frac{E_{f1}h_{f1}}{E_c h_c + E_{f2}h_{f2}}, \tag{12.22b}$$

therefore (12.22a) becomes

$$P_{d1}\frac{a}{E_{f1}h_{f1}} - P_{s1}\frac{a}{E_c h_c + E_{f2}h_{f2}} = \frac{[D_b\lambda_b\cot\lambda_b(L-a) + D_d\lambda_d\cot\lambda_d a + D_s\lambda_s\cot\lambda_s a]\,a}{E_{f1}h_{f1}(\frac{h_{f1}}{2}+\frac{h_c}{2}+e_b)}. \tag{12.23}$$

By comparing (12.18f) and (12.23), we can see that the left-hand side of (12.23) can be eliminated. Thus, we obtain the following characteristic equation:

$$\frac{[D_b\lambda_b\cot\lambda_b(L-a) + D_d\lambda_d\cot\lambda_d a + D_s\lambda_s\cot\lambda_s a]\,a}{E_{f1}h_{f1}(\frac{h_{f1}}{2}+\frac{h_c}{2}+e_b)} + \left(e_s+\frac{h_c}{2}+\frac{h_{f1}}{2}\right) = 0. \tag{12.24}$$

Equation (12.24) is a nonlinear algebraic equation which can be solved numerically for the critical strain ε_0 (or critical load from (12.16)). In the numerical procedure, a solution is sought near the Euler buckling strain of the debonded face, $\varepsilon_0 = \pi^2 h_{f1}^2/(12a^2)$.

Example. To offer insight into the debond buckling behavior, let us consider a symmetric sandwich plate with unidirectional carbon/epoxy faces and hexagonal glass/phenolic honeycomb core. The material data are given in Section 5.1.4. The thickness of each face sheet is $h_{f1} = h_{f1} = 2$ mm and the core thickness, $h_c = 16$ mm, thus the total thickness is $h = 20$ mm. A total length ($2L$) of 10 times the total thickness is assumed, thus $L = 5h$.

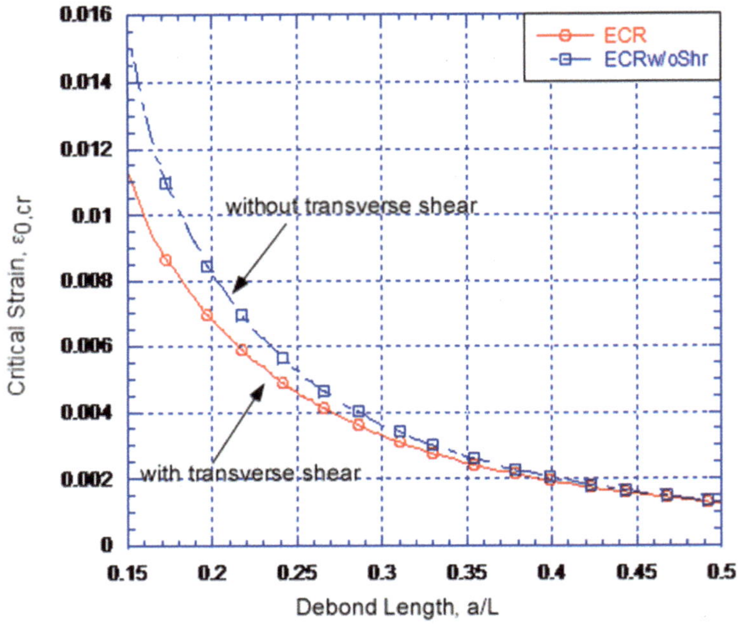

Figure 12.3 Critical strain vs. debond length.

A range of debond lengths ($2a$) between 15 to 50% of the total length is considered, i.e. a ranges between 0.15 and $0.50L$.

Figure 12.3 shows the critical strain, $\varepsilon_{0,\mathrm{cr}}$ plotted vs. debond length (a/L). In the results presented, the case of no transverse shear effect corresponds to setting $\beta_i = 0$. It is seen that the critical strain decreases with debond length, as expected, but also interesting is the effect of transverse shear, which reduces the critical strain. This effect is most pronounced for short debonds. For example, for $a = 0.20L$, transverse shear reduces the critical strain by 20% but at $a = 0.30L$ the reduction is only 10%.

12.3 Intitial Post-Buckling Behavior of a Face/Core Debond in a Wide Plate/Beam

12.3.1 Second-Order Equations, $O(\xi^2)$

From (12.3), (12.11) and (12.12), we obtain the following second-order differential equation:

$$D_i \frac{d^2\theta_{i2}(x_i)}{dx_i^2} + \left(\frac{\beta_i P_{i0}^2}{A_i \bar{G}_i} + P_{i0}\right)\theta_{i2}(x_i)$$
$$= -\left(\frac{2\beta_i P_{i0} P_{i1}}{A_i \bar{G}_i} + P_{i1}\right)\theta_{i1}(x_i), \quad i = b, d, s \tag{12.25}$$

and from (12.4), (12.14) and (12.15):

$$\theta_{i2}(0) = 0, \quad i = b, d, s, \tag{12.26a}$$

$$\theta_{b2}(L-a) = \theta_{d2}(-a) = \theta_{s2}(-a) = 0. \tag{12.26b}$$

The second-order moment equilibrium from (12.7) is

$$D_b \frac{d\theta_{b2}}{dx_b}\bigg|_{x_b = L-a} - D_d \frac{d\theta_{d2}}{dx_d}\bigg|_{x_d = -a} - D_s \frac{d\theta_{s2}}{dx_s}\bigg|_{x_s = -a}$$
$$- P_{d2}\left(\frac{h_{f1}}{2} + \frac{h_c}{2} + e_b\right) + P_{s2}(e_s - e_b) = 0, \tag{12.27}$$

and the second-order force equilibrium

$$P_{d2} + P_{s2} = P_{b2}. \tag{12.28}$$

Finally, the second-order displacement compatibility from (12.8)–(12.12) is

$$\frac{1}{2}\int_{-a}^{0} \theta_{s1}^2(x_s)dx_s + \frac{P_{s2}a}{E_c h_c + E_{f2} h_{f2}}$$
$$= \frac{1}{2}\int_{-a}^{0} \theta_{d1}^2(x_d)dx_d + \frac{P_{d2}a}{E_{f1} h_{f1}}. \tag{12.29}$$

The general solution for the second-order differential equation (12.25) is

$$\theta_{i2}(x_i) = C_{i2}\sin\lambda_i x_i + B_{i2}\cos\lambda_i x_i + \frac{P_{i1}}{2\lambda_i D_i}\left(\frac{2\beta_i P_{i0}}{A_i \bar{G}_i} + 1\right) C_{i1} x_i \cos\lambda_i x_i. \tag{12.30}$$

The constants B_{i2} are zero due to the boundary conditions (12.26a)

$$B_{i2} = 0, \quad i = b, d, s. \tag{12.31}$$

Applying the conditions (12.26b), we can find the constants C_{i2} as

$$C_{b2} = -\frac{P_{b1}}{2\lambda_b D_b} C_{b1}\left(\frac{2\beta_b P_{b0}}{A_b \bar{G}_b} + 1\right)(L-a)\cot\lambda_b(L-a), \tag{12.32a}$$

and

$$C_{i2} = -\frac{P_{i1}}{2\lambda_i D_i} C_{i1} \left(\frac{2\beta_i P_{i0}}{A_i \bar{G}_i} + 1 \right) a \cot \lambda_i a, \quad i = d, s. \tag{12.32b}$$

Now the displacement compatibility equation (12.29) becomes

$$P_{d2} \frac{a}{E_{f1} h_{f1}} - P_{s2} \frac{a}{E_c h_c + E_{f2} h_{f2}} = \frac{1}{4} \left[C_{s1}^2 \left(a - \frac{\sin 2\lambda_s a}{2\lambda_s} \right) - C_{d1}^2 \left(a - \frac{\sin 2\lambda_d a}{2\lambda_d} \right) \right]. \tag{12.33}$$

The moment equilibrium (12.27), obtained by substituting the second-order deflections (12.30) and the relationship for the neutral axes of the substrate and the base part (12.22b), becomes

$$\begin{aligned} &\left[P_{d2} \frac{a}{E_{f1} h_{f1}} - P_{s2} \frac{a}{E_c h_c + E_{f2} h_{f2}} \right] \frac{E_{f1} h_{f1} (\frac{h_{f1}}{2} + \frac{h_c}{2} + e_b)}{a} \\ &= D_b \Bigg\{ C_{b2} \lambda_b \cos \lambda_b (L-a) + \frac{C_{b1} P_{b1}}{2\lambda_b D_b} \left(\frac{2\beta_b P_{b0}}{A_b \bar{G}_b} + 1 \right) \\ &\times [\cos \lambda_b (L-a) - (L-a) \lambda_b \sin \lambda_b (L-a)] \Bigg\} \\ &- \sum_{i=d,s} D_i \Bigg[C_{i2} \lambda_i \cos \lambda_i a + \frac{C_{i1} P_{i1}}{2\lambda_i D_i} \left(\frac{2\beta_i P_{i0}}{A_i \bar{G}_i} + 1 \right) \\ &\times (\cos \lambda_i a - a \lambda_i \sin \lambda_i a) \Bigg], \end{aligned} \tag{12.34}$$

Comparing (12.34) and (12.33), we can eliminate the left-hand side of the latter equation, which contains the second-order forces, and, by using also (12.32), an equation for the first-order forces is obtained, i.e.

$$\begin{aligned} B_d P_{d1} + B_s P_{s1} = \frac{1}{4} &\left[C_{s1}^2 \left(a - \frac{\sin 2\lambda_s a}{2\lambda_s} \right) - C_{d1}^2 \left(a - \frac{\sin 2\lambda_d a}{2\lambda_d} \right) \right] \\ &\times \frac{E_{f1} h_{f1} (\frac{h_{f1}}{2} + \frac{h_c}{2} + e_b)}{a}, \end{aligned} \tag{12.35}$$

where

$$\begin{aligned} B_i = &\frac{C_{b1}}{2\lambda_b} \left(\frac{2\beta_b P_{b0}}{A_b \bar{G}_b} + 1 \right) \left[\cos \lambda_b (L-a) - \frac{(L-a)\lambda_b}{\sin \lambda_b (L-a)} \right] \\ &+ \frac{C_{i1}}{2\lambda_i} \left(\frac{2\beta_i P_{i0}}{A_i \bar{G}_i} + 1 \right) \left(\frac{a\lambda_i}{\sin \lambda_i a} - \cos \lambda_i a \right), \quad i = d, s. \end{aligned} \tag{12.36}$$

The second equation for the first-order forces is the first-order compatibility equation (12.18f). The system of the two linear equations (12.35) and (12.18f) can be solved for the first-order forces, P_{d1} and P_{s1}.

12.3.2 Third-Order Equations, $O(\xi^3)$

Collecting terms $O(\xi^3)$ from (12.3), (12.11) and (12.12) gives the following differential equations for the three parts, $i = b$ (base), d (debonded), and s (substrate):

$$D_i \frac{d^2\theta_{i3}(x_i)}{dx_i^2} + D_i\lambda_i^2\theta_{i3}(x_i) = -\left(\frac{2\beta_i P_{i0} P_{i1}}{A_i\bar{G}_i} + P_{i1}\right)\theta_{i2}(x)$$
$$-\left(\frac{\beta_i P_{i1}^2}{A_i\bar{G}_i} + \frac{2\beta_i P_{i0} P_{i2}}{A_i\bar{G}_i} + P_{i2}\right)\theta_{i1}(x_i) + \left(\frac{2\beta_i P_{i0}^2}{3A_i\bar{G}_i} + \frac{P_{i0}}{6}\right)\theta_{i1}^3(x_i). \tag{12.37}$$

Substituting (12.20) and (12.30) for $\theta_{i1}(x)$ and $\theta_{i2}(x)$, respectively, gives the solution

$$\theta_{i3}(x_i) = C_{i3}\sin\lambda_i x_i + B_{i3}\cos\lambda_i x_i + \left(q_{1i} + \frac{k_{0i}C_{i1}}{2\lambda_i}P_{i2}\right) x_i\cos\lambda_i x_i$$
$$-\frac{k_{0i}^2}{8\lambda_i^2}P_{i1}^2 C_{i1} x_i^2 \sin\lambda_i x_i + \frac{q_{0i}}{32\lambda_i^2}C_{i1}^3\sin 3\lambda_i x_i, \tag{12.38a}$$

where

$$k_{0i} = \frac{1}{D_i}\left(\frac{2\beta_i P_{i0}}{A_i\bar{G}_i} + 1\right); \quad k_{1i} = \frac{\beta_i P_{i1}^2}{D_i A_i\bar{G}_i}; \quad q_{0i} = \frac{1}{D_i}\left(\frac{2\beta_i P_{i0}^2}{3A_i\bar{G}_i} + \frac{P_{i0}}{6}\right), \tag{12.38b}$$

$$q_{1i} = \frac{k_{0i}C_{i2}}{2\lambda_i}P_{i1} + \frac{k_{1i}C_{i1}}{2\lambda_i} - \frac{3q_{0i}C_{i1}^3}{8\lambda_i} - \frac{k_{0i}^2 C_{i1}}{8\lambda_i^3}P_{i1}^2, \tag{12.38c}$$

From (12.4):

$$\theta_{i3}(0) = 0, \tag{12.39a}$$

which gives

$$B_{i3} = 0. \tag{12.39b}$$

From (12.5):

$$\theta_{i3}(-a) = 0; \quad i = d, s \text{ (debonded and substrate parts)} \tag{12.39c}$$

and

$$\theta_{b3}(L-a)=0 \text{ (base part)}. \tag{12.39d}$$

Equations (12.39c, d) give C_{i3} as follows:

$$C_{i3}=q_{2i}+k_{2i}P_{i2}, \tag{12.39e}$$

where for the debonded and substrate parts, $i=d,b$,

$$k_{2i}=-\frac{k_{0i}C_{i1}}{2\lambda_i}a\cot\lambda_i a, \tag{12.39f}$$

$$q_{2i}=-q_{1i}a\cot\lambda_i a+\frac{k_{0i}^2C_{i1}}{8\lambda_i^2}P_{i1}^2a^2-\frac{q_{0i}C_{i1}^3}{32\lambda_i^2}\frac{\sin 3\lambda_i a}{\sin\lambda_i a} \tag{12.39g}$$

and for the base part

$$k_{2b}=-\frac{k_{0b}C_{b1}}{2\lambda_b}(L-a)\cot\lambda_b(L-a), \tag{12.39h}$$

$$q_{2b}=-q_{1b}(L-a)\cot\lambda_b(L-a)+\frac{k_{0b}^2C_{b1}}{8\lambda_b^2}P_{b1}^2(L-a)^2$$
$$-\frac{q_{0b}C_{b1}^3}{32\lambda_b^2}\frac{\sin 3\lambda_b(L-a)}{\sin\lambda_b(L-a)}. \tag{12.39i}$$

The third-order moment equilibrium from (12.7) is

$$D_b\frac{d\theta_{b3}}{dx_b}\bigg|_{x_b=L-a}-D_d\frac{d\theta_{d3}}{dx_d}\bigg|_{x_d=-a}-D_s\frac{d\theta_{s3}}{dx_s}\bigg|_{x_s=-a}$$
$$-P_{d3}\left(\frac{h_{f1}}{2}+\frac{h_c}{2}+e_b\right)+P_{s3}(e_s-e_b)=0, \tag{12.40a}$$

and the third-order force equilibrium

$$P_{d3}+P_{s3}=P_{b3}. \tag{12.40b}$$

Using the relations for the neutral axes of the substrate and the base parts (Equations (12.22b), we can write Equation (12.40a) as

$$D_b\frac{d\theta_{b3}}{dx_b}\bigg|_{x_b=L-a}-D_d\frac{d\theta_{d3}}{dx_d}\bigg|_{x_d=-a}-D_s\frac{d\theta_{s3}}{dx_s}\bigg|_{x_s=-a}$$
$$=\left(\frac{P_{d3}a}{E_{f1}h_{f1}}-\frac{P_{s3}a}{E_ch_c+E_{f2}h_{f2}}\right)\frac{E_{f1}h_{f1}}{a}\left(\frac{h_{f1}}{2}+\frac{h_c}{2}+e_b\right). \tag{12.40c}$$

Now, the third-order displacement compatibility from (12.8)–(12.13) is

$$\int_{-a}^{0} \theta_{d1}\theta_{d2}dx_d + \frac{P_{d3}a}{E_{f1}h_{f1}} = \int_{-a}^{0} \theta_{s1}\theta_{s2}dx_s + \frac{P_{s3}a}{E_c h_c + E_{f2}h_{f2}}, \tag{12.41a}$$

which can be written as

$$\int_{-a}^{0} \theta_{s1}\theta_{s2}dx_s - \int_{-a}^{0} \theta_{d1}\theta_{d2}dx_d = \frac{P_{d3}a}{E_{f1}h_{f1}} - \frac{P_{s3}a}{E_c h_c + E_{f2}h_{f2}}. \tag{12.41b}$$

Comparing (12.40c) and (12.41b), we can eliminate the right-hand side, which contains the third-order forces

$$D_b \frac{d\theta_{b3}}{dx_b}\bigg|_{x_b=L-a} - D_d \frac{d\theta_{d3}}{dx_d}\bigg|_{x_d=-a} - D_s \frac{d\theta_{s3}}{dx_s}\bigg|_{x_s=-a} = \left[\int_{-a}^{0} \theta_{s1}\theta_{s2}dx_s - \int_{-a}^{0} \theta_{d1}\theta_{d2}dx_d\right] \frac{E_{f1}h_{f1}}{a}\left(\frac{h_{f1}}{2} + \frac{h_c}{2} + e_b\right). \tag{12.41c}$$

Substituting the definitions for θ_{i1} and θ_{i2}, $i = d, s$ gives

$$\int_{-a}^{0} \theta_{s1}\theta_{s2}dx_s - \int_{-a}^{0} \theta_{d1}\theta_{d2}dx_d = U_s - U_d, \tag{12.41d}$$

where

$$U_i = C_{i1}C_{i2}\left(\frac{a}{2} - \frac{\sin 2\lambda_i a}{4\lambda_i}\right) + C_{i1}^2 \frac{k_{0i}}{4\lambda_i} P_{i1}\left(\frac{\sin 2\lambda_i a}{4\lambda_i^2} - \frac{a\cos 2\lambda_i a}{2\lambda_i}\right), \quad i = d, s. \tag{12.41e}$$

Moreover, substituting the third-order solutions into (12.38a), gives

$$\frac{d\theta_{i3}}{dx_i} = Q_i(x) + R_i(x)P_{i2}, \tag{12.42a}$$

where the functions $Q_i(x)$ and $R_i(x)$ are defined as

$$Q_i(x) = \left[q_{2i}\lambda_i + q_{1i} - \frac{k_{0i}^2}{8\lambda_i}P_{i1}^2 C_{i1}x^2 + \frac{3q_{0i}}{32\lambda_i}C_{i1}^3\right]\cos\lambda_i x - \left[q_{1i}\lambda_i + \frac{k_{0i}^2}{4\lambda_i^2}P_{i1}^2 C_{i1}\right]x\sin\lambda_i x, \tag{12.42b}$$

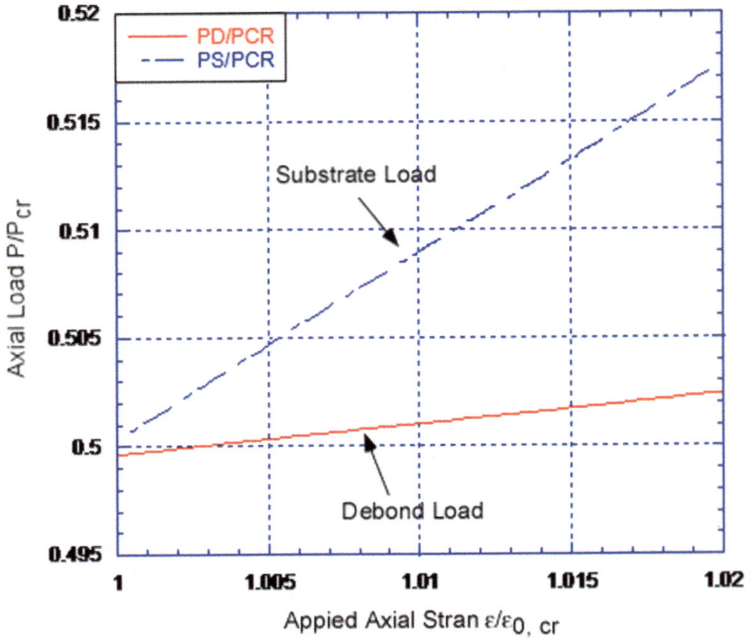

Figure 12.4 Debond and substrate axial load at the initial post-buckling phase.

$$R_i(x) = \left(k_{2i}\lambda_i + \frac{k_{0i}}{2\lambda_i}C_{i1}\right)\cos\lambda_i x - \frac{k_{0i}}{2}C_{i1}x\sin\lambda_i x. \qquad (12.42c)$$

Therefore, Equations (12.41c)–(12.42c) and (12.28) give the following equation for the second-order forces P_{d2} and P_{s2}:

$$\begin{aligned}
&[D_b R_b(L-a) - D_d R_d(-a)]\, P_{d2} + [D_b R_b(L-a) - D_s R_s(-a)]\, P_{s2} \\
&\quad = D_d Q_d(-a) + D_s Q_s(-a) - D_b Q_b(L-a) \\
&\quad\quad + (U_s - U_d)\frac{E_{f1}h_{f1}}{a}\left(\frac{h_{f1}}{2} + \frac{h_c}{2} + e_b\right), \qquad (12.43)
\end{aligned}$$

The second equation for the second-order forces is the second-order compatibility equation (12.33). The system of the two linear equations (12.43) and (12.33) can be solved for the second-order forces P_{d2} and P_{s2}. The first-order applied load P_{b2} is in turn found from the second-order force equilibrium (Equation (12.28). The solution for the higher-order terms can proceed in the same fashion although the math become increasingly more involved.

Notice that from (12.11), since $P_{b0} = P_{cr}$, the perturbation parameter ξ can be found from the applied external load, $\bar{P}$, by solving the quadratic equation in terms of ξ:

$$\xi = -\frac{P_{b1} \pm \sqrt{\Delta}}{2P_{b2}}; \quad \Delta = P_{b1}^2 - 4P_{b2}\,(P_{b0} - P)\,. \tag{12.44}$$

This, of course, presumes that we account up to the second-order load terms.

Example. The initial post-buckling results that follow are produced for the sandwich material system and geometry of the previous section and for a debond length $a = L/4$. Figure 12.4 shows the debond and substrate axial loads (normalized with the critical load) in the initial phase of post-buckling. Both loads increase with applied strain but the substrate load increases at a higher rate, which means that a redistribution of load occurs. These results are with the first-order loads (second term of the perturbation series) only. Extending further into the post-buckling phase would require additional terms to be included, thus additional analysis to be carried out. More direct results could be obtained with application of the elastica theory, properly modified to include transverse shear. This has been done for a delaminated monolithic composite (without transverse shear effects) by Kardomateas (1993) but not yet for the sandwich case with a face sheet/core debond.

12.3.3 Deflections

The deflections of the debonded substrate and base elements can be found by integrating the relationship (Huang and Kardomateas, 2002):

$$\frac{dy_i}{dx_i} = \sin\theta_i + \frac{\beta_i P_i}{2A_i\bar{G}_i}\sin 2\theta_i\,. \tag{12.45a}$$

Introducing the asymptotic expansions (12.11) and (12.12) and the first and second-order expressions (12.20) and (12.30), gives

$$\begin{aligned}\frac{dy_i}{dx_i} &= \xi\left(1 + \frac{\beta_i P_{i0}}{\bar{G}_i A_i}\right)\theta_{i1} + \xi^2\left[\left(1 + \frac{\beta_i P_{i0}}{\bar{G}_i A_i}\right)\theta_{i2} + \frac{\beta_i P_{i1}}{\bar{G}_i A_i}\theta_{i1}\right] \\ &\quad + \xi^3\left[\left(1 + \frac{\beta_i P_{i0}}{\bar{G}_i A_i}\right)\theta_{i3} + \frac{\beta_i P_{i1}}{\bar{G}_i A_i}\theta_{i2} + \frac{\beta_i P_{i2}}{\bar{G}_i A_i}\theta_{i1} - \left(1 + \frac{4\beta_i P_{i0}}{\bar{G}_i A_i}\right)\frac{\theta_{i1}^3}{6}\right] \\ &\quad + O(\xi^4),\end{aligned} \tag{12.45b}$$

This can be easily integrated and give the deflections in the form

$$y_i = \xi y_{i1} + \xi^2 y_{i2} + \xi^3 y_{i3} + \cdots, \tag{12.46}$$

$$y_{i1} = \left(1 + \frac{\beta_i P_{i0}}{\bar{G}_i A_i}\right) I_{i1}(x), \tag{12.47a}$$

$$y_{i2} = \left(1 + \frac{\beta_i P_{i0}}{\bar{G}_i A_i}\right) I_{i2}(x) + \frac{\beta_i P_{i1}}{\bar{G}_i A_i} I_{i1}(x), \tag{12.47b}$$

and

$$\begin{aligned} y_{i3} = {} & \left(1 + \frac{\beta_i P_{i0}}{\bar{G}_i A_i}\right) I_{i3}(x) + \frac{\beta_i P_{i1}}{\bar{G}_i A_i} I_{i2}(x) + \frac{\beta_i P_{i2}}{\bar{G}_i A_i} I_{i1}(x) \\ & - \left(1 + \frac{4\beta_i P_{i0}}{\bar{G}_i A_i}\right) \frac{C_{i1}^3}{18\lambda_i}(2 + \cos^3 \lambda_i x - 3\cos \lambda_i x), \end{aligned} \tag{12.47c}$$

where

$$I_{i1}(x) = \int_0^x \theta_{i1}(x)dx = \frac{C_{i1}}{\lambda_i}\left(1 - \cos \lambda_i x\right), \tag{12.48a}$$

$$\begin{aligned} I_{i2}(x) = \int_0^x \theta_{i2}(x)dx = {} & \left[\frac{C_{i2}}{\lambda_i} - \frac{P_{i1}C_{i1}}{2\lambda_i^3 D_i}\left(1 + \frac{2\beta_i P_{i0}}{\bar{G}_i A_i}\right)\right](1 - \cos \lambda_i x) \\ & + \frac{P_{i1}C_{i1}}{2\lambda_i^2 D_i}\left(1 + \frac{2\beta_i P_{i0}}{\bar{G}_i A_i}\right) x \sin \lambda_i x, \end{aligned} \tag{12.48b}$$

and

$$\begin{aligned} I_{i3}(x) = {} & \int_0^x \theta_{i3}(x)dx \\ = {} & \left[\frac{C_{i3}}{\lambda_i} - \left(\frac{q_{1i}}{\lambda_i^2} + \frac{k_{0i}C_{i1}}{2\lambda_i^3} P_{i2}\right) + \frac{k_{0i}^2 C_{i1}}{4\lambda_i^5} P_{i1}^2\right](1 - \cos \lambda_i x) \\ & + \left[\left(\frac{q_{1i}}{\lambda_i} + \frac{k_{0i}C_{i1}}{2\lambda_i^2} P_{i2}\right) - \frac{k_{0i}^2 C_{i1}}{4\lambda_i^4} P_{i1}^2\right] x \sin \lambda_i x \\ & + \frac{k_{0i}^2 C_{i1}}{8\lambda_i^3} P_{i1}^2 x^2 \cos \lambda_i x + \frac{q_{0i}}{96\lambda_i^3} C_{i1}^3 \left(1 - \cos 3\lambda_i x\right). \end{aligned} \tag{12.48c}$$

Notice that the deflections calculated from the above equations are referred to a local coordinate system for each part, as shown in Figure 12.1. Thus, the mid-point debond deflection would be $y_b(L - a) - y_d(-a)$.

The initial post-buckling behavior derived in this section is a pre-requisite to the fracture mechanics analysis since the stress intensity factors and energy release rate at the post-buckling phase are a function of the displacements, forces, and moments at the debond tip section.

Table 12.1 Face and core mechanical properties for sandwich specimens with carbon/epoxy face sheets and Nomex honeycomb core, E = Young's modulus, X_c = compression strength.

Material	Density, kg/m^3	E, MPa	X_c, Pa
Face	N/A	53,000	531×10^6
Core	28.8	2.30	18.9×10^3
Core	48	3.86	42.0×10^3
Core	96	7.72	157×10^3

12.4 Experimental Studies on Buckling and Growth of Face/Core Debonds

Experimental studies of buckling and growth of face/core debonds in sandwich columns and panels will be discussed. In addition, finite element analysis results for debond buckling and growth in compression loaded sandwich panels with face/core debonds will be presented.

12.4.1 Column Testing

Avery and Sankar (2000) and Sankar and Narayan (2001) examined a large set of sandwich specimens consisting of plain weave carbon/epoxy face sheets and various Nomex honeycomb cores. The faces consisted of 1, 3, 5, or 7 plies, each ply nominally 0.221 mm thick. The core thicknesses examined were 6.35, 9.53, and 12.7 mm. The face sheets were bonded to the core using vacuum bagging and an autoclave cure procedure. To define a face/core debond, a thin Teflon strip of 12.7, 25.4, 38.1 and 50.8 mm width was introduced between one of the face sheets and the core before cure. The specimen were 102 mm long and 50.8 mm wide. Properties of the face sheets and cores are listed in Table 12.1.

Compression testing of the columns revealed that the specimens failed in a variety of failure modes, schematically illustrated in Figure 12.5, depending on the specific combination of face sheet thickness, core thickness and density, and debond size.

The failure modes (Figure 12.5) were classified as "GS", GA", "LS" and "LA". These modes are called the "global symmetric" (GS) and "global asymmetric modes" (GA). Here, symmetry refers to the buckling shape with respect to a horizontal axis through the center of the specimen. "Local sym-

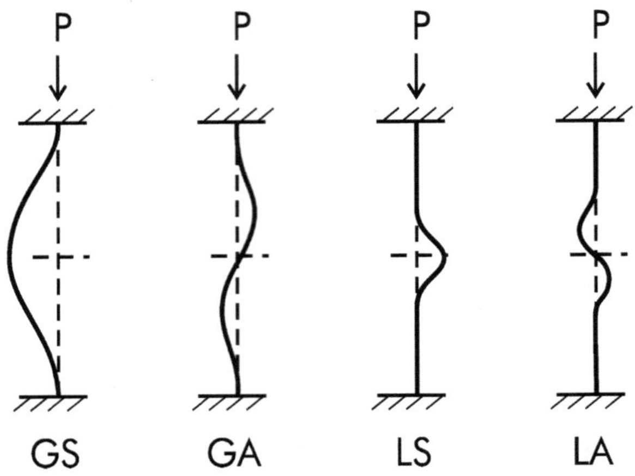

Figure 12.5 Failure modes observed in compression testing of debonded sandwich column specimens.

metric buckling" (LS) means that the debond buckled in its fundamental load. The "local asymmetric" (LA) failure mode refers to local buckling of the column in a mode shape that is asymmetric.

The specimen configuration and test program was designed according to an "experiment design approach" which provides a large range of empirical data. This approach is quite different from a more traditional verification program where one parameter at a time is symmetrically varied. Unfortunately, the local buckling loads were not measured. Only the failure modes and ultimate loads are reported. It should be pointed out that, due to the thin face sheets used, none of the specimens failed through debond propagation.

Specific details on the specimen configurations and test results are summarized in Table 12.2. Only one specimen (#1) was observed to fail in the local asymmetric mode. This specimen has very thin face sheets (0.221 mm) and thin core which would promote such a failure mode. All other specimens with thin face sheets (#2–9) failed in a local symmetric (LS) mode. As an example, Figure 12.6 details the progression of collapse of specimen #4 that failed in the LS mode. Buckling of the debonded face in this case occurred symmetrically in the "LS" failure mode. The load vs. end shortening graph is shown in Figure 12.7.

It is noted that the local buckling of the extremely thin face sheet occurred at quite small loads (point b in Figure 12.7) well before the ultimate load was reached. Specimens with thicker face sheets, $h_f = 1.11$ and 1.55 mm, failed by global buckling (GS or GA modes). These columns failed at much higher

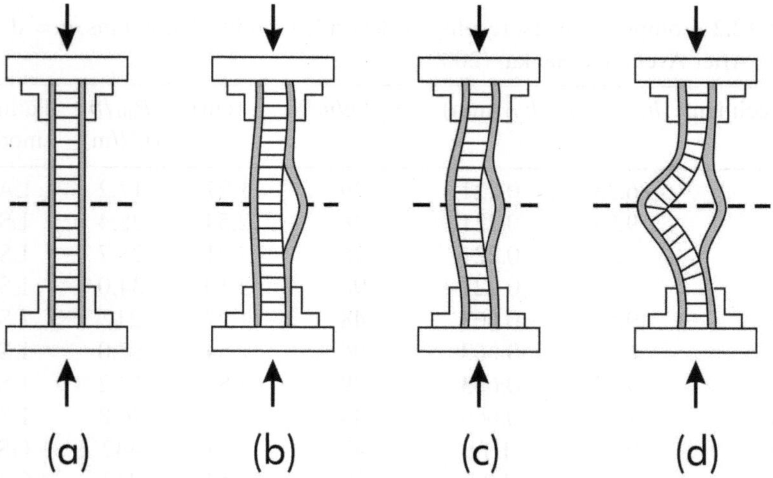

Figure 12.6 Progression of compression failure (LS mode) in sandwich column #4.

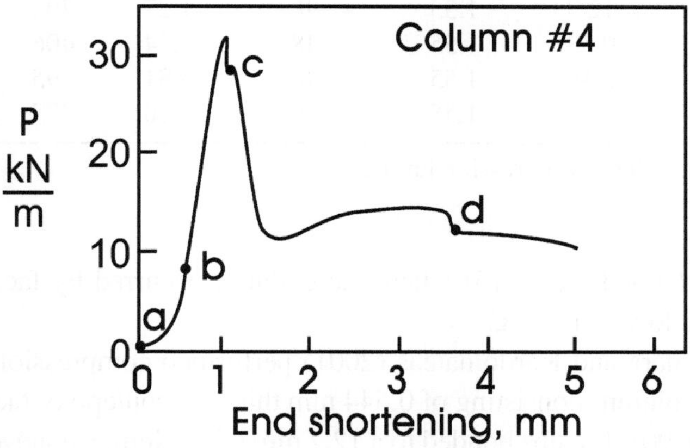

Figure 12.7 Load vs. end shortening curve for specimen #4. The letters a–d refer to Figure 12.6.

loads than columns with thin faces as a result of their greater loading area and buckling resistance, see the data in Table 12.2.

The buckling load, P_{cr}, of each column specimen was calculated from linear finite element analysis. Results are displayed vs. the measured collapse load in Figure 12.8. It is noted that the calculated local buckling load is a reasonable predictor of the experimental collapse load for most specimens, except for those with very thin faces (#1–4), where the measured strength exceeds the calculated local buckling load and for specimens with thick faces

Table 12.2 Compression test results for debonded sandwich columns, a = debond length. After Avery and Sankar (2000).

Specimen	h_c (mm)	h_f (mm)	ρ_c (kg/m^3)	a (cm)	P_{ult}/b (kN/m)	Failure mode
1	6.35	0.221	29	1.27	17.2	LA
2	9.53	0.221	48	2.54	28.4	LS
3	12.7	0.221	48	3.81	28.7	LS
4	9.53	0.221	96	5.10	34.0	LS
5	9.53	0.663	48	1.27	212	LS
6	6.35	0.663	96	2.54	87.0	LS
7	9.53	0.663	29	3.81	63.2	LS
8	12.7	0.663	48	5.10	76.8	LS
9	9.53	1.11	48	1.27	442	GS
10	12.7	1.11	29	2.54	213	GA
11	9.53	1.11	96	3.81	242	GA
12	6.35	1.11	29	5.10	156	GS
13	12.7	1.55	96	1.27	792	FC*
14	9.53	1.55	48	2.54	406	GA
15	6.35	1.55	48	3.81	295	GS
16	9.53	1.55	29	5.10	277	GA

*FC = face sheet compression failure.

and short debonds (#13–15) where the collapse occurred by face compression before local buckling.

La Saponara and Kardomateas (2001) performed compression testing of sandwich columns consisting of 0.544 mm thick carbon/epoxy face sheets of $[0]_4$ and $[0/90]_s$ lay-ups bonded to a 12.7 mm thick Nomex honeycomb core. They implanted a thin Teflon film of length 3.4 cm at one face/core interface to define an initial debond. The specimens were 139 mm long, 25.4 mm wide and 13.8 mm thick. The debond was placed at the center of the column specimen. Compression testing was conducted in displacement control at a rate of 2.54 mm/min. Figure 12.9 shows compression load vs. axial strain response for columns with $[0]_4$ and $[0/90]_s$ face sheets.

The testing, conducted at a slow rate, allowed detailed visual observation of the specimens. It was observed that the debond opened up, indicating local buckling. The local buckling of the debond in the $[0/90]_s$ specimen occurred at a load of about 679 N, well before the maximum load. The $[0]_4$ specimen similarly buckled well before the ultimate load at a load of about 1.18 kN. These observations are consistent with those reported for similar thin faced column specimens examined by Avery and Sankar (2000), see

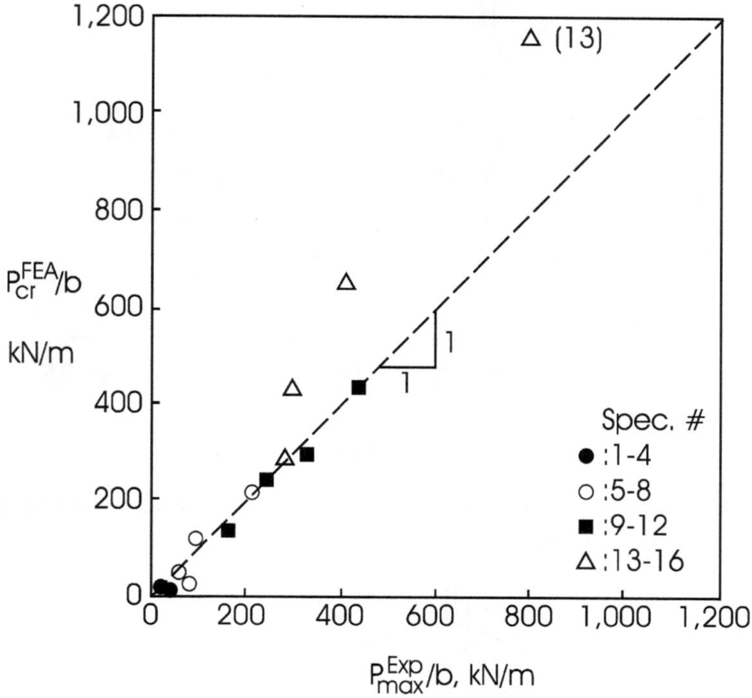

Figure 12.8 Buckling load calculated for the debonded face sheet vs. measured collapse load.

earlier part of this section. Ultimate failure occurred by through-thickness crack propagation in the core. The ultimate failure loads per unit width were about 46 and 67 kN/m for the $[0/90]_s$ and $[0]_4$ specimens, respectively. These load levels compare reasonably well with those determined by Avery and Sankar (2000) for similar thin-faced sandwich specimens, see Table 12.2. Observation of the load vs. strain response curves for the column specimens (Figure 12.9) reveals very little post-failure strength of the columns.

Vadakke and Carlsson (2004) conducted compressive testing of sandwich columns containing a one-sided face/core debond. The sandwich specimens consisted of 2 mm thick glass/vinylester face laminates bonded to 50 mm thick H45, H80 and H100 PVC foam core materials. The face elastic modulus was $E_f = 20.6$ GPa and the core moduli were $E_c = 42$, 80 and 105 MPa, for the H45, H80 and H100 cores, respectively. Debond lengths of 25 and 50 mm were examined. Compression testing of the debonded sandwich columns utilized a test rig shown in Figure 12.10.

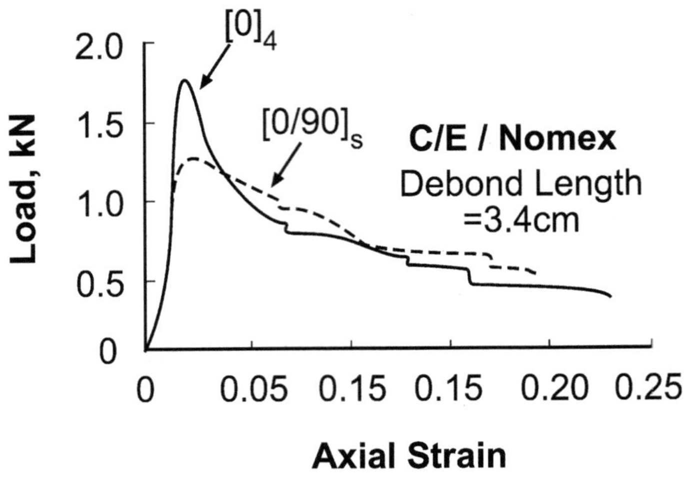

Figure 12.9 Compression load vs. strain response for sandwich column specimens with a 3.4 cm face/core debond.

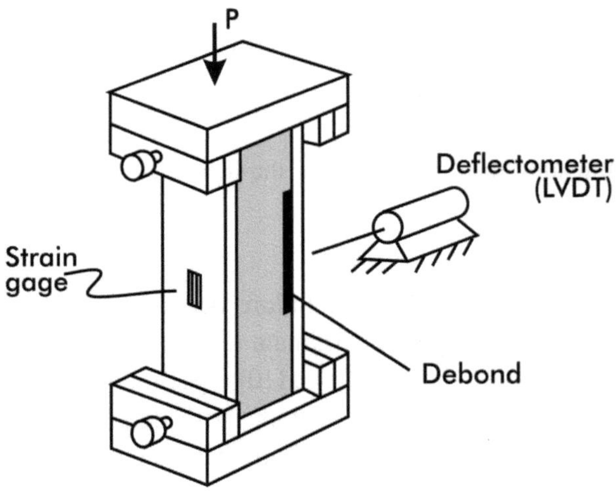

Figure 12.10 Compression end-loading test fixture.

The sandwich specimen is end-loaded, which requires that the fixture and ends of the column are properly aligned and parallel. Rectangular steel clamps were used to position the specimen at the center of the fixture and clamp the sandwich column. The load is applied to the end surfaces of the sandwich column specimen through steel platens attached to the crosshead and base of the test machine. Strain gages were glued on both face sheets at the center of the gage length to monitor the axial strain response during

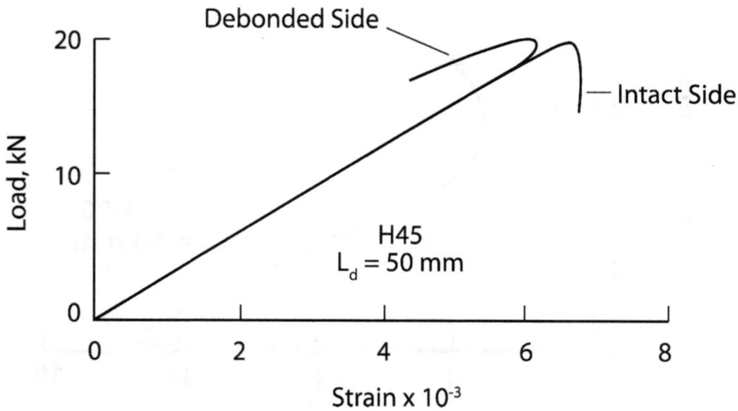

Figure 12.11 Load vs. strain response for a sandwich column with H45 core and a 50 mm debond.

loading and detect local or global bucking. Additionally, a deflectometer was put in contact with the center of the debond to monitor lateral (out-of-plane) deflections. The total specimen length was 150 mm. The width was 37.5 mm. Each end was clamped over a distance of 25.4 mm, resulting in an unsupported length of about 100 mm, short enough to suppress wrinkling and global buckling failures, but long enough to allow local buckling of the 25 and 50 mm long debonds.

Figure 12.11 shows the load vs. axial strain response of a specimen with H45 core and a 50 mm debond. Near the maximum load (19 kN), a strain reversal is observed for the debonded face indicating local buckling of the debonded face. Once the face buckled, the debond rapidly propagated to the clamped ends of the specimen causing total collapse of the column. Hence, local buckling of these column specimens caused catastrophic failure.

The out-of-plane deflection vs. load response is displayed in Figure 12.12 for a column with H80 core and a 50 mm debond. The results show that the out-of-plane deflection of the debonded face sheet remains small until local buckling occurred as indicated by the rapid increase of the deflection. The load decrease is attributed to failure of the column. Both fronts of the face/core debond propagated until the crack fronts were arrested by the clamps in the test fixture.

To illustrate the strength reduction due to a face/core debond, the collapse load of the debonded column was normalized with the measured compression failure load P_0 for columns without a debond. Figure 12.13 shows that the compression strength is substantially reduced by the presence of a debond and more so for longer debonds as they buckle at lower loads.

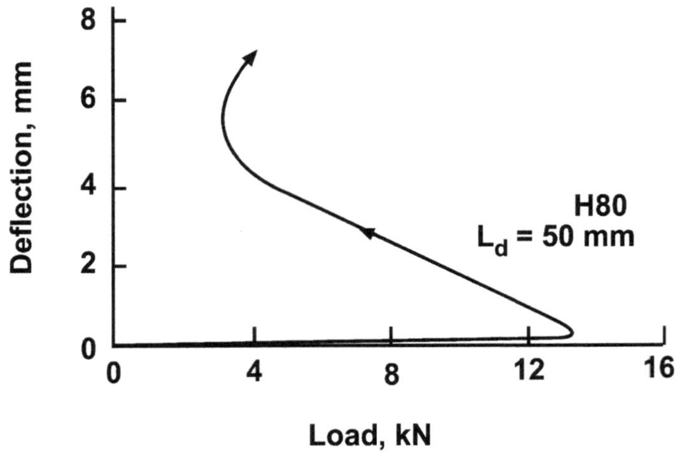

Figure 12.12 Out-of-plane deflection of debonded face sheet vs. load for a sandwich column with H80 core and a 50 mm debond.

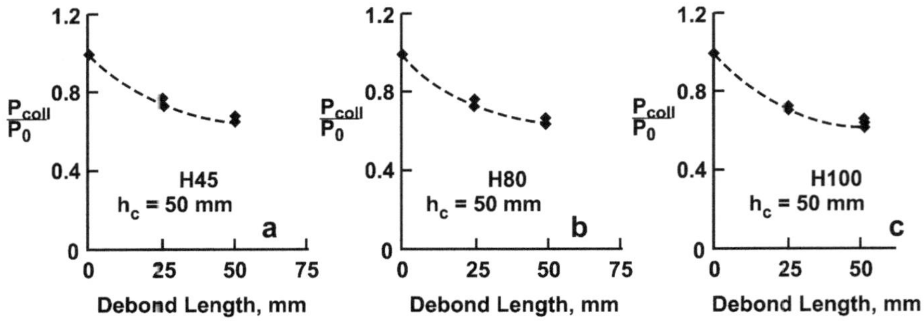

Figure 12.13 Compression strength of debonded sandwich columns: (a) H45 core, (b) H80 core, and (c) H100 core.

12.4.2 Panel Testing

An experimental study of local buckling and failure behavior of compression-loaded sandwich panels containing one-sided, centrally located circular face/core debonds was conducted by Aviles and Carlsson (2006a). Such a configuration should be much more representative for a sandwich structure containing a localized face/core debond than the previously considered single load-path debonded columns. Sandwich panels with 2 mm thick glass/epoxy face sheets and 25 mm thick PVC foam cores were prepared. The panels were 15 cm wide and 20 cm long. The face modulus and Poisson ratio were: $E_f = 20.6$ GPa and $\nu_f = 0.42$. The foams were assumed

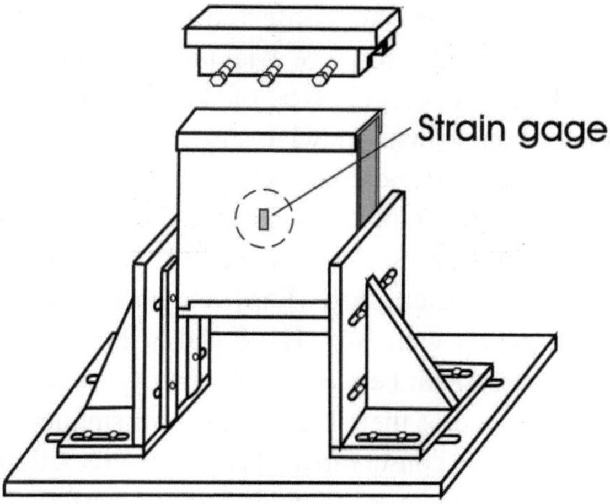

Figure 12.14 Compression test fixture for sandwich panels with a face/core debond.

to be isotropic with a Poisson ratio $\nu_c = 0.32$, and Young's moduli of 40, 100 and 200 MPa for the H45, H100 and H200 foams. A circular face/core debond was defined in each panel by implanting a thin (30 μm thick) non-stick Teflon film between one face sheet and the core, at the center of the panel. The debonds were 5, 6.3, 7.5, and 10 cm in diameter.

A special test fixture was developed for uniaxial compression loading of the test panels (see Figure 12.14). The test fixture was designed for a panel of nominal size of 20 × 15 × 3 (cm) (length × width × thickness) with some flexibility to accommodate other dimensions. Rounded metal edge supports were used to constrain out-of-plane deflections of the vertical panel edges. The lower panel edge was resting on a horizontal steel platen that constrains translation and rotation. A load was introduced as the upper edge using a steel platen with a central groove (adjustable to the panel thickness) and hinged at the center to promote uniform loading across the panel width. This type of load introduction constrains rotation of the upper edge of the panel. Compressive displacement of the upper steel platen was applied using the movable crosshead of a universal (displacement controlled) test machine. Strain gages were bonded on both face sheets at the panel center to enable monitoring of the strains during loading. Three replicate panel specimens were tested.

Typical test results are shown in Figure 12.15 for a panel with H45 core and a 7.5 cm diameter debond. The local buckling load was determined from the bifurcation point in the load-strain response. At the point of local (face) instability, a strain reversal is observed for the debonded face sheets. The rapid decrease in load after face buckling is indicative of a lack of post-buckling strength. Hence, this type of damage can be very serious in compression-loaded panels in a sandwich structure.

Experimentally measured local buckling loads for the panels with H45, H100, and H200 cores and a range of debond sizes were compared to predictions from finite element analysis (Aviles and Carlsson, 2006a). Figure 12.16 presents experimentally determined (Exp) and predicted (FEA) buckling loads for sandwich panels with H45, H100 and H200 cores. For the panels with H45 core, the agreement between FEA and experiments is quite favorable. For higher density cores there is agreement for smaller debonds, but for larger debonds the FEA is conservative. The under-prediction of the critical load for large debonds is believed to be due to non-perfect definition of the artificially generated debond in the experimental study. As mentioned earlier, the face/core debond was defined by implanting a Teflon film in the panels. It has been observed that such films may adhere to the resin and core after resin infusion resulting in tractions preventing the opening of the debond. The tendency for such problems was more pronounced for large debonds.

Experimental observations of compression loaded panels with a face/core debond revealed that failure was initiated by local buckling of the debond. After buckling, it was observed that the debond rapidly propagated perpendicular to the loading direction followed by the collapse of the panel. Figure 12.17 is a schematic illustration of the failure mechanism.

To analyze the face/core debond propagation, a fracture mechanics analysis was conducted by Aviles and Carlsson (2007b). As discussed in Chapter 9, a face/core debond may be considered as a crack-like fault located between the face and core. The solution for the near tip crack flank displacements derived by Suo (1990), provided in Equation (9.3b), is utilized to extract the stress intensity factors K_{I} and K_{II} from the relative opening and sliding displacements of the crack flanks, δ_{I} and δ_{II} (Figure 9.4)

$$(K_{\mathrm{I}}, K_{\mathrm{II}}) = \sqrt{\frac{\pi}{2H_{11}H_{22}x}}(\delta_{\mathrm{I}}, \delta_{\mathrm{II}}). \tag{12.49}$$

H_{11} and H_{22} are defined in Equations (9.4). Two locations at the debond front shown in Figure 12.18 were specifically examined: location (L), corresponding to debond propagation along the direction of loading (y axis), and location (T), corresponding to debond propagation perpendicular to the direction of loading. The displacements of the (initially coincident) nodes next to crack tip at the "L" and "T" locations were recorded for each load increment in order to determine the stress intensity factors.

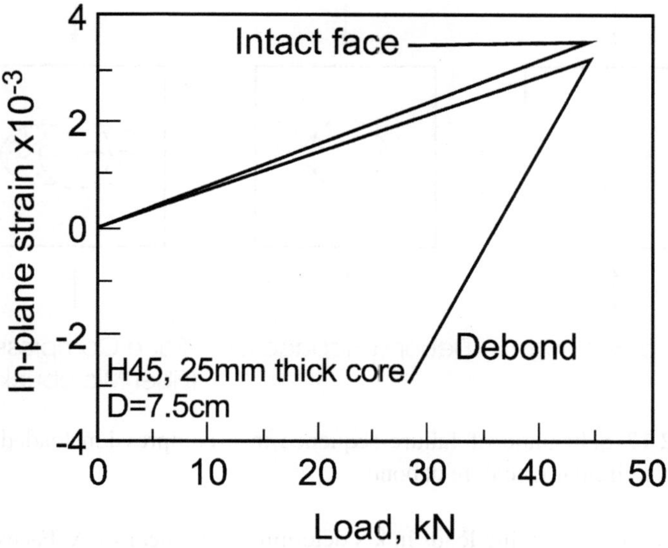

Figure 12.15 Strain vs. applied load for a sandwich panel with a 7.5 cm diameter debond (H45 core).

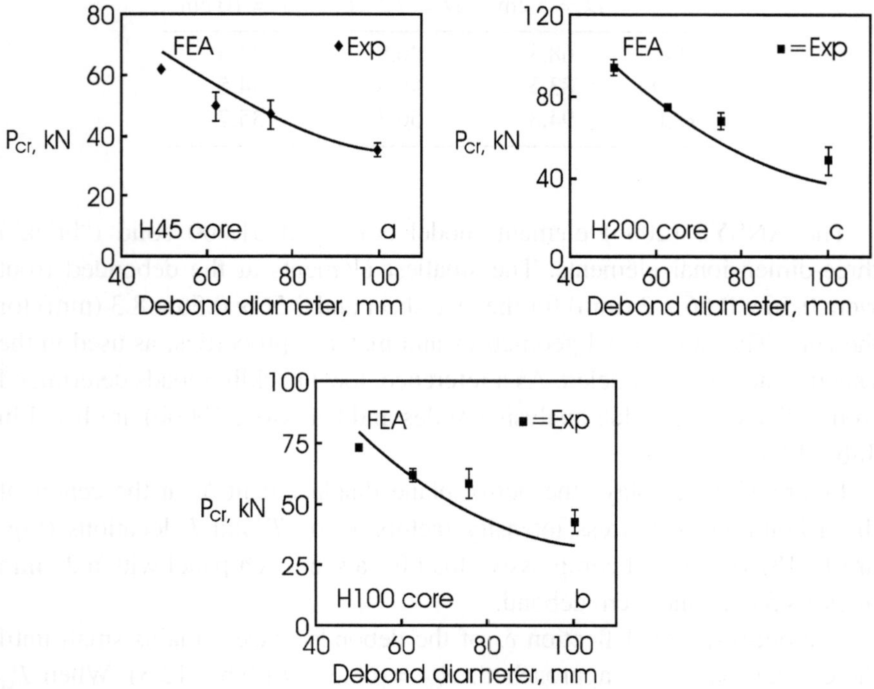

Figure 12.16 Local buckling loads determined from FEA and experimental testing for sandwich panels with circular debonds: (a) H45 core, (b) H100 core, and (c) H200 core.

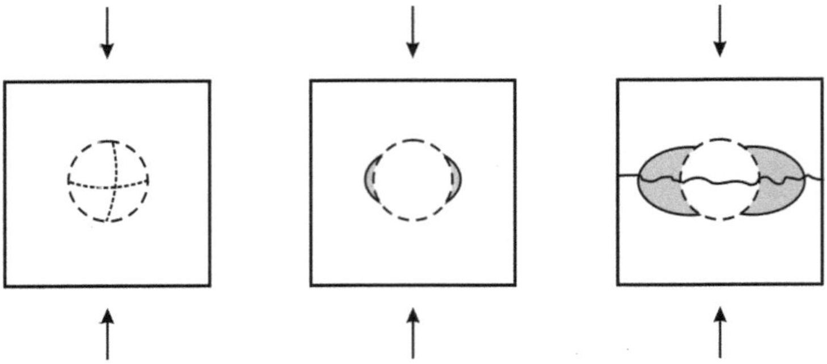

Figure 12.17 Schematic of failure sequence in a compression loaded sandwich panel with a circular face/core debond.

Table 12.3 Local buckling loads in kN determined by linear FEA. Faces are 2 mm thick S2-glass/epoxy (plain weave). Core thickness is 25 mm.

Core	Buckling load, kN		
	$D = 5$ cm	$D = 7.5$ cm	$D = 10$ cm
H45	68.3	46.8	33.5
H100	83.5	49.8	34.5
H200	94.8	56.3	35.7

The ANSYS (2006) element model employed 31,000 solid ("brick") three-dimensional elements. The smallest elements at the debonded front were $0.56 \times 0.56 \times 2$ (mm) for the face sheet and $0.56 \times 0.56 \times 8.3$ (mm) for the core. The same panel geometries and material properties, as used in the experiments, were modeled. As a reference, local buckling loads determined from a linear eigenvalue analysis (Aviles and Carlsson, 2006b) are listed in Table 12.3.

Figure 12.19 displays the out-of-plane displacement δ_c at the center of the debond and the stress intensity factors at the T and L locations (Figure 12.18) vs. applied compressive load for a sandwich panel with a 25 mm thick H45 core and 5 cm debond.

The out-of-plane deflection δ_c of the debonded face remains small until the critical load P_{cr} is approached, $P_{\text{cr}} = 68.3$ kN (Table 12.3). When P_{cr} is approached, δ_c increases rapidly and peaks at a load of about 68 kN after which the debond starts to close. The stress intensity factors K_{I} and K_{II} follow the same trend as the deflection, increasing drastically as P_{cr} is ap-

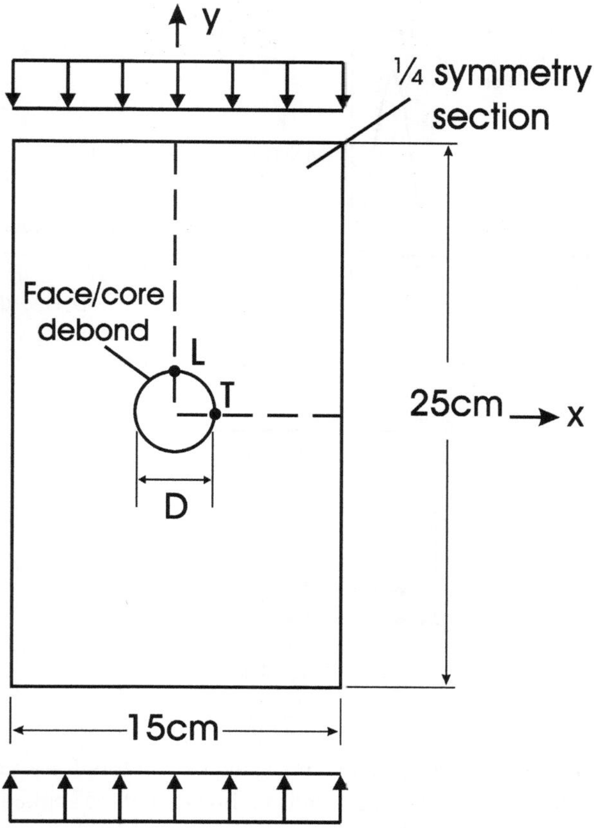

Figure 12.18 Definition of L and T locations in a sandwich panel.

proached. The maximum stress intensity factor at the T location is K_{I}. This result indicates that crack growth would occur under mode I-dominated conditions. The energy release rate, G, was calculated from the stress intensity factors K_{I} and K_{II} using Equation (9.8). Figure 12.20 shows G as a function of the applied load at the L and T locations for sandwich panels with a 5 cm debond diameter. G is much larger at the T location than at the L location. Thus, if the condition for crack propagation is met, i.e. $G = G_c$, crack propagation should occur perpendicular to the applied load. This is consistent with experimental observations of the crack propagation, see Figure 12.17. Results for panels with different debond sizes and foam cores are qualitatively similar to those for the panel discussed at some detail above. More details can be found in Aviles and Carlsson (2007b).

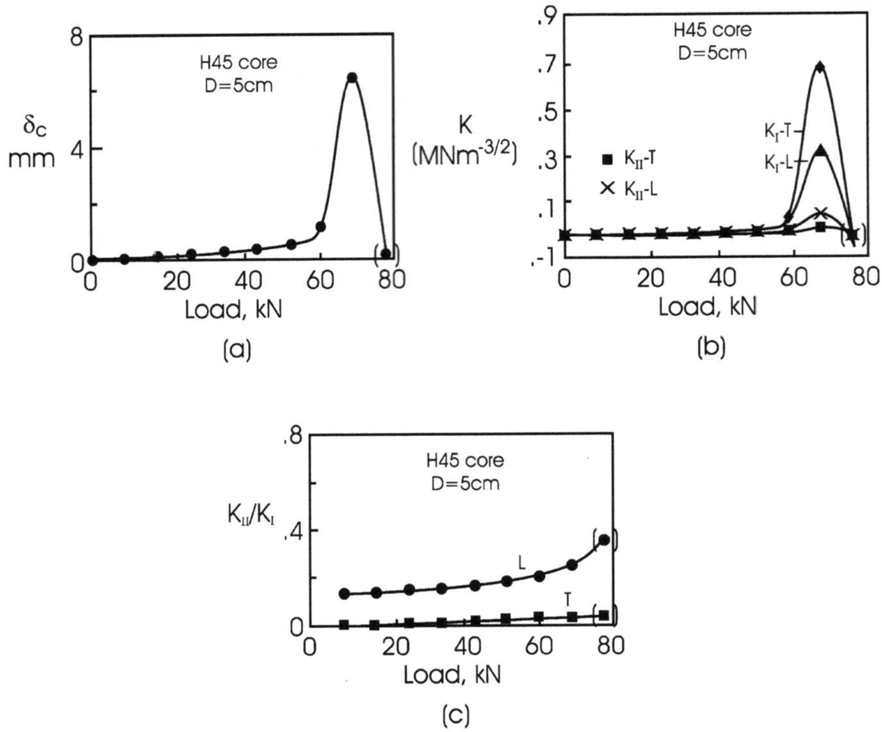

Figure 12.19 Debond deflection and stress intensity factors for a sandwich panel with H45 core and a 5 cm diameter debond as a function of the applied load. (a) Out-of-plane displacement of the debonded interface, (b) K_{I} and K_{II} at the L and T locations, and (c) mode ratio ($K_{\mathrm{II}}/K_{\mathrm{I}}$).

To predict the critical load corresponding to debond propagation, G was compared to the experimentally determined debond fracture toughness G_c. Such data were obtained from the sandwich debond (TSD) test discussed in Chapters 10 and 11. For the panels with H45, H100 and H200 PVC foam cores, $G_c = 0.18, 0.36$ and 0.51 kJ/m^2 (Viana and Carlsson, 2003). As shown in Figure 12.20, G reaches its critical value G_c well before the peak magnitude is reached. Hence, the peak value of G does not govern the fracture process of the sandwich panels examined herein.

Table 12.4 summarizes predicted and measured critical loads for debond propagation in sandwich panels with 5, 7.5, and 10 cm debonds. Very good agreements between measured and predicted loads is observed for panels with the smallest (5 cm) debond. For the panels with H100 core with debonds larger than 5 cm, the predicted debond propagation loads fall below the experimentally measured values. The difference between measured

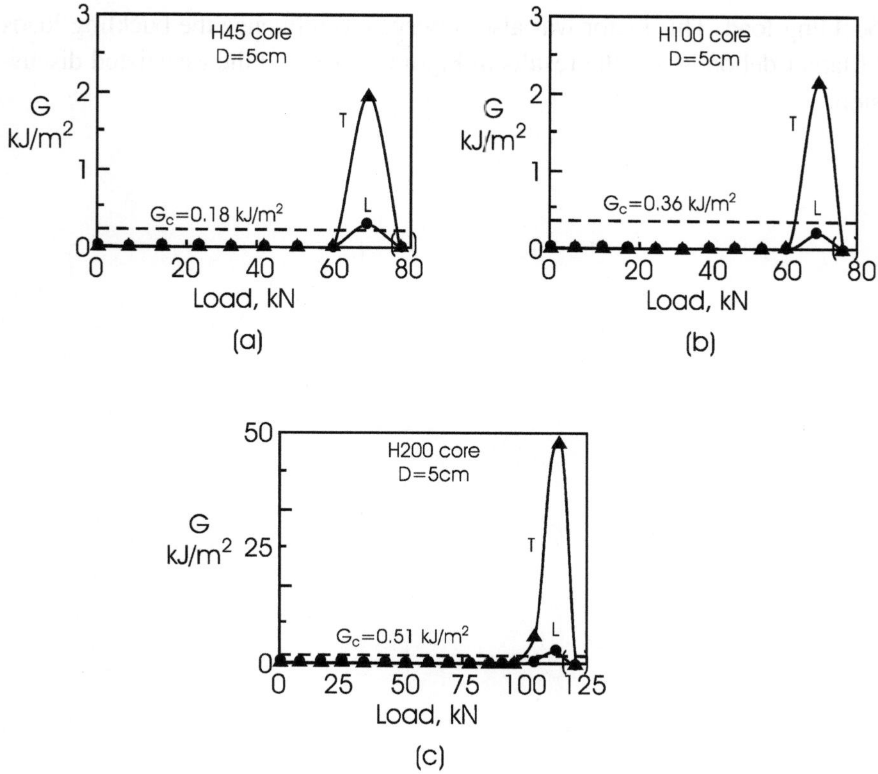

Figure 12.20 Energy release rate at the L and T locations for sandwich panels with different foam cores and a 5 cm diameter debond. (a) H45, (b) H100, and (c) H200.

Table 12.4 Debond propagation loads, for panels with circular debonds predicted by nonlinear FEA and measured experimentally, D = debond diameter.

Core	Debond propagation load (kN)					
	$D = 5$ cm		$D = 7.5$ cm		$D = 10$ cm	
	NL-FEA	Exp.	NL-FEA	Exp.	NL-FEA	Exp.
H45	61.9	62.2*	N/A	47.1 ± 3.6	N/A	34.8 ± 1.1
H100	79.3	75.7*	45.7	60.5 ± 5.4	28.1	43.8 ± 5.2
H200	94.1	93.5 ± 4.2	N/A	67.2 ± 4.3	N/A	47.4 ± 7.2

*Test results are for one panel only.

and predicted loads for large debonds may possibly be due to sticking of the debonded face sheet to the core, which would restrict the opening of the debond, reduce the stress intensity factors, and elevate the measured local buckling load. This factor was also observed to influence the buckling loads of larger debonds; see the results in Figure 12.16, and the associated discussion.

Appendix A
Stress-Strain Relations for On-Axis and Off-Axis Composite Elements

A.1 On-Axis System

Consider an element of a unidirectional on-axis composite, i.e., a composite where the principal material axes (1, 2, 3) are aligned with the coordinate system, see Figure A.1.

Figure A.1 also shows the definition of stress components associated with the material coordinate system, 1, 2, 3, where the stresses are volume averages over the fiber and matrix domains. The normal stresses are σ_1, σ_2, and σ_3, while the shear stresses are τ_{12}, τ_{13}, and τ_{23}. Corresponding normal strains are ε_1, ε_2, and ε_3, and the engineering shear strains are γ_{12}, γ_{13}, and γ_{23}, see Hyer (1998) for additional discussion.

In thin, sheet-like structures such as a ply in a laminate, it is common to assume a state of plane stress by setting

$$\sigma_3 = \tau_{13} = \tau_{23} = 0. \tag{A.1}$$

It may be shown that such a state of stress leads to vanishing of the out-of-plane shear strains, i.e.

$$\gamma_{13} = \gamma_{23} = 0. \tag{A.2}$$

The out-of-plane extensional strain, ε_3, does not vanish but becomes coupled to the in-plane stresses σ_1 and σ_2 and does not remain an independent quantity. The stress-strain relation for plane stress becomes

$$\begin{bmatrix} \sigma_1 \\ \sigma_2 \\ \tau_{12} \end{bmatrix} = \begin{bmatrix} Q_{11} & Q_{12} & 0 \\ Q_{12} & Q_{22} & 0 \\ 0 & 0 & Q_{66} \end{bmatrix} \begin{bmatrix} \varepsilon_1 \\ \varepsilon_2 \\ \gamma_{12} \end{bmatrix}, \tag{A.3}$$

where the so-called reduced stiffnesses, Q_{ij}, can be expressed in terms of engineering constants as

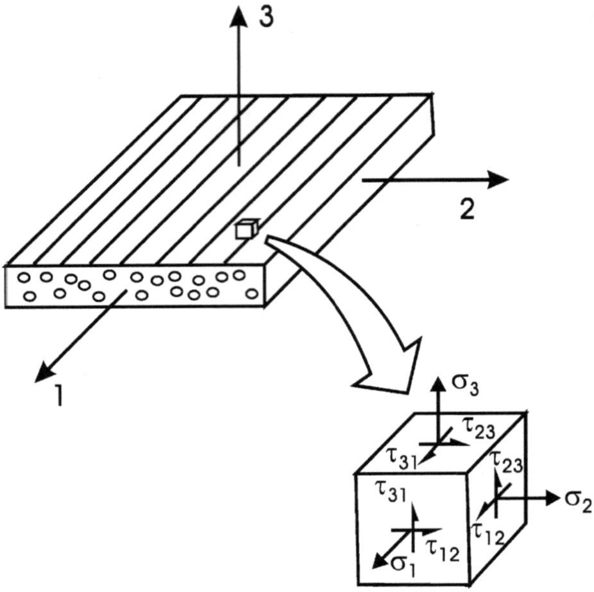

Figure A.1 Element of on-axis composite and definition of normal and shear stress components.

$$Q_{11} = \frac{E_1}{1 - \nu_{12}\nu_{21}}, \tag{A.4a}$$

$$Q_{12} = \frac{\nu_{12}E_2}{1 - \nu_{12}\nu_{21}}, \tag{A.4b}$$

$$Q_{22} = \frac{E_2}{1 - \nu_{12}\nu_{21}}, \tag{A.4c}$$

$$Q_{66} = G_{12}, \tag{A.4d}$$

where E_1 and E_2 are the principal Young's moduli, and G_{12} is the in-plane shear modulus. ν_{12} and ν_{21} are the principal (major and minor) Poisson ratios.

A common approach to achieve a set of more balanced mechanical properties is to utilize woven fabric composites, where tows of several thousand fibers are arranged in a specific pattern, such as plain weave fabrics consisting of "one fiber bundle over-one under", see Figure A.2.

In addition to the plain weave fiber pattern, there are several other weave patterns. The mechanical response of a plain weave composite layer consisting of fibers in the 0 and 90° directions can also be represented by Equation (A.3).

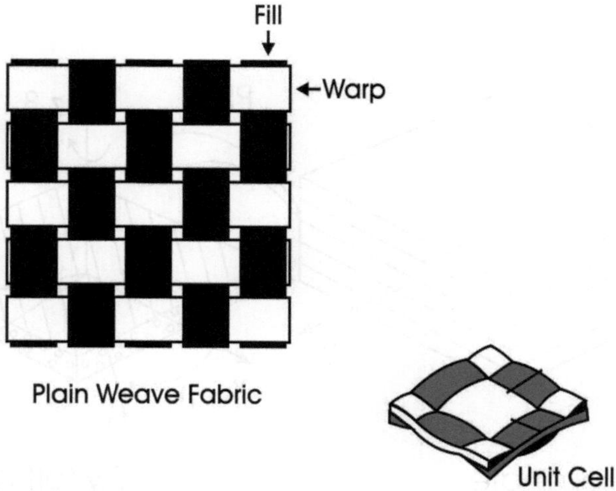

Figure A.2 Plain weave fabric structure.

A.2 Off-Axis System

Unidirectional composites, such as shown in Figure A.1, are rarely utilized in actual structures because of their extremely anisotropic properties and weak failure planes. Most composite structures are multi-directional laminates, where unidirectional plies are oriented in different directions θ, e.g., $\theta = 0°, 45°, -45°$, and 90° (see Figure A.3). It is observed that each ply orientation θ is obtained by a suitable rotation of the ply around the 3 axis.

For an off-axis ply (Figure A.3b) the stresses, strains and stiffness elements must be transformed

$$\begin{bmatrix} \sigma_x \\ \sigma_y \\ \tau_{xy} \end{bmatrix} = \begin{bmatrix} \overline{Q}_{11} & \overline{Q}_{12} & \overline{Q}_{16} \\ \overline{Q}_{12} & \overline{Q}_{22} & \overline{Q}_{26} \\ \overline{Q}_{16} & \overline{Q}_{26} & \overline{Q}_{66} \end{bmatrix} \begin{bmatrix} \varepsilon_x \\ \varepsilon_y \\ \gamma_{xy} \end{bmatrix}, \qquad \text{(A.5)}$$

where the overbars denote transformed properties.

The transformed (off-axis) stiffnesses $\overline{Q}_{ij}$ are calculated from the on-axis stiffnesses, Q_{ij}, defined in Equations (A.4) (Hyer, 1998).

$$\overline{Q}_{11} = m^4 Q_{11} + 2m^2n^2 (Q_{12} + 2Q_{66}) + n^4 Q_{22},$$

$$\overline{Q}_{12} = m^2n^2 (Q_{11} + Q_{22} - 4Q_{66}) + \left(m^4 + n^4\right) Q_{12},$$

$$\overline{Q}_{22} = n^4 Q_{11} + 2m^2n^2 (Q_{12} + 2Q_{66}) + m^4 Q_{22},$$

$$\overline{Q}_{16} = m^3n (Q_{11} - Q_{12}) + mn^3 (Q_{12} - Q_{22}) - 2mn \left(m^2 - n^2\right) Q_{66},$$

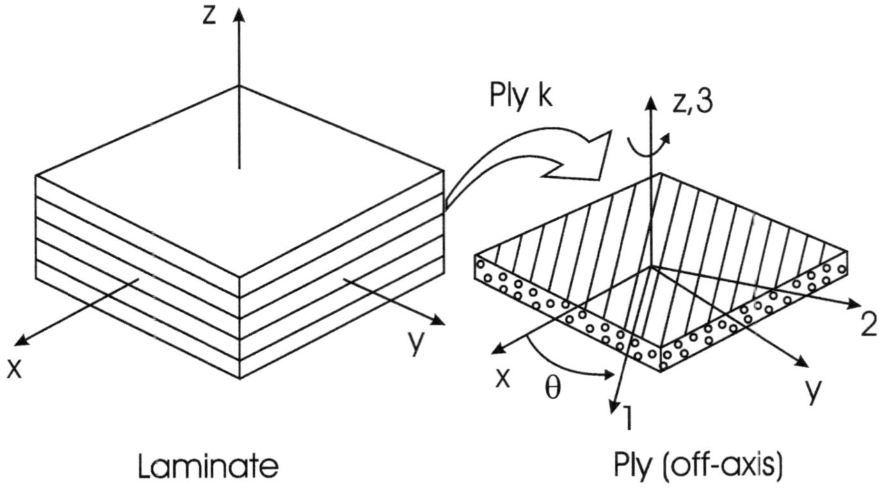

Figure A.3 (a) A multidirectional laminate is built by a stack of unidirectional composites bonded together. (b) Definition of ply orientation angle θ.

$$\overline{Q}_{26} = mn^3 (Q_{11} - Q_{12}) + m^3 n (Q_{12} - Q_{22}) + 2mn \left(m^2 - n^2\right) Q_{66},$$
$$\overline{Q}_{66} = m^2 n^2 (Q_{11} + Q_{22} - 2Q_{12} - 2Q_{66}) + \left(m^4 - n^4\right) Q_{66}, \qquad \text{(A.6)}$$

where $m = \cos\theta$ and $n = \sin\theta$.

Appendix B
Calculation of Compliance Matrices for Sandwich Panels

The stiffness form of the laminate constitutive equations (3.18) and (3.19) for a sandwich panel may be expressed in compressed form as

$$\begin{bmatrix} N \\ M \end{bmatrix} = \begin{bmatrix} A & B \\ C & D \end{bmatrix} \begin{bmatrix} \varepsilon^{\circ} \\ \kappa \end{bmatrix}. \tag{B.1}$$

Sometimes it is desirable to express the core mid-plane strains and curvatures in terms of force and moment resultants and this is achieved by inversion of the 6×6 ABCD matrix in Equation (B.1)

$$\begin{bmatrix} \varepsilon^{0} \\ \kappa \end{bmatrix} = \begin{bmatrix} a & b \\ c & d \end{bmatrix} \begin{bmatrix} N \\ M \end{bmatrix}, \tag{B.2}$$

where the 3×3 compliance matrices $[a]$, $[b]$, $[c]$ and $[d]$ are given by

$$[a] = [A^*] - [B^*][D^*]^{-1}[C^*], \tag{B.3a}$$

$$[b] = [B^*][D^*]^{-1}, \tag{B.3b}$$

$$[c] = -[D^*]^{-1}[C^*], \tag{B.3c}$$

$$[d] = [D^*]^{-1}, \tag{B.3d}$$

where

$$[A^*] = [A^*]^{-1}, \tag{B.4a}$$

$$[B^*] = -[A^*]^{-1}[B], \tag{B.4b}$$

$$[C^*] = [C][A]^{-1}, \tag{B.4c}$$

$$[D^*] = [D] - [C][A^*]^{-1}[B]. \tag{B.4d}$$

For the special case considered in Section 3.2, where the faces are treated as homogeneous orthotropic materials, it was shown in that $B_{ij} = C_{ij}$, which yields

$$[A^*] = [A]^{-1}, \tag{B.5a}$$

$$[B^*] = -[A]^{-1}[B], \tag{B.5b}$$

$$[C^*] = [B][A]^{-1}, \tag{B.5c}$$

$$[D^*] = [D] - [B][A]^{-1}[B]. \tag{B.5d}$$

Furthermore, for the special case of a symmetric sandwich (Section 3.2), i.e., where the face sheets are identical and laid-up with their mid-plane as a mirror plane,

$$[B] = [C] = [0]. \tag{B.6}$$

In Equations (B.3) and (B.4) this results in

$$[B^*] = [C^*] = [0]. \tag{B.7}$$

Substitution into Equations (B.3) gives

$$[a] = [A]^{-1}, \tag{B.8a}$$

$$[b] = [c] = [0], \tag{B.8b}$$

$$[d] = [D]^{-1}. \tag{B.8c}$$

For a sandwich with zero D_{16} and D_{26} terms, it may furthermore be shown that

$$d_{11} = \frac{D_{22}}{D_{11}D_{22} - D_{12}^2}, \tag{B.9a}$$

$$d_{12} = \frac{-D_{12}}{D_{11}D_{22} - D_{11}^2}, \tag{B.9b}$$

$$d_{22} = \frac{D_{11}}{D_{11}D_{22} - D_{12}^2}, \tag{B.9c}$$

$$d_{66} = \frac{1}{D_{66}}. \tag{B.9d}$$

Appendix C
Southwell Method

The theory of Euler indicates that an initially straight and centrally loaded column will remain straight under increasing load until the load attains a value, called the critical load P_{cr}, where the straight shape ceases to be stable and the column bends into the form of a single bow. This shape may be maintained by the end forces alone. If, however, the column is not perfectly straight initially, but bowed, the bifurcation from a straight to a bent shape will not occur; see Figure C.1, which illustrates load vs. out-of-plane deflection of the center of the column. When the amplitude of the initial im-

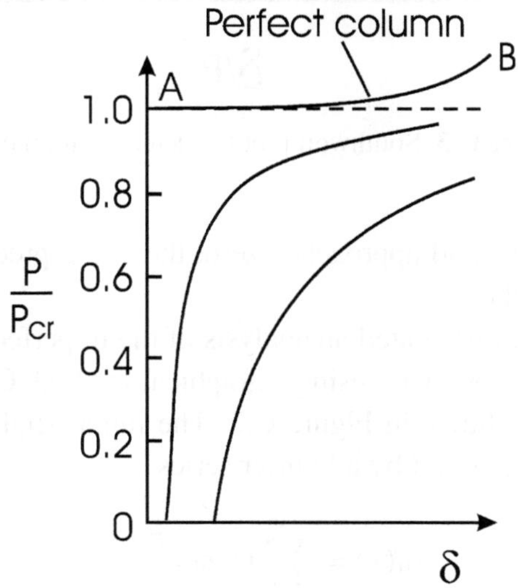

Figure C.1 Load vs. out-of-plane deflection for elastic columns. The intersection of the curves with the δ axis is the initial imperfection amplitude.

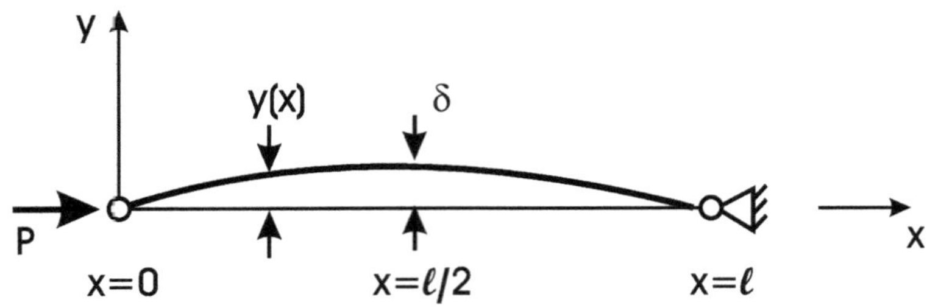

Figure C.2 Compression loading of a column with initial bow.

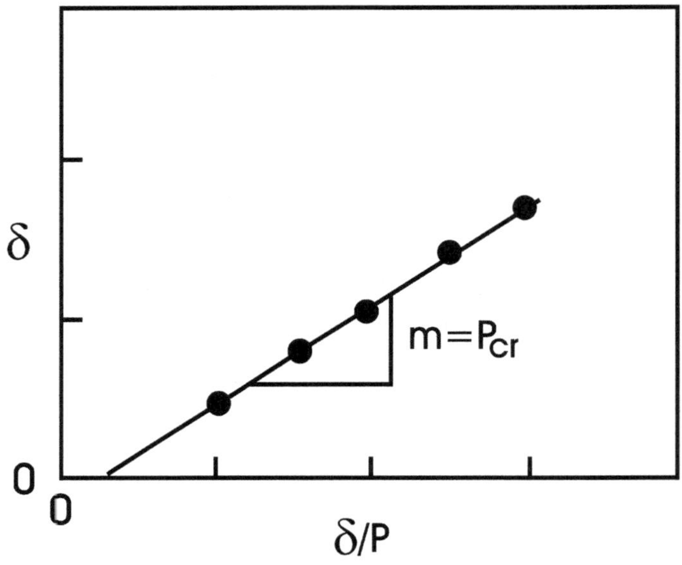

Figure C.3 Southwell plot for determination of P_{cr}.

perfection decreases and approaches zero, the curve predicted by the Euler analysis is approached.

Southwell (1932) presented an analysis of the imperfect column allowing a simple determination of P_{cr} using a graphical method. Consider the simply supported column shown in Figure C.2. The initial slight deflection of the column $y_0(x)$ is expressed by a Fourier series

$$y_0(x) = \sum_{n=1}^{\infty} \bar{w}_n \sin \frac{n\pi x}{l}, \tag{C.1}$$

where $\bar{w}_n$ represents the amplitude of each half sine wave representing each mode (n) of the initial imperfection, and l is the column length. x is a coordinate measured along the line of force action between the two pinned ends of the column. Southwell showed tat the total deflection δ_T at the center of the column ($x = l/2$) may be approximated by

$$\delta_T = \frac{\bar{w}_1}{1 - P/P_{\text{cr}}}, \tag{C.2}$$

where $\bar{w}_1$ is the amplitude of the fundamental mode ($n = 1$) of the initial imperfection (Equation (C.1)), and P_{cr} is the buckling load corresponding to the fundamental buckling mode. Equation (C.2) approximates a rectangular hyperbola in P vs. δ_T space with P_{cr} as the horizontal asymptote, and an intersection of the x axis ($P = 0$) at $\delta_T = \bar{w}_1$. Southwell recognized that the load vs. the *measured* deflection at the center of a column δ (Figure C.2) is also expected to follow a rectangular hyperbola that is passing through the origin ($P = 0$, $\delta = 0$). For $P > 0$, the load and deflection are connected by

$$\delta - P_{\text{cr}}\frac{\delta}{P} + \alpha = 0, \tag{C.3}$$

where α is a measure of the initial deflection. Hence, by plotting δ vs. δ/P, the data should fall on a straight line with P_{cr} as the slope of the line, see Figure C.3.

References

Adams, D.F. and Welsh, J.S., 1997, The Wyoming Combined Loading Compression (CLC) Test Method, *J. Compos. Tech. Res.*, **19**(3), pp. 123–133.

Adams, D.F., Carlsson, L.A., and Pipes, R.B., 2003, *Experimental Characterization of Advanced Composite Materials*, 3rd ed., CRC Press, Boca Raton.

Advani, S.G. and Sozer, E.M., 2002, *Process Modeling in Composites Manufacturing*, Marcel Dekker, New York.

Alfredsson, K.S., Bogetti, T.A., Carlsson, L.A., Gillespie, J.W., and Yiournas, A., 2008, Flexure of Beams with an Interlayer-Symmetric Beams with Orthotropic Adherends, *J. Mech. Mater. Struct.*, **3**, pp. 45–62.

Alif, N. and Carlsson, L.A., 1997, Failure Mechanisms of Woven Carbon and Glass Composites, in *ASTM STP 1285*, pp. 471–493.

Allen, H.G., 1969, *Analysis and Design of Structural Sandwich Panels*, Pergamon Press, Oxford.

ANSYS 8.1, 2006, *Swanson Analysis Systems*, Houston, PA.

Argon, A.S., 2000, Fracture, Strength and Toughness Mechanisms, in *Comprehensive Composite Materials*, A. Kelly and C. Zweben (Eds.), Vol. 1, Chapter 24, pp. 763–802, Elsevier, Oxford.

Armanios, E.A. (Ed.), 1996, *Fracture of Composties*, Key Eng. Mater., pp. 120–121, Trans Tech Publications, Zurich.

Ashby, M.F., 1999, *Materials Selection in Mechanical Design*, 2nd ed., Butterworth-Heinemann, Woburn.

Ashby, M.F., Evans, A., Fleck, N.J., Gibson, L.J., Hutchinson, J.W., and Wadley, H.N.G., 2000, *Metal Foams – A Design Guide*, Butterworth-Heinemann, Woburn.

ASTM Annual Book of Standards, 1998, Vol. 15.03, Space Simulation and Applications of Space Technology; Aerospace and Aircraft; Composite Materials, American Society for Testing and Materials, West Conshohocken, PA.

ASTM C273, 2000, *Standard Test Method for Shear Properties of Sandwich Core Materials*, American Society for Testing and Materials, West Conshohocken, PA.

ASTM C297/C297, 2004, *Standard Test Method for Flatwise Tensile Strength of Sandwich Construction*, American Society for Testing and Materials, West Conshohocken, PA.

ASTM C365, 2003, *Standard Test Method for Flatwise Compressive Properties of Sandwich Cores*, American Society for Testing and Materials, West Conshohocken, PA.

ASTM C393, 2000, *Standard Test Method for Flexural Properties of Sandwich Constructions*, American Society for Testing and Materials, West Conshohocken, PA.

ASTM D790, 2002, *Test Methods for Flexural Properties of Unreinforced and Reinforced Plastics and Electrical Insulating Materials*, American Society for Testing and Materials, West Conshohocken, PA.

ASTM D3039, 2000, *Standard Test Method for Tensile Properties of Polymer Matrix Composite Materials*, American Society for Testing and Materials, West Conshohocken, PA.

ASTM D3044-94, 2000, *Standard Test Method for Shear Modulus of Wood-Based Structural Panels*, American Society for Testing and Materials, West Conshohocken, PA.

ASTM D3410, 2003, *Standard Test Method for Compressive Properties of Polymer Matrix Composite Materials with Unsupported Gage Section by Shear Loading*, American Society for Testing and Materials, West Conshohocken, PA.

ASTM D4255-01, 2002, *Test Method for In-Plane Shear Properties of Polymer Matrix Composites Materials by the Rail Shear Method*, American Society for Testing and Materials, West Conshohocken, PA.

ASTM D5379-98, 2001, *Test Method for Shear Properties of Composite Materials by the V-Notched Beam Method*, American Society for Testing and Materials, West Conshohocken, PA.

ASTM D5528-94a, 2001, *Test Method for Mode I Interlaminar Fracture Toughness of Unidirectional Fiber-Reinforced Polymer Matrix Composites*, American Society for Testing and Materials, West Consohocken, PA.

ASTM D6641-01, 2001, *Determining the Compressive Properties of Polymer Matrix Composite Laminates Using a Combined Loading Compression (CLC) Test Fixture*, American Society for Testing and Materials, West Conshohocken, PA.

ASTM E8, 1998, *Standard Test Method for Tension Testing of Metallic Materials*, American Society for Testing and Materials, West Conshohocken, PA.

Avery, J.L. and Sankar, B.V., 2000, Compressive Failure of Sandwich Beam with Debonded Face-Sheets, *J. Compos. Mater.*, **34**, pp. 1176–1199.

Aviles, F., 2005, Local Buckling and Debond Propagation in Sandwich Columns and Panels, Ph.D. Dissertation, Florida Atlantic University.

Aviles, F. and Carlsson, L.A., 2006a, Experimental Study of Debonded Sandwich Panels Loaded in Compression, *J. Sandwich Struct. Mater.*, **8**, pp. 7–31.

Aviles, F. and Carlsson, L.A., 2006b, Three-Dimensional Finite Element Buckling Analysis of Debonded Sandwich Panels, *J. Compos. Mater.*, **40**, pp. 993–1008.

Aviles, F. and Carlsson, L.A., 2007a, Analysis of the Sandwich DCB Specimen for Debond Characterization, *Eng. Fracture Mech.*, **75**, pp. 153–168.

Aviles, F. and Carlsson, L.A., 2007b, Post-Buckling and Debond Propagation in Sandwich Panels Subject to In-plane Loading, *Eng. Fract. Mech.*, **74**, pp. 794–806.

Aviles, F., Carlsson, L.A., Browning, G., and Millay, K., 2009, Investigation of the Sandwich Plate Twist Test, *Exper. Mech.*, **49**, pp. 813–822.

Bao, G. and Suo, Z., 1992, Remarks on Crack-Bridging Concepts, *Appl. Mech. Rev.*, **45**, pp. 355–361.

Barrett, J.D. and Foschi, R.O., 1977, Mode II Stress Intensity Factors for Cracked Wood Beams, *Eng. Fracture Mech.*, **9**, pp. 371–378.

Bau-Madsen, N.K, Svendsen, K.H., and Kildegaard, A., 1992, Finite Deformation of Sandwich Plates, in *Proceedings of 2nd International Conference on Sandwich Construction*, EMAS, London, pp. 401–415.

Bazant, Z.P. and Beghini, A., 2004, Sandwich Buckling Formulas and Applicability of Standard Computational Algorithm for Finite Strain, *Composites: Part B*, **35**, pp. 573–581.

Bazant, Z.P. and Beghini, A., 2006, Stability and Finite Strain of Homogenized Structures Soft in Shear: Sandwich or Fiber Composites, and Layered Bodies, *Int. J. Solids Struct.*, **43**, pp. 1571–1593.

Bazant, Z.P. and Cedolin, L., 1991, *Stability of Structures*, Oxford University Press, New York, pp. 30–35.

Berggreen, C., 2004, Damage Tolerance of Debonded Sandwich Structures, Ph.D. Thesis, Department of Mechanical Engineering, Technical University of Denmark, Lyngby, Denmark.

Berry, J.P., 1963, Determination of Fracture Energies by the Cleavage Technique, *J. Appl. Phys.*, **34**, pp. 62–68.

Blaas, C. et al., 1984, Local Instability for Sandwich Panels, Fokker report TR-N-84-CSE-061.

Bogetti, T.A., Hoppel, C.P.R., Harik, V.M., Newill, J.F., and Burns, B.P., 2004. Predicting the Nonlinear Response and Progressive Failure of Composite Laminates, *Compos. Sci. Tech.*, **64**, pp. 329–342.

Budianky, B. and Rice, J.R., 1973, Conservation Laws and Energy Release Rates, *J. Appl. Mech.*, **x**, pp. 201–203.

Burman, M., 1998, Fatigue Crack Initiation and Propagation in Sandwich Structures, Doctoral Thesis, Report 98-29, Department of Aeronautics, KTH, Stockholm, Sweden.

Cantwell, W.J. and Davies, P., 1994, A Test Technique for Assessing Core-Skin Adhesion in Composite Sandwich Structures, *J. Mater. Sci. Lett.*, **13**, pp. 203–205.

Cantwell, W.J. and Davies, P., 1996, A Study of Skin-Core Adhesion in Glass Fiber Reinforced Sandwich Materials, *Appl. Compos. Mater.*, **3**, pp. 407–420.

Cantwell, W.J., Scudamore, R., Ratcliffe, J., and Davies, P., 1999, Interfacial Fracture in Sandwich Laminates, *Compos. Sci. Tech.*, **59**, pp. 2079–2085.

Carlsson, L.A., 1991, On the Design of the Cracked Sandwich Beam (CSB) Specimen, *J. Reinf. Plast. Compos.*, **10**, pp. 434–444.

Carlsson, L.A., Sendlein, L.S., and Merry, S.L., 1991, Characterization of Face Sheet/Core Shear Fracture of Composite Sandwich Beam, *J. Compos. Mater.*, **25**, pp. 101–116.

Carlsson, L.A., Matteson, R.C., Aviles, F., and Loup, D.C., 2006, Crack Path in Foam Cored DCB Sandwich Specimens, *Compos. Sci. Tech.*, **65**, pp. 2612–2621.

Chau, F.W., 1987, Discussion of Paper by Minguez (1986), *Exper. Mech.*, **27**, pp. 388–389.

Chou, P.C., Coreleone, J., and Hsu, C.M., 1972, Elastic Constants of Layered Media, *J. Compos. Mater.*, **6**, pp. 80–93.

Chow, T.S., 1971, On the Propagation of Flexural Waves in an Orthotropic Laminated Plate and Its Response to an Impulsive Load, *J. Compos. Mater.*, **5**, pp. 306–319.

Cox, H.L., 1933, Buckling of Thin Plates in Compression, British Air Ministry, Air Research Committee Reports and Memoranda, No. 1554.

Daniel, I.M. and Ishai, O., 1994, *Engineering Mechanics of Composite Materials*, Oxford University Press, New York.

Daniel, I.M. and Ishai, O., 2006, *Engineering Mechanics of Composite Materials*, 2nd ed., Oxford University Press, New York.

Donnell, L.H., 1938, On the Application of Southwell's Method for the Analysis of Buckling Tests, in *Timoshenko 60th Anniversary Volume*, McGraw-Hill, New York.

Dundurs, J., 1969, Edge-Bonded Dissimilar Orthogonal Wedges, *J. Appl. Mech.*, **36**, pp. 650–652.

Eshelby, J.D., Read, W.T., and Shockley, W., 1953, Anistropic Elasticity with Appplications to Dislocation Theory, *Acta Metal.*, **1**, pp. 251–259.

Engesser, F., 1891, Die Knickfestigheit gerader Stabe, *Zentralblatt der Bauverwaltung*, **11**, pp. 483–486.

England, A.H., 1965, A Crack between Dissimilar Media, *J. Appl. Mech.*, **32**, pp. 400–402.

Erdogan, F., 1965, Stress Distributions in Bonded Dissimilar Materials with Cracks, *J. Appl. Mech.*, **32**, 403–410.

Erdogan, F. and Sih, G.C., 1963, On the Crack Extension in Plates under Plane Loading and Transverse Shear, *J. Basic Eng.*, **85**, pp. 519–527.

Ewalds, H.L. and Wanhill, R.J.H., 1989, *Fracture Mechanics*, Edward Arnold, London.

Farris, R. and Filippov, A., 1982, Experimental Method of Determining the Uniaxial-Compression Characteristics of Thin-Walled Structures, *Exper. Mech.*, **22**(2), pp. 49–51.

Friedrich, K. (Ed.), 1989, *Application of Fracture Mechanics to Composite Materials*, Elsevier, Amsterdam.

Frostig, Y., 2010, On Wrinkling of a Sandwich Panel with a Compliant Core and Self-Equilibrated Loads, in *Proceedings of 9th International Conference of Sandwich Structure (ICSS9)*, G. Ravichandran (Ed.), June 2010, Pasadena, USA.

Frostig, Y., Baruch, M., Vilnay, O., and Sheinman, I., 1992, High-Order Theory for Sandwich-Beam Behavior with Transversely Flexible Core, *J. Eng. Mech.*, **118**, pp. 1026–1043.

Gere, J.M., 2004, *Mechanics of Materials*, 6th ed., Brooks/Cole, Belmont.

Gibson, L.J. and Ashby, M.F., 1997, *Cellular Solids: Structure and Properties*, 2nd ed., Cambridge University Press, Cambridge.

Grangard, H. and Rudstrom, L., 1970, A Theoretical Consideration of the Compressive Strength of a Carton, Report B-30, Swedish Forest Products Laboratory, Stockholm, Sweden.

Haringx, J.A., 1948, On Highly Compressible Helical Springs and Rubber Rods, and Their Application for Vibration-Free Mountings, I, *Phillips Research Reports, Eindhoven, the Netherlands*, **3**, pp. 401–449.

Haringx, J.A., 1949, On Highly Compressible Helical Springs and Rubber Rods, and Their Application for Vibration-Free Mountings, II, *Phillips Research Reports, Eindhoven, the Netherlands*, **4**, pp. 49–80.

Hayman, B., Wiese, M., Davies, P., Choquese, D., Hoyning, B., and Mitusch, P., 1998, Foam Cored Sandwich Panels Under Static Pressure Loading: Some New Tests and Analyses, in *Proceedings of the 4th International Conference on Sandwich Construction*, KTH, Stockholm, Sweden, 9–11 June, 1998, EMAS, London, pp. 623–634.

He, M.Y. and Hutchinson, J.W., 1989, Kinking of a Crack out of an Interface, *J. Appl. Mech.*, **56**, pp. 270–278.

Helm, J.D., McNeill, S.R., and Sutton, M.A., 1996, Improved Three-Dimensional Image Correlation for Surface Displacement Measurements, *Opt. Eng.*, **35**, pp. 1911–1920.

High Performance Test Fixtures, 2000, Product Catalog No. 106, Wyoming Test Fixtures, Inc., Laramie, WY (see also www.wyomingtestfixtures.com).

Hoff, N.J., 1950, Bending and Buckling of Rectangular Sandwich Plates, NACA TN 2225.

Hoff, N.J. and Mautner, S.F., 1945, The Buckling of Sandwich-Type Panels, *J. Aeron. Sci.*, **12**, pp. 285–297.

Hohe, J., Librescu, L. and Oh, S.Y., 2006, Dynamic Buckling of Flat and Curved Sandwich Panels with Transversely Compressible Core, *Compos. Struct.*, **74**, pp. 10–24.

Huang, H. and Kardomateas, G.A., 2001, Mixed-Mode Stress Intensity Factors for Cracks Located at or Parallel to the Interface in Bimaterial Half Planes, *Int. J. Solids Struct.*, **38**, pp. 3719–3734.

Huang, H. and Kardomateas, G.A., 2002, Buckling and Initial Postbuckling Behavior of Sandwich Beams Including Transverse Shear, *AIAA J.*, **40**, pp. 2331–2335.

Hutchinson, J.W., 1990, Mixed Mode Fracture Mechanics of Interfaces, in *Metal-Ceramic Interfaces*, M. Rühle et al. (Eds.), Pergamon Press, New York, pp. 295–306.

Hutchinson, J.W. and Suo, Z., 1992, Mixed Mode Cracking in Layered Materials, *Adv. Appl. Mech.*, **29**, pp. 63–191.

Hyer, M.W., 1998, *Stress Analysis of Fiber-Reinforced Composite Materials*, WCB/McGraw-Hill, Boston.

Irwin, G.R., 1958, *Fracture, in Handbuch der Physik*, 6 (Flugge ed.), Springer, Berlin, pp. 551–590.

Jones, R.M., 1999, *Mechanics of Composite Materials*, Taylor & Francis, Philadelphia.

Kanninen, M.F., 1973, An Augmented Double Cantilever Beam Model for Studying Crack Propagation and Arrest, *Int. J. Fracture*, **9**, pp. 83–91.

Kardomateas, G.A., 1993, The Initial Postbuckling and Growth Behavior of Internal Delaminations in Composite Plates, *J. Appl. Mech.*, **60**, pp. 903–910.

Kardomateas, G.A., 2001, Elasticity Solutions for a Sandwich Orthotropic Cylindrical Shell under External Pressure, Internal Pressure and Axial Force, *AIAA J.*, **39**, pp. 713–719.

Kardomateas, G.A., 2005, Wrinkling of Wide Sandwich Panels/Beams with Orthotropic Phases by an Elasticity Approach, *J. Appl. Mech.*, **72**, pp. 818–825.

Kardomateas G.A., 2009, Three-Dimensional Elasticity Solution for Sandwich Plates with Orthotropic Phases: the Positive Discriminant Case, *J. Appl. Mech.*, **76**, pp. 014505-1–014505-4.

Kardomateas, G.A., 2010, An Elasticity Solution for the Global Buckling of Sandwich Beams/Wide Panels with Orthotropic Phases, *J. Appl. Mech.*, **77**, pp. 021015-1–021015-7.

Kardomateas, G.A. and Huang, H., 2003, The Initial Postbuckling Behavior of Face-Sheet Delaminations in Sandwich Composites, *J. Appl. Mech.*, 70, pp. 191–199.

Kardomateas, G.A. and Phan, C.N., 2010, Three Dimensional Elasticity Solution for Sandwich Beams/Wide Plates with Orthotropic Phases: The Negative Discriminant Case, *Journal of Sandwich Structures and Materials*, in press.

Kardomateas, G.A., Berggreen, C. and Carlsson, L.A., 2010, Energy Release Rate and Mode Mixity for a General Asymmetric Debonded Sandwich Specimen, in *Proceedings of 9th International Conference on Sandwich Structures (ICSS-9)*, Caltech, Pasadena, CA, 14–16 June 2010.

Kelly, A. and Zweben, C. (Eds.), 2000, *Comprehensive Composite Materials*, Vol. 2: Polymer Matrix Composites, Elsevier Science, Oxford.

Khot, N.S. and Bauld Jr., N.R., 1983, Further Comparison of the Numerical and Experimental Buckling Behaviors of Composite Panels, *Comput. & Struct.*, **17**, pp. 61–68.

Kinloch, A.J. and Young, R.J., 1983, *Fracture Behavior of Polyumers*, Elsevier, Amsterdam.

Krueger, R., 2004, Virtual Crack Closure Technique: History, Approach and Applications, *Appl. Mech. Rev.*, **57**, pp. 109–143.

La Saponara, V. and Kardomateas, G.A., 2001, Tests on the Compression Behavior of Debonds in Sandwich Beams, Paper presented at ASME International Conference, New York, NY, November 11–16.

Lekhnitskii, S.G., 1963, *Theory of Elasticity of an Anisotropic Elastic Body*, Holden-Day, San Francisco.

Lekhnitskii, S.G., 1968, *Anisotropic Plates*, Gordon and Breach, New York.

Li, R. and Kardomateas, G.A., 2006, A Solution to the Thermo-elastic Interface Crack Branching in Dissimilar Anisotropic Bi-material Media, *Int. J. Solids Struct.*, **43**, pp. 913–942.

Li, R. and Kardomateas, G.A., 2008, A Nonlinear High Order Core Theory for Sandwich Plates with Orthotropic Phases, *AIAA J.*, **46**, pp. 2926–2934.

Li, S., Wang, J., and Thouless, M.D., 2004, The Effects of Shear on Delamination in Layered Materials, *J Mech. Phys. Solids*, **52**, pp. 193–204.

Li, V.C. and Ward, R.J., 1989, A Novel Testing Technique for Post-Peak Tensile Behavior of Cementitious Materials, in *Fracture Toughness and Fracture Energy – Testing Methods for Concrete and Rocks*, H. Mihashi et al. (Eds.), A.A. Balkema Publishers, Rotterdam, pp. 183–195.

Li, X. and Carlsson, L.A., 1999, The Tilted Sandwich Debond (TSD) Specimen for Face/Core Interface Fracture Characterization, *J. Sandwich Struct. Mater.*, **1**, pp. 60–75.

Li, X. and Carlsson, L.A., 2000, Elastic Foundation Analysis of Tilted Sandwich Debond (TSD) Specimen, *J. Sandwich Struct. Mater.*, **2**, pp. 3–32.

Li, X. and Carlsson, L.A., 2001, Fracture Mechanics Analysis of Tilted Sandwich Debond (TSD) Specimen, *J. Compos. Mater.*, **35**, pp. 2145–2168.

Libove, C. and Batdorf, S.B., 1948, A General Small Deflection Theory for Flat Sandwich Plates, NACA TN 1526 Report.

Libove, C. and Hubka, R.E., 1951, Elastic Constants for Corrugated Core Sandwich Plates, NACA TN 2289 Report.

Liechti, K.M. and Chai, Y.S., 1992, Asymmetric Shielding in Interfacial Fracture under In-Plane Shear, *J. Appl. Mech.*, **59**, pp. 295–304.

Liu, L., Kardomateas, G.A., and Holmes, J.W., 2004, Mixed-Mode Stress Intensity Factors for a Crack in an Anisotropic Bi-Material Strip, *Int. J. Solids Struct.*, **41**, pp. 3095–3107.

Love, A.E.H., 1927, *A Treatise on the Mathematical Theory of Elasticity*, 4th ed., Dover.

Lundsgaard-Larsen, C., Sorensen, B.F., Berggreen, C., and Ostergaard, R.C., 2008, A Modified DCB Sandwich Specimen for Measuring Mixed-Mode Cohesive Laws, *Eng. Fract. Mech.*, **75**, pp. 2514–2530.

Majumdar, P., Srinivasagupta, D., Mahfuz, H., Joseph, B., Thomas, M., and Christensen, S., 2003, Effect of Processing Conditions and Material Properties on the Debond Fracture Toughness of Foam-Core Sandwich Composites: Experimental Optimization, *Composites: Part A*, **34**, pp. 1097–1104.

Matthews, M.J. and Swanson, S.R., 2005, A Numerical Approach to Separate the Modes of Fracture in Interface Crack Propagation, *J. Compos. Mater.*, **39**, pp. 247–264.

McKee, R.C., Gander, J.W., and Wachuta, J.R., 1963, Compression Strength Formula for Corrugated Boxes, *Paperboard Packaging*, **48**, pp. 149–159.

Mindlin, R.D., 1951, Influence of Rotary Inertia and Shear on Flexural Motions of Isotropic, Elastic Plates, *J. Appl. Mech.*, **18**, pp. 336–343.

Minguez, J.M., 1986, An Experimental Investigation of How Accurate, Simply Supported Boundary Conditions Can Be Achieved in Compression Testing of Panels, *Exper. Mech.*, **26**, pp. 238–244.

Moyer, E.T., Amir, G.G., Olsson, K.A. and Hellbratt, S.E., 1992, Response of GRP Sandwich Structures Subjected to Shock Loading, in *Proceedings of 2nd International Conference on Sandwich Construction*, EMAS, London, pp. 49–65.

Mure, M., 1986, Corrugated Board – A New Method: Anticlastic Rigidity, *Revue A.T.I.P.*, **40**, pp. 325–330.

Nairn, J.A., 2009, Analytical and Numerical Modeling of R-Curves for Cracks with Bridging Zones, *Int. J. Fracture*, **5**, pp. 167–181.

Nordstrand, T., 2003, Basic Testing and Strength Design of Corrugated Board and Containers, Doctoral Thesis, Lund University, Lund, Sweden.

Nordstrand, T. and Carlsson, L.A., 1997, Evaluation of Transverse Shear Stiffness of Structural Core Sandwich Plates, *Compos. Struct.*, **37**, pp. 145–153.

Nordstrand, T., Carlsson, L.A., and Allen, H.G., 1994, Transverse Shear Stiffness of Structural Core Sandwich, *Compos. Struct.*, **27**, pp. 317–329.

Norris, C.B., 1942, *Technique of Plywood*, I.F. Haucks, Seattle.

Norris, C.B. and Krommer, W.J., 1950, Short-Column Compressive Strength of Sandwich Constructions as Affected by the Size of the Cells of Honeycomb-Core Materials, Forest Products Laboratory Report No. 1817, Madison, Wisconsin.

Norton, R.L., 2003, *Machine Design – An Integrated Approach*, 3rd ed., Pearson Prentice Hall, Upper Saddle River.

Ostergaard, R.C., 2007, Interface Fracture in Composite Materials and Structures, DCAMM Special Report No. S102, DTU, Lyngby, Denmark.

Ostlund, S., 1995, Fracture Modeling of Brittle-Matrix Composites with Spatially Dependent Crack Bridging, *Fatigue Fracture Eng. Mater. Struct.*, **18**, pp. 1213–1230.

Pagano, N.J., 1969, Exact Solutions for Composite Laminates in Cylindrical Bending, *J. Compos. Mater.*, **3**, pp. 398–411.

Pagano, N.J., 1970a, Exact Solutions for Rectangular Bidirectional Composites and Sandwich Plates, *J. Compos. Mater.*, **4**, pp. 20–34,

Pagano, N.J., 1970b, Influence of Shear Coupling in Cylindrical Bending of Anisotropic Laminates, *J. Compos. Mater.*, **4**, pp. 330–343.

Patel, P., Nordstrand, T., and Carlsson, L.A., 1997, Local Buckling and Collapse of Corrugated Board Under Biaxial Stress, *Compos. Struct.*, **39**, pp. 93–110.

Petras, A. and Sutcliffe, M.P.F., 1999, Failure Mode Maps for Honeycomb Sandwich Panels, *Compos. Struct.*, **44**, pp. 237–252.

Pfund, B., 2005, Core Installation, *Professional Boat Builder*, **94**, April/May, pp. 48–69.

Phan, C.N., Frostig, Y., and Kardomateas, G.A., 2010, Analysis of Sandwich Panels with a Compliant Core and with In-Plane Rigidity – Extended High-Order Sandwich Panel Theory versus Elasticity, submitted for publication.

Plantema, F.J., 1966, *Sandwich Construction*, John Wiley, New York.

Prasad, S. and Carlsson, L.A., 1994a, Debonding and Crack Kinking in Foam Core Sandwich Beams – I: Analysis of Fracture Specimens, *Eng. Fracture Mech.*, **47**, pp. 813–824.

Prasad, S. and Carlsson, L.A., 1994b, Debonding and Crack Kinking in Foam Core Sandwich Beam – II: Experimental Investigation, *Eng. Fracture Mech.*, **47**, pp. 825–841.

Quispitupa, A., Berggreen, C., and Carlsson, L.A., 2009, On the Analysis of a Mixed Mode Bending (MMB) Sandwich Specimen for Debond Fracture Characterization, *Eng. Fracture Mech.*, **76**, pp. 594–613.

Raju, I.S., Crews Jr., J.H., and Aminpour, M.A., 1988, Convergence of Strain Energy Release Rate Components for Edge-Delaminated Composite Laminates, *Eng. Fracture Mech.*, **30**, pp. 383–396.

Reeder, J.R. and Crews Jr., J.H., 1990, Mixed Mode Bending Method for Delamination Testing, *AIAA J.*, **28**, pp. 1270–1276.

Reissner, E., 1945, The Effect of Transverse Shear Deformation on the Bending of Elastic Plates, *J. Appl. Mech.*, **12**, pp. 69–77.

Rhodes, J. and Harvey, J.M., 1977, Examination of the Post-Buckling Behavior, *J. Eng. Mech. Div.*, **103**, pp. 461–478.

Riber, H.J., 1997, Non-linear Analytical Solutions for Laterally Loaded Sandwich Plates, *Compos. Struct.*, **39**, pp. 63–83.

Rice, J.R., 1968, A Path-Independent Integral and the Approximate Analysis of Strain Concentration by Notches and Cracks, *J. Appl. Mech.*, **35**, pp. 376–386.

Rice, J.R. and Sih, G.C., 1965, Plane Problems of Cracks in Dissimilar Media, *J. Appl. Mech.*, **32**, pp. 418–423.

Rothschild, Y., Berge, J., and Moen, A.J., 1992, Experiments and Numerical Analyses of Hydrostatically Loaded Sandwich Plates, Report 92-13, Department of Lightweight Structures, KTH, Stockholm, Sweden.

Russell, A.J. and Street, K.N., 1982, Factors Affecting the Interlaminar Fracture Energy of Graphite/Epoxy Laminates, in *Proceedings of the 4th International Conference on Composite Materials (ICCM-4)*, Japan Society for Composite Materials, pp. 279–286.

Rybicki, E.F. and Kanninen, M.F., 1977, A Finite Element Calculation of Stress Intensity Factors by a Modified Crack Closure Integral, *Eng. Fracture Mech.*, **9**, pp. 931–938.

Sankar, B.V. and Narayanan, M., 2001, Finite Element Analysis of Debonded Sandwich Beams under Axial Compression, *J. Sandwich Struct. Mater.*, **3**, pp. 197–219.

SAS Users Manual, 2003, 100 SAS Campus Drive, Cary, North Carolina.

Satapathy, N.R. and Vinson, J.R., 2000, Sandwich Beams with Mid-Plane Asymmetry Subjected to Lateral Loads, *J. Sandwich Struct. Mater.*, **2**, pp. 379–390.

Sciammarella, C.A., 1982, The Moire Method – A Review, *Exper. Mech.*, **22**(11), pp. 418–433.

Seeman, W.H., 1990, U.S. Patent 49002215.

Sendlein, L.S. and Carlsson, L.A., 1990, Experimental and Numerical Evaluation of the Cracked Sandwich Beam (CSB) Specimen, presented at the 3rd International Conference on Marine Applications of Composite Materials, FIT, Melbourne, FL, March 19–21, 1990.

Shipsa, A., Burman, M., and Zenkert, D., 1999, Interfacial Fatique Crack Growth in Foam Core Sandwich Structures, *Fatigue Fract. Eng. Mater. Struct.*, **22**, pp. 123–131.

Shivakumar, K.N. and Smith, S.A., 2004, In-Situ Fracture Toughness Testing of Core Materials in Sandwich Panels, *J. Compos. Mater.*, **38**, pp. 655–668.

Shivakumar, K.N., Chen, H., and Smith, S.A., 2005, An Evaluation of Data Reduction Methods for Opening Mode Fracture Toughness of Sandwhich Panels, *J. Sandwich Struct. Mater.*, **7**, pp. 77–90.

Sihn, S. and Rice, B.P., 2002, Characterization of Compressive and Shear Properties of Carbon Foam and Its Core-Application in Sandwich Construction, in *Proceedings from the 17th ASC Composites Conference*, Purdue University, West Lafayette, Indiana, October 21–23.

Sorensen, B.F. and Jacobsen, T.K., 1998, Large-Scale Bridging in Composites: R-Curves and Bridging Laws, *Composites, Part A*, **29**, pp. 1443–1451.

Sorensen, B.F. and Kirkegaard, P., 2006, Determination of Mixed-Mode Cohesive Laws, *Eng. Fract. Mech.*, **73**, pp. 2642–2646.

Sorensen, B.F., Jorgensen, K., Jacobsen, T.K., and Ostergaard, R.C., 2006, DCB-Specimen Loaded with Uneven Bending Moments, *Int. J. Fract.*, **141**, pp. 163–176.

Southwell, R., 1932, On the Analysis of Experimental Observations of Problems of Elastic Stability, *Proc. Royal Soc., London*, pp. 601–616.

Souza, M.A., Kck, W.C., and Walker, A.C., 1983, Review of Experimental Techniques for Thin-Walled Structures Liable to Buckling, *Exper. Tech.*, **7**(9), pp. 21–25, and **7**(10), pp. 36–39.

Spencer, H.H. and Walker, A.C., 1975, Critique of Southwell Plots with Proposals for Alternative Methods, *Exper. Mech.*, **15**, 1975, pp. 303–310.

Steeves, C.A. and Fleck, N., 2004a, Collapse Mechanism of Sandwich Beams with Composite Faces and Foam Core, Loaded in Three-Point Bending, Part I: Analytical Models and Minimum Weight Design, *Int. J. Mech. Sci.*, **46**, pp. 561–583.

Steeves, C.A. and Fleck, N., 2004b, Collapse Mechanism of Sandwich Beams with Composite Faces and Foam Core, Loaded in Three-Point Bending, Part II: Experimental Investigation and Numerical Modeling, *Int. J. Mech. Sci.*, **46**, pp. 585–608.

Stroh, A.N., 1958, Dislocations and Cracks in Anisotropic Elasticity, *Phil. Mag.*, **7**, pp. 625–646.

Suo, Z., 1990, Singularities, Interfaces and Cracks in Dissimilar Anisotropic Media, *Proc. R. Soc. Lond.*, **A427**, pp. 331–358.

Suo, Z. and Hutchinson, J.W., 1990, Interface Crack between Two Elastic Materials, *Int. J. Fracture*, **43**, pp. 1–18.

Thomsen, O.T., 1977, Localized Loads, in D. Zenkert (Ed.), *An Introduction to Sandwich Construction*, Chapter 12, EMAS, London.

Thouless, M.D., 2009, The Effects of Transverse Shear on the Delamination of Edge-Notch Flexure and 3-Point Bend Geometries, *Compos. Part B*, **40**, pp. 305–312.

Timoshenko, S., 1936, *Theory of Elastic Stability*, McGraw-Hill, New York.

Timoshenko, S. and Woinowsky-Krieger, S., 1959, *Theory of Plates and Shells*, McGraw-Hill, New York.

Ting, T., 1986, Explicit Solution and Invariance of the Singularities at an Interface Crack in Anisotropic Composites, *Int. J. Solids Struct.*, **22**, pp. 965–983.

Triantafillou, T.C. and Gibson, L.J., 1987, Failure Mode Maps for Foam Core Sandwich Beams, *Mater. Sci. Eng.*, **95**, pp 37–53.

Tsai, S.W., 1965, Experimental Determination of the Elastic Behavior of Orthotropic Plates, *J. Eng. Ind.*, **63**, pp. 315–318.

Tsai, S.W. and Wu, E.M., 1971, A General Theory of Strength for Anisotropic Materials, *J. Compos. Mater.*, **5**, pp. 58–80.

Vadakke, V. and Carlsson, L.A., 2004, Experimental Investigation of Compression Failure of Sandwich Specimens with Face/Core Debond, *Composites, Part B*, **35**, pp. 583–590.

Viana, G.M., 2001, Factors Influencing Face/Core Crack Propagation in PVC Foam Core Sandwich, Master Thesis, Florida Atlantic University.

Viana, G.M. and Carlsson, L.A., 2002a, Mechanical Properties and Fracture Characterization of Cross-Linked PVC Foams, *J. Sandwich Struct. Mater.*, **4**, pp. 99–113.

Viana, G.M. and Carlsson, L.A., 2002b, Mode Mixity and Crack Tip Yield Zones in TSD Sandwich Specimens with PVC Foam Core, J. Sandwich Struct. Mater., 4, pp. 141-155.

Viana, G.M. and Carlsson, L.A., 2003, Influences of Foam Density and Core Thickness on Debond Toughness of Sandwich Specimens with PVC Foam Core, *J. Sandwich Struct. Mater.*, **5**, pp. 103–118.

Vinson, J.R., 1999, *The Behavior of Sandwich Structures of Isotropic and Composite Materials,* Technomic, Lancaster.

von Kármán, T., Sechler, E.E., and Donnell, L.H.J., 1932, The Strength of Thin Plates in Compression, *J. Appl. Mech.*, **54**, pp. 53–58.

Weikel, R.C. and Kobayashi, A.S., 1959, On the Local Elastic Stability of Honeycomb Face Plate Subjected to Uniaxial Compression, *J. Aero/Space Sci.*, pp. 672–674.

Wennhage, P. and Zenkert, D., 1998, Testing of Sandwich Panels under Uniform Pressure, *J. Testing and Evaluation*, **26**, pp. 101–108.

Westerlind, B.S. and Carlsson, L.A., 1992, Compressive Response of Corrugated Board, *Tappi J.*, **75**, pp. 145–154.

Whitney, J.M., 1972, Stress Analysis of Thick Laminated Composite and Sandwich Plates, *J. Compos. Mater.*, **6**, pp. 426–440.

Whitney, J.M., 1987, *Structural Analysis of Laminated Anisotropic Plates*, Technomic, Lancaster.

Whitney, J.M., Browning, C.E., and Hoogsteden, W., 1982, A Double Cantilever Beam Test for Characterizing Mode I Delamination of Composite Materials, *J. Reinf. Plast. Compos.*, **1**, pp. 297–313.

Williams, M.L., 1959, The Stresses around a Fault or Crack in Dissimilar Media, *Bull. Seismol. Soc. Amer.*, **49**, pp. 199–204.

Yen, K.T., Gunturkun, S., and Pohle, F.V., 1951, Deflections of a Simply Supported Rectangular Sandwich Plate Subjected to Transverse Loads, NACA TN 2581.

Yin, W.L. and Wang, J.T.S., 1984, The Energy Release Rate in the Growth of a One-Dimensional Delamination, *J. Appl. Mech.*, **51**, pp. 939–941.

Zenkert, D., 1997, *An Introduction to Sandwich Construction*, EMAS, London.

Subject Index

FSC
www.fsc.org
MIX
Papier aus verantwortungsvollen Quellen
Paper from responsible sources
FSC® C105338

If you have any concerns about our products,
you can contact us on
ProductSafety@springernature.com

In case Publisher is established outside the EU,
the EU authorized representative is:
Springer Nature Customer Service Center GmbH
Europaplatz 3, 69115 Heidelberg, Germany

Printed by Libri Plureos GmbH
in Hamburg, Germany